T0325696

Rare Diseases and Orphan Drugs

Keys to Understanding and Treating the Common Diseases

Rare Diseases and Orphan Drugs

Keys to Understanding and Treating the Common Diseases

Jules J. Berman, Ph.D., M.D.

AMSTERDAM • BOSTON • HEIDELBERG • LONDON
NEW YORK • OXFORD • PARIS • SAN DIEGO
SAN FRANCISCO • SINGAPORE • SYDNEY • TOKYO

Academic Press is an imprint of Elsevier

Academic Press is an imprint of Elsevier
32 Jamestown Road, London NW1 7BY, UK
225 Wyman Street, Waltham, MA 02451, USA
525 B Street, Suite 1800, San Diego, CA 92101-4495, USA

Medicine is an ever-changing field. Standard safety precautions must be followed, but as new research and clinical experience broaden our knowledge, changes in treatment and drug therapy may become necessary or appropriate. Readers are advised to check the most current product information provided by the manufacturer of each drug to be administered to verify the recommended dose, the method and duration of administrations, and contraindications. It is the responsibility of the treating physician, relying on experience and knowledge of the patient, to determine dosages and the best treatment for each individual patient. Neither the publisher nor the authors assume any liability for any injury and/or damage to persons or property arising from this publication.

Notice

No responsibility is assumed by the publisher for any injury and/or damage to persons or property as a matter of products liability, negligence or otherwise, or from any use or operation of any methods, products, instructions or ideas contained in the material herein. Because of rapid advances in the medical sciences, in particular, independent verification of diagnoses and drug dosages should be made

British Library Cataloguing-in-Publication Data
A catalogue record for this book is available from the British Library

Library of Congress Cataloging-in-Publication Data
A catalog record for this book is available from the Library of Congress

ISBN: 978-0-12-419988-0

For information on all Academic Press publications
visit our website at elsevierdirect.com

Typeset by Scientific Publishing Services
www.sps.co.in

Printed and bound in United States of America

14 15 16 17 10 9 8 7 6 5 4 3 2 1

Dedication

For my mother, Ida

Contents

PART II
Rare Lessons for Common Diseases

PART III

Fundamental Relationships between Rare and Common Diseases

Acknowledgments

It is impossible to deeply understand the rare diseases, or their relationships to the common diseases, without access to the OMIM data set. OMIM, the Online Mendelian Inheritance in Man, is almost certainly the largest, best-curated, and longest-running collection of information on Mendelian disorders and disease genes. OMIM began simply as MIM in the early 1960s, a creation of Dr. Victor A. McKusick. Starting in 1966, it was printed in annual volumes. By 1998, the print version was heavier than most people could safely lift. Currently, it is available as a query engine online, and as a file that can be downloaded at no cost, and studied as a stand-alone plain-text document. The current length of OMIM is about 175 megabytes, and is curated at the McKusick-Nathans Institute of Genetic Medicine at the Johns Hopkins University School of Medicine, under the direction of Dr. Ada Hamosh. Over my career, I have spent many hundreds of hours, possibly thousands of hours, reading the OMIM file. Without this remarkable resource, I could not have written this book.

I would like to thank the entire staff at the Office of Rare Diseases at the National Institutes of Health. On two separate occasions, these defenders of orphan diseases chose to fund my projects on GIST and on borderline ovarian tumors. This support from the Office of Rare Diseases, received at a time in my career when I was vulnerable to criticism for my preoccupation with orphaned conditions, boosted my resolve and inspired me to follow a path less traveled.

Special thanks go to Mara Conner and Jeffrey Rossetti (Editorial Department), and Caroline Johnson (Production), at Elsevier, for their extraordinary care and effort on this complex book project.

Foreword

Dr. Berman's book, *Rare Diseases and Orphan Drugs: Keys to Understanding and Treating the Common Diseases*, addresses a topic of great importance at this particular moment in research history. Recent advances in the molecular biology of disease have taught us that the genetic changes in the common diseases are complex and that there is remarkable variation among affected individuals in the clinical presentation and in the genetic signature of common diseases. Research scientists are beginning to recognize that the common diseases are best conceived as aggregates of many different rare diseases. To benefit from our newly acquired knowledge of the genetics of common diseases, we will need to understand how treatments for the rare diseases will apply to subsets of the common diseases.

Breakthrough discoveries among the rare diseases are now viewed as opportunities to understand and treat the common diseases. Hence, there is an increasing emphasis, coming from government, academia, and private research organizations, to increase funding for rare diseases research and orphan drug development.

Worldwide, orphan drugs are being developed and approved at a rapid rate. In the United States, expedited programs adopted by the FDA should continue to move potential products through the research and development continuum toward approval for safe and effective products. Streamlined programs such as Fast Track, Breakthrough Therapy, Accelerated Approval, and Priority Review bring optimism to patients and their families for the quick approval of new products. Additional regulatory approaches and incentives have been expanded and include the rare pediatric and tropical diseases priority review vouchers. Repurposed products now qualify for orphan product incentives at the FDA. Compounds included in the Best Pharmaceuticals for Children Act program are eligible for a 6-month extension to existing exclusivity as an incentive to expand the indications for use from the adults to the pediatric population. For antibiotics, a newer incentive program, Generating Antibiotic Incentives Now (GAIN), and authorizing legislation add 5 years to existing exclusivity for products considered new chemical entities and those included under the Orphan Drug Act. The FDA now uses expert consultants to facilitate applications for orphan drugs while still in the pre-approval stage. The success of clinical trials for orphan drugs can be credited, in no small part, to the willingness of individuals with rare diseases to participate in clinical trials.

In the U.S., more than 2,900 active Orphan Product Designations have been made, and 50 additional designations have been provided thus far in 2014. There were 258 Orphan Product Designations in 2013. Obtaining the Orphan Product

Designation from the Office of Orphan Products Development at the FDA provides incentives such as 7-year marketing exclusivity, eligibility for research grants, along with exemption from filing fees for some qualifying applications. Pharmaceutical Research and Manufacturers of America reported more than 450 compounds in development for rare diseases. Current activity levels indicate a continued emphasis on rare diseases. Many pharmaceutical companies have established programs to make available the needed orphan products for rare diseases regardless of the patients' ability to pay for the product.

The U.S. NIH Clinical Center research portfolio contains more than 860 research protocols for approximately 520 rare diseases. In fiscal year 2013, NIH provided $3.456 billion for rare diseases research projects and included approximately $764 million for orphan product research projects.

Our new-generation orphan drugs are designed to target disease-causing pathways that operate in the rare diseases. It is everyone's hope that these new drugs will prove to be effective against common diseases that share disease pathways with rare diseases. If so, then we can begin to design clinical trials that include common and rare diseases in the same trials.

Rare Diseases and Orphan Drugs: Keys to Understanding and Treating the Common Diseases bridges our understanding of the common diseases and the rare diseases. This unique and much-needed book provides an insightful glimpse of how biomedical research will play out as the rare diseases take an increasing role in the way we understand and treat the common diseases. Healthcare professionals, students, biomedical researchers, and advocates for rare disease research will find that this book applies common sense to a rare subject.

April, 2014

Stephen G. Groft, Pharm.D.

Note added by editor: Dr. Groft retired from the NIH in 2014, where he served as Director of the Office of Rare Diseases Research for more than two decades. His work in the field of rare diseases and orphan drugs began in 1982 at the FDA Office of Orphan Products, a division dedicated to advancing the evaluation and development of therapeutics for the diagnosis and treatment of rare diseases. Dr. Groft served as the Executive Director of the National Commission on Orphan Diseases, from 1987 to 1989. In all, Dr. Groft has played a prominent role in the field of rare diseases and orphan drugs for more than 30 years.

Preface

"All interest in disease and death is only another expression of interest in life."

—Thomas Mann

For a few decades now, I have been interested in writing a book that treats the rare diseases as a separate specialty within medicine. Most of my colleagues were not particularly receptive to the idea. Here is a sample of their advice, paraphrased: "Don't waste your time on the rare diseases. There are about 7000 rare diseases that are known to modern medicine. The busiest physician, over the length of a long career, will encounter only a tiny fraction of the total number of rare diseases. Surely, an attempt to learn them all would be silly; an exercise in purposeless scholarship. Furthermore, each rare disease accounts for so few people, it is impractical to devote much research funding to these medical outliers. To get the most bang for our bucks, we should concentrate our research efforts on the most common diseases: heart disease, cancer, diabetes, Alzheimer's disease, and so on."

Other colleagues questioned whether rare diseases are a legitimate area of study: "Rare diseases do not comprise a biologically meaningful class of diseases. They are simply an arbitrary construction, differing from common diseases by a numeric accident. A disease does not become scientifically interesting just by being rare." For some of my colleagues, the rare diseases are mere aberrations, best ignored.

In biology, there are no outliers; no circumstances that are rare enough to be ignored. Every disease, no matter how rare, operates under the same biological principles that pertain to common diseases. In 1657, William Harvey, the noted physiologist, wrote: "Nature is nowhere accustomed more openly to display her secret mysteries than in cases where she shows tracings of her workings apart from the beaten paths; nor is there any better way to advance the proper practice of medicine than to give our minds to the discovery of the usual law of nature, by careful investigation of cases of rarer forms of disease."

We shall see that the rare diseases are much simpler, genetically, than the common diseases. The rare disease can be conceived as controlled experiments of nature, in which everything is identical in the diseased and the normal organisms, except for one single factor that is the root cause of the ensuing disease. By studying the rare diseases, we can begin to piece together the more complex parts of common diseases.

The book has five large themes that emerge, in one form or another, in every chapter.

1. In the past two decades, there have been enormous advances in the diagnosis and treatment of the rare diseases. In the same period, progress in the common diseases has stagnated. Advances in the rare diseases have profoundly influenced the theory and the practice of modern medicine.
2. The molecular pathways that are operative in the rare diseases contribute to the pathogenesis of the common diseases. Hence, the rare diseases are not the exceptions to the general rules that apply to common diseases; the rare diseases are the exceptions upon which the general rules of common diseases are based.
3. Research into the genetics of common diseases indicates that these diseases are much more complex than we had anticipated. Many rare diseases have simple genetics, wherein a mutation in a single gene accounts for a clinical outcome. The same simple pathways found in the rare diseases serve as components of the common diseases. If the common diseases are the puzzles that modern medical researchers are mandated to solve, then the rare diseases are the pieces of the puzzles.
4. If we fail to study the rare diseases in a comprehensive way, we lose the opportunity to see the important biological relationships among diseases consigned to non-overlapping subdisciplines of medicine.
5. Every scientific field must have a set of fundamental principles that describes, explains, or predicts its own operation. Rare diseases operate under a set of principles, and these principles can be inferred from well-documented pathologic, clinical, and epidemiologic observations.

Today, there is no recognized field of medicine devoted to the study of rare diseases; but there should be.

CONTENT AND ORGANIZATION OF THE BOOK

There are three parts to the book. In Part I (Understanding the Problem), we discuss the differences between the rare and the common diseases, and why it is crucial to understand these differences. To stir your interest, here are just a few of the most striking differences: (1) most of the rare diseases occur in early childhood, while most of the common diseases occur in adulthood; (2) the genetic determinants of most rare diseases have a simple Mendelian pattern, dependent on whether the disease trait occurs in the father, or mother, or both. Genetic influences in the common diseases seldom display Mendelian inheritance; (3) rare diseases often occur as syndromes involving multiple organs through seemingly unrelated pathological processes. Common diseases usually involve a single organ or involve multiple organs involved by a common pathologic process.

The most common pathological conditions of humans are aging, metabolic diseases (including diabetes, hypertension, and obesity), diseases of the heart and vessels, infectious diseases, and cancer. Each of these disorders is characterized by pathologic processes that bear some relation to the processes that operate in rare diseases. In Part II (Rare Lessons for Common Diseases), we discuss the rare diseases that have helped us understand the common diseases. Emphasis is placed on the enormous value of rare disease research. We begin to ask and answer some of the fundamental questions raised in Part I. Specifically, how is it possible for two diseases to share the same pathologic mechanisms without sharing similar genetic alterations? Why are the common diseases often caused, in no small part, by environmental (i.e., non-genetic) influences, while the rare disease counterparts are driven by single genetic flaws? Why are the rare diseases often syndromic (i.e., involving multiple organs with multiple types of abnormalities and dysfunctions), while the so-called complex common diseases often manifest in a single pathological process? In Part II, we will discuss a variety of pathologic mechanisms that apply to classes of rare diseases. We will also see how these same mechanisms operate in the common diseases. We will explore the relationship between genotype and phenotype, and we will address one of the most important questions in modern disease biology: "How is it possible that complex and variable disease genotypes operating in unique individuals will converge to produce one disease with the same biological features from individual to individual?"

In Part III (Fundamental Relationships between Rare and Common Diseases), we answer the as-yet unanswered questions from Part I, plus the new questions raised in Part II. The reasons why rare diseases are different from common diseases are explained. The convergence of pathologic mechanisms and clinical outcome observed in rare diseases and common diseases, as it relates to the prevention, diagnosis, and treatment of both types of diseases, is described in detail.

The book includes a scientific rationale for funding research in the rare diseases. Currently, there is a vigorous lobbying effort, launched by coalitions of rare disease organizations, to attract research funding and donations. Funding for the rare diseases has always been small, relative to the common diseases. Funding agencies find it impractical to devote large portions of their research budget to the rare diseases, while so many people are suffering from the common diseases. As it turns out, direct funding of the common diseases has not been particularly cost effective. It is time for funders to re-evaluate their goals and priorities.

Laypersons advocating for rare disease research almost always appeal to our charitable instincts, hoping that prospective donors will respond to the plight of a few individuals. Readers will learn that such supplications are unnecessary and misdirect attention from more practical arguments. When rare diseases are funded, everyone benefits. We will see that it is much easier to find effective targeted treatments for the rare diseases than for common diseases. Furthermore, treatments that are effective against rare diseases will almost always find a place in the treatment of one or more common diseases. This assertion is not based on wishful thinking, and is not based on extrapolation from a few past triumphs wherein some treatment overlap has been found in rare and common diseases. The assertion is based on the observation that rare diseases encapsulate the many biological pathways that drive, in the aggregate, our common diseases. This simple theme is described and justified throughout the book. Society will benefit when we increase funding for the rare diseases with the primary goal of curing the common diseases. The final chapter of this book discusses promising new approaches to rare disease research.

RULES GOVERNING THE RARE DISEASES

What is the value of learning a lot of facts about the rare diseases if this information cannot deepen our understanding of medicine? The genetics of human disease is incredibly complex. As we learn more and more about the human genome, we find ourselves less able to cope with all the incoming information. We need to have some way of relating intangible and invisible molecular complexities to the stark, clinical reality of human diseases. A good way to understand the complex data is by building generalizations. When we generalize, we force ourselves to think about biological relationships and their clinical consequences. Suddenly, we are no longer passive collectors of information; we become innovators, creators, and puzzle solvers. Facts that were formerly too esoteric to recall are burned into our memories as vital clues in a vast biological mystery. For the clinically minded, generalizations drive down the complexity of genetics and molecular pathology. For the research minded, generalizations are testable hypotheses; inspirational fodder for the next grant application.

The text is sprinkled with general rules that can be inferred from the chapter contents. **The term "rule" herein means observations that are generally true; not natural laws. In many cases, counter-examples and constraints are also provided. The rules are primarily intended to encourage readers to think critically about the subject matter.** Readers will find that the disease descriptions in the chapters will have greater meaning if the disease can be associated with a biological rule. Every rule appearing in the text is listed again in Appendix II, where they are numbered by chapter and section. The reader is encouraged to browse through the list. When a provocative rule is encountered, the reader can easily refer back to the chapter to read a full discussion.

WHO SHOULD READ THIS BOOK?

This book is written primarily as a text for healthcare students, professionals, and for biomedical researchers. Advocates for the rare diseases will find that this book provides a practical scientific rationale to support increased funding and new initiatives into the rare diseases and orphan drug development. For funders and administrators who have poured vast resources into genetic infrastructure, such as the Human Genome Project, they will find that the rare diseases are the bridge leading from genome databases to practical clinical innovations.

My hope is that the book will reach non-biologists working on large, multidisciplinary biomedical projects; so-called Big Science. Systems biologists, computational biologists, biomedical computer scientists, data modelers, biostatisticians, bioinformaticists, and biomedical informaticians often sit on the sidelines of biomedical research. Too often, brilliant professionals serve in complex biomedical projects without fully realizing the potential of their personal contributions. One of the purposes of this book is to provide a practical perspective of modern disease research; one that clarifies the relationships between genes, pathogenesis, and clinical phenotype.

Readers will encounter specialized terminology from the fields of genetics, pathology, microbiology, cellular physiology, and anatomy. Rather than devote chapter space to defining terms, a large glossary is provided. Glossary terms appearing for the first time within the text are labeled. In addition to defining terms, glossary items are provided with a detailed explanation of their relevance to the themes developed within the book. The glossary can be enjoyed as a stand-alone text.

HOW TO READ THIS BOOK

Because this book attempts to establish the general biological rules that govern the rare diseases, it is necessary to provide examples of diseases to which those rules apply. This book contains descriptions of the genetic and clinical features of hundreds of rare diseases. Laypersons reading this book should take solace in knowing that the most seasoned medical professionals will be unfamiliar with many of the disease entities described herein. To facilitate the reader's understanding of fundamental principles, I have written the book in such a way that many of the burdensome technicalities can be compartmentalized and saved for later reading after the main points of the book are absorbed. Here is how it works. The book contains about 130 biological rules, with each rule followed by a brief, non-technical rationale that explains why the rule makes sense. Rules and rationales are indented and displayed in bold font easily distinguished from the surrounding text. Each rule and rationale is followed by a detailed discussion with examples. Readers are encouraged to read through the rules, dwelling on the full discussions that have particular relevance to their own interests. Those readers who seek an in-depth treatment of the book's subject are welcome to study the text cover to cover. In addition, the text introduces specialized terminology that may be unfamiliar to many readers. Throughout the book, short definitions of terms are provided as parenthesized comments. Terms that require in-depth explanations are discussed in the glossary. Non-biologists should not be intimidated by the highly specialized nature of the topics included here. You may need to consult a dictionary from time to time, but if you can read the science articles in *The New York Times*, then you will be able to read and understand every chapter in this book.

Part I

Understanding the Problem

Chapter 1

What are the Rare Diseases, and Why do we Care?

1.1 THE DEFINITION OF RARE DISEASE

"The beginnings and endings of all human undertakings are untidy."

—John Galsworthy

In the U.S., Public Law 107-280, the Rare Diseases Act of 2002 states: "Rare diseases and disorders are those which affect small patient populations, typically populations smaller than 200,000 individuals in the United States" [1]. Since the population of the U.S. is about 314 million (in 2013), this comes to about one case for every 1570 persons. This is not too far from the definition recommended by the European Commission on Public Health; fewer than one in 2000 people. It is important to have numeric criteria for the rare diseases, because special laws exist in the U.S. and in Europe to stimulate research and drug development for diseases that meet the criteria for being "rare" (see Section 14.2). Unfortunately, it is very difficult to know, with any certainty, the specific prevalence or incidence of any of the rare diseases (see Glossary items, Prevalence, Incidence). A certain percentage of the cases will go unreported, or undiagnosed, or misdiagnosed. Though it is impossible to obtain accurate and up-to-date prevalence data on every rare disease, in the U.S. the National Institutes of Health has estimated that rare diseases affect, in aggregate, 25–30 million Americans [2].

There seems to be a growing consensus that there are about 7000 rare diseases [3]. Depending on how you choose to count diseases, this may be a gross underestimate. There are several thousand inherited conditions with a Mendelian inheritance pattern [4]. To these, we must add the different types of cancer. Every cancer, other than the top five or ten most common cancers, occurs with an incidence much less than 200,000 and would qualify as a rare disease. There are more than 3000 named types of cancer, and many of these cancers have well-defined subtypes, with their own morphologic, clinical, or genetic characteristics. Including defined subtypes, there are well over 6000 rare types of cancer [5–8]. Regarding the rare infectious diseases, well over 1400 different infectious organisms have been reported in the literature [9].

Rare Diseases and Orphan Drugs. http://dx.doi.org/10.1016/B978-0-12-419988-0.00001-8

A single infectious organism may manifest as several different named conditions, each with its own distinctive clinical features. For example, leishmaniasis, an infectious disease that is common in Africa but rare in Europe, may present in one of four different forms (cutaneous, visceral, diffuse cutaneous, and mucocutaneous). When we add in the many rare nutritional, toxic, and degenerative diseases that occur in humans, the consensus estimate of the number of rare diseases seems woefully inadequate. Nonetheless, the low-ball "7000" number tells us that there are many rare diseases; way too many for any individual to fully comprehend.

The rare diseases are sometimes referred to as orphan diseases. The term is apt for several reasons. First, the term "orphan" applies to children, and it happens that neonates, infants, and children are at highest risk for the most devastating rare diseases. Second, the concept of an "orphan disease" implies a lack of stewardship. For far too long, the rare diseases were neglected by clinicians, medical researchers, the pharmaceutical industry, and society in general (see Glossary item, Neglected disease). The rare diseases manifested as strange and often disfiguring maladies that occurred without any obvious cause. Primitive and not-so-primitive cultures have attributed a supernatural origin for the rare diseases of childhood. It was common for children with disfiguring diseases to be confined in homes or institutions and hidden from society. Over the past 40 years, these conditions have changed drastically, and for the better. A confluence of political, social, and scientific enlightenments has led to stunning advances in the field of rare diseases, and these advances have spilled over into the common diseases. If the rare diseases are orphans, then orphans have been adopted by caring and competent guardians.

Today, there are effective treatments for many of the rare diseases. Hence, it is crucial to make correct diagnoses, at early stages of disease, before irreversible organ damage develops.

1.1.1 Rule—Rare diseases are easily misdiagnosed, and are often mistaken for a common disease or for some other rare disease.

Brief Rationale—It is impossible for any physician to attain clinical experience with more than a small fraction of the total number of rare diseases. When it comes to rare diseases, every doctor is a dilettante.

In 1993, Reggie Lewis was the 27-year-old captain of the Boston Celtics basketball team. Mr. Lewis enjoyed good health until the moment when he collapsed during a basketball game. Mr. Lewis' collapse attracted the attention of cardiologists across the nation. A medical team assembled by the New England Baptist Hospital opined that Mr. Lewis had cardiomyopathy, a life-threatening condition that would require Mr. Lewis to retire from basketball immediately. A second team of experts, assembled at the Brigham and Women's Hospital, disagreed. They rendered a diagnosis of vaso-vagal fainting, a benign condition. A third team of experts, from St. John's Hospital in Santa Monica, California, was non-committal. The Santa Monica team suggested that Mr. Lewis play basketball, but with a heart monitor attached to his body. With three discordant diagnoses, Mr. Lewis decided to take his chances, continuing his athletic career. Soon thereafter, Lewis died, quite suddenly, from cardiomyopathy, while playing basketball [10].

A few dozen common diseases account for the majority of ailments encountered in the typical medical practice. When a physician encounters a rare disease for the first time, he or she may be no more capable than a medical student to reach a correct diagnosis. The presenting symptoms of many rare diseases are disarmingly pedestrian (e.g., failure to thrive, weakness, fatigability, etc.) and the first reaction of any physician might be to make a tentative diagnosis of a common disease. Only after treatment fails, and symptoms do not resolve, are alternate diagnoses considered. It is not unusual for an accurate diagnosis to follow numerous visits to several physicians [11]. In the interim, the disease worsens, the medical bills grow, and the emotional distress builds.

1.2 REMARKABLE PROGRESS IN THE RARE DISEASES

"Most [rare diseases] result from a dysfunction of a single pathway due to a defective gene. Understanding the impact of a single defect may therefore yield insights into the more complex pathways involved in common diseases which are generally multifactorial."

—Segolene Ayme and Virginie Hivert, from Orphanet [12].

Excluding genes causing rare cancers, more than 2000 genes have been linked to 2000 rare diseases [12]. In most cases, these links are presumed to be causal (i.e., mutations in the gene lead to the development of the disease). Virtually every gene known to cause a rare disease was discovered within the past half century. The diseases whose underlying causes were known, prior to about 1960, numbered in the hundreds, and the majority of these well-understood diseases were caused by infectious organisms (see Glossary item, Infectious disease).

Progress in the genetic diseases greatly accelerated in the 1960s, and the earliest advances came to the group of diseases known as inborn errors of metabolism. Treatments consisted of avoidance of substances that could not be metabolized in affected individuals or supplementations for missing metabolites (e.g., avoidance of phenylalanine in newborns with phenylketonuria, supplements of thyroid hormone in congenital hypothyroidism, avoidance of galactose in newborns with galactosemia, supplementation with biotin in newborns with biotinidase deficiency, specially formulated low protein diets for newborns with maple syrup urine disease, and so on).

Some of the groundbreaking advances in rare disease research include the 1956 discovery of the specific molecular alteration in hemoglobin that causes sickle cell disease [13,14]; and the identification of the cystic fibrosis gene in 1989 [15]. In 2007, Leber congenital amaurosis, a form of inherited blindness, was the first disease to be treated, with some clinical improvement, using genetic engineering. The mutated RPE65 gene was replaced with a functioning gene [16]. Partial vision was obtained in individuals who were previously blind. It remains to be seen whether genetic engineering will ever restore adequate and long-term vision to individuals with Leber congenital amaurosis [17]. It is noteworthy that the test case was made on an extremely rare form of blindness, not a common form such as macular degeneration. The reason why rare diseases are superior to common diseases, when developing innovative treatment methods, is a topic that will be discussed in Chapter 14.

Currently, drug development for the rare diseases is far exceeding anything seen in the common diseases. Since 1983, more than 350 drugs have been approved to treat rare diseases [18]. By 2011, the U.S. Food and Drug Administration had designated over 2300 medicines as orphan drugs (see Glossary item, Orphan drug). That same year, 460 drugs were in development to treat or prevent the rare diseases [18]. Meanwhile, in Europe, 20% of the innovative products with marketing authorization were developed for rare diseases [12].

As we shall discuss in later chapters, many factors have contributed to the remarkable advances in the rare diseases. The upshot of these advances is that we know much more about the rare diseases, in terms of pathogenesis and treatment, than we know about the common diseases (see Glossary item, Pathogenesis). At this point, there is every expectation that the greatest breakthroughs in understanding the general mechanisms of disease processes will come from research on the rare diseases [19].

Let us briefly examine a few general statements that will be developed in ensuing chapters.

1.2.1 Rule—Rare diseases are not the exceptions to the general rules of disease biology; they are the exceptions upon which the general rules are based.

Brief Rationale—All biological systems must follow the same rules. If a rare disease is the basis for a general assertion about the biology of disease, then the rule must apply to the common diseases.

Every rare disease tells us something about the normal functions of organisms. When we study a rare hemoglobinopathy, we learn something about the consequences that befall when normal hemoglobin is replaced with an abnormal hemoglobin. This information leads us to a deeper understanding of the normal role of hemoglobin. Likewise, rare urea cycle disorders, coagulation disorders, metabolic disorders, and endocrine disorders have taught us how these functional pathways operate under normal conditions (see Glossary item, Pathway) [19].

1.2.2 Rule—Every common disease is a collection of different diseases that happen to have the same clinical phenotype (see Glossary item, Phenotype).

Brief Rationale—Numerous causes and pathways may lead to the same biological outcome.

Consider the heart attack; its risk of occurrence is elevated by many factors. Obesity, poor diet, smoking, stress, lack of exercise, hypertension, diabetes, disorders of blood lipid metabolism, infections, male gender; they all contribute to heart attacks. Regardless of the contributing factors, a common event precedes and causes the heart attack; the blockage of a coronary artery. Blockage is often caused by an atherosclerotic plaque. Consequently, rare inherited conditions that produce atherosclerotic plaques can produce the common heart attack (e.g., inherited disorders of lipid metabolism). We infer that for every common disease, there are rare, inherited diseases that account for a small subset of cases. This topic will be revisited and expanded in Section 12.2.

1.2.3 Rule—Rare diseases inform us how to treat common diseases.

Brief Rationale—When we encounter a common disease, we look to see what pathways are dysfunctional, and we develop a rational approach to prevention, diagnosis, and treatment based on experiences drawn from the rare diseases that are driven by the same dysfunctional pathways.

Many heart attacks are caused by atherosclerotic plaque blocking a coronary artery. Many conditions produce atherosclerotic plaque, but a rare condition known as familial hypercholesterolemia is associated with some cases of coronary atherosclerosis that occur in young individuals. Studies on familial hypercholesterolemia led to the finding that statins inhibit the rate-limiting enzyme in cholesterol synthesis (hydroxymethylglutaryl coenzyme A), thus reducing the blood levels of cholesterol and blocking the formation of plaque. The treatment of a pathway operative in a rare form of hypercholesterolemia has become the most effective treatment for commonly occurring forms of hypercholesterolemia, and a mainstay in the prevention of the common heart attack [20]. This topic will be revisited and expanded in Section 13.2.

REFERENCES

1. Rare Diseases Act of 2002, Public Law 107-280, 107th U.S. Congress, November 6, 2002.
2. FAQ About Rare Diseases. National Center for Advancing Translational Sciences, National Institutes of Health. http://www.ncats.nih.gov/about/faq/rare/rare-faq.html, viewed on October 24, 2013.
3. M.J. Field, T. Boat, Rare diseases and orphan products: accelerating research and development. Institute of Medicine (US) Committee on Accelerating Rare Diseases Research and Orphan Product Development, 2010. The National Academies Press, Washington, DC. Available from: http://www.ncbi.nlm.nih.gov/books/NBK56189/.
4. OMIM. Online Mendelian Inheritance in Man. Available from: http://omim.org/downloads, viewed June 20, 2013.
5. Berman JJ. Modern classification of neoplasms: reconciling differences between morphologic and molecular approaches. BMC Cancer 5:100, 2005. Available from: http://www.biomedcentral.com/1471-2407/5/100.
6. Berman JJ. Tumor taxonomy for the developmental lineage classification of neoplasms. BMC Cancer 4:88, 2004.

7. Berman JJ. Tumor classification: molecular analysis meets Aristotle. BMC Cancer 4:10, 2004. Available from: http://www.biomedcentral.com/1471-2407/4/10.
8. Berman JJ. Neoplasms: Principles of Development and Diversity. Jones & Bartlett, Sudbury, 2009.
9. Berman JJ. Taxonomic Guide to Infections Diseases: Understanding the Biologic Classes of Pathogenic Organisms. Academic Press, Waltham, 2012.
10. Altman LK. After a highly publicized death, second-guessing second opinions. The New York Times, August 3, 1993.
11. Rare Diseases and Scientific Inquiry. Developed by BSCS Under a Contract from the National Institutes of Health, Office of Rare Diseases Research, 2011.
12. Report on Rare Disease Research its Determinants in Europe and the Way Forward. Ayme S, Hivert V, eds. INSERM, May 2011. Available from: http://asso.orpha.net/RDPlatform/upload/file/RDPlatform_final_report.pdf, viewed February 26, 2013.
13. Pauling L, Itano HA, Singer SJ, Wells IC. Sickle cell anemia, a molecular disease. Science 110:543–548, 1949.
14. Ingram VM. A specific chemical difference between globins of normal and sickle-cell anemia hemoglobins. Nature 178:792–794, 1956.
15. Riordan JR, Rommens JM, Kerem B, Alon N, Rozmahel R, Grzelczak Z, et al. Identification of the cystic fibrosis gene: cloning and characterization of complementary DNA. Science 245:1066–1073, 1989.
16. Hauswirth WW, Aleman TS, Kaushal S, Cideciyan AV, Schwartz SB, Wang L, et al. Treatment of leber congenital amaurosis due to RPE65 mutations by ocular subretinal injection of adeno-associated virus gene vector: short-term results of a phase I trial. Hum Gene Ther 19:979–990, 2008.
17. Cideciyan AV, Jacobson SG, Beltran WA, Sumaroka A, Swider M, Iwabe S, et al. Human retinal gene therapy for Leber congenital amaurosis shows advancing retinal degeneration despite enduring visual improvement. Proc Natl Acad Sci USA 110:E517–E525, 2013.
18. Orphan Drugs in Development for Rare Diseases. 2011 Report. America's Biopharmaceutical Research Companies. Available from: http://www.phrma.org/sites/default/files/pdf/rarediseases2011.pdf, viewed July 14, 2013.
19. Wizemann T, Robinson S, Giffin R. Breakthrough Business Models: Drug Development for Rare and Neglected Diseases and Individualized Therapies Workshop Summary. National Academy of Sciences, 2009.
20. Stossel TP. The discovery of statins. Cell 134:903–905, 2008.

Chapter 2

What are the Common Diseases?

2.1 THE COMMON DISEASES OF HUMANS, A SHORT BUT TERRIFYING LIST

"Not everything that counts can be counted, and not everything that can be counted counts."

—William Bruce Cameron

There are about 7 billion humans living in the world today, with about 57 million people dying each year [1,2]. There are about 312 million persons residing in the U.S. [3,1]. The U.S. Central Intelligence Agency estimates that U.S. crude death rate is 8.36 per 1000 and the world crude death rate is 8.12 per 1000 [4]. This translates to 2.6 million people dying in 2011 in the U.S. These figures are just a tad higher than the total U.S. deaths calculated independently from the 2003 National Vital Statistics Report [5]. Authoritative death statistics correlate surprisingly well with the widely used rule of thumb that 1% of the human population dies every year. What diseases account for all of these deaths?

Let us take a look at diseases that cause the greatest number of human deaths worldwide.

Worldwide deaths in 2008, from the World Health Organization [2]:

TOTAL DEATHS WORLDWIDE	56,888,289
1. Cardiovascular diseases	17,326,646
2. Infectious and parasitic diseases	8,721,166
3. Malignant neoplasms	7,583,252
4. Respiratory infections	3,533,652
5. Diabetes mellitus	1,255,585
6. Alzheimer and other dementias	539,948
7. Other neoplasms	188,227

Rare Diseases and Orphan Drugs. http://dx.doi.org/10.1016/B978-0-12-419988-0.00002-X

U.S. deaths in 2003, from National Vital Statistics Report [5]:

TOTAL DEATHS IN THE U.S. (calculated from table)	2,512,873
1. Diseases of heart	596,339
2. Malignant neoplasms	575,313
3. Chronic lower respiratory diseases	143,382
4. Cerebrovascular diseases	128,931
5. Accidents (unintentional injuries)	122,777
6. Alzheimer's disease	84,691
7. Diabetes mellitus	73,282

There is much to be learned from these two short lists. We see that although there are thousands of human diseases, many of which are capable of causing death, only a few diseases account for the bulk of death occurring in populations. For both U.S. deaths and worldwide deaths, the first three conditions on each list account for more than 50% of the total number of deaths. The top seven conditions account for 70% of the total number of deaths worldwide.

2.1.1 Rule—A small number of diseases account for most instances of morbidity or mortality.

Brief Rationale—Pareto's principle applies to biological systems.

Pareto's principle, also known as the 80/20 rule, holds that a small number of causes will account for the vast majority of observed instances of real-world distributions (see Glossary item, Pareto's principle). For example, a small number of rich people account for the majority of wealth. A few troublemakers in a classroom may draw the bulk of a teacher's attention. Just two countries, India and China, account for 37% of the world population. Within most countries, a small number of provinces or geographic areas contain the majority of the population of a country (e.g., east and west coastlines of the U.S.). A small number of books, compared with the total number of published books, account for the majority of book sales.

In the realm of medicine, a small number of diseases account for the bulk of human morbidity and mortality. For example, two common types of cancer, basal cell carcinoma of skin and squamous cell carcinoma of skin, account for about 1 million new cases of cancer each year in the U.S. This is approximately the sum total of all other types of cancer combined.

Sets of data that follow Pareto's principle are often said to follow a Zipf distribution, or a power law distribution (see Glossary item, Zipf distribution). These types of distributions are not tractable by standard statistical descriptors because they do not produce a symmetric distribution around a central peak. Simple measurements such as average and standard deviation have very little practical meaning when applied to Zipf distributions. Furthermore, none of the statistical inferences built upon an assumption of a normal or Gaussian distribution will apply to data sets that observe Pareto's principle.

2.1.2 Rule—Funding for disease research adheres to Pareto's principle.

Brief Rationale—The diseases that kill the greatest number of individuals receive the highest levels of funding, in the simple-minded expectation that advances against common diseases will provide the greatest benefit to society.

National Institutes of Health (NIH) spending, by institute, for the budget year 2010, based on data from the American Association for the Advancement of Science [6], is:

National Cancer Institute	$5.295 billion
National Institute of Heart, Lung, and Blood	$3.213 billion
National Institute of Allergy and Infectious Diseases	$4.690 billion
Total NIH budget	$32.127 billion

The NIH comprises 27 institutes and centers. The top three institutes account for 41% of NIH budget.

Let us look at cancer funding, based on cancer incidence.

Cancer funding from the National Cancer Institute in millions of dollars, and listed in decreasing order of cancer incidence, for 2010 [7], is:

Lung	281.9
Prostate	300.5
Breast	631.2
Colorectal	270.4
Bladder	22.6
Melanoma	102.3
Non-Hodgkin lymphoma	122.4
Kidney	44.6
Thyroid	15.6
Endometrial (uterine)	14.2

Once again, Pareto's principle applies. The top four sites of cancer occurrence are the top four recipients of funding, accounting for 82% of the funding provided to the top 10 cancer sites.

Whenever we look at death rates, we find that a few common diseases have a disproportionate effect on human mortality. For example, using WHO data, a drop in the cardiovascular death rate of a mere 3% would be equivalent, in terms of lives saved, to eliminating all deaths due to Alzheimer's disease plus all other dementias.

Is there any wonder that the bulk of research spending at NIH is directed toward cancer, cardiovascular diseases, and infectious diseases? Actually, there are reasons that weigh against spending the bulk of research funds on common diseases. These reasons will be discussed throughout this book. For now, let us consider how the argument for investing in common diseases considerably weakens when we consider the effect of age-at-diagnosis on life expectancy.

2.1.3 Rule—The cancers that account for the majority of cancer deaths occur in elderly individuals.

Brief Rationale—Common diseases are caused by cellular events that accumulate over time or that arise over time. Hence, the chance of developing a common disease increases steadily as individuals age.

For example, let us compare the incidence of cancer in children aged 4 and under compared with the incidence of cancer in adults aged 85 and older. In England, males and females 4 and under have a cancer incidence of 19.3/100,000 population and 17.4/100,000 population, respectively [8]. We see here that females have a lower rate of cancer than males for this age group. In the same statistical survey, the incidence of cancer in males and females 85 and older is 3393.5/100,000 and 2095.3/100,000 [8]. Males and females 85 and older have a cancer incidence 176 and 120 times that seen with male and female children 4 years and under, respectively.

2.1.4 Rule—The most common causes of death, if eliminated entirely, will not greatly increase human life expectancy.

Brief Rationale—Elderly individuals who do not die from one common disease will likely die from some other common disease.

In 1978, Tsai and coworkers calculated the increase in life expectancy that would occur if cancer was eliminated as a human disease. They predicted that the elimination of cancer would extend human life by no more than 2.5 years [9].

Readers may be surprised to know that if we were to finally win the war against cancer, the increase in life expectancy would only equal 2.5 years. Although we would all appreciate having an average of 2.5 years added to our lifespans, we must understand that differences of life expectancy of 2.5 years are found among populations living in different countries. For example, life expectancy in the U.S. is 78.6 years. The life expectancies in Canada and Italy are 81.6 and 81.9 years, and life expectancies in Australia and Japan are 82.0 and 84.2 years [10]. Had we been born in these countries, or in any of the developed European countries, our life expectancies would be extended by about as much as we might expect to gain by eliminating cancer.

Why are the benefits so small? It comes down to Pareto's principle. Most of the deaths from cancer occur from a few common diseases, and these diseases occur almost exclusively in elderly patients. The rare cancers receive a small portion of NIH funding, but they strike children in disproportionate numbers.

2.2 THE RECENT DECLINE IN PROGRESS AGAINST COMMON DISEASES

"Despite large public investments in genome-wide association studies of common human diseases, so far, few gene discoveries have led to applications for clinical medicine or public health."

—Idris Guessous, Marta Gwinn and Muin J. Khoury in 2009 [11]

We like to think that we are living in an era of rapid scientific advancement; more rapid than any prior era in human history. This is nonsense. In the field of medicine, the 50-year progress between 1913 and 1963 greatly exceeded progress between 1964 and 2014. By 1921, we had insulin. Over the next four decades, we developed antibiotics effective against an enormous range of infectious diseases, including tuberculosis. Civil engineers prevented a wide range of common diseases by providing a clean water supply and improved waste management. Safe methods to preserve food, such as canning, refrigeration, and freezing, saved countless lives. In 1941, Papanicolaou introduced the smear technique to screen for precancerous cervical lesions, resulting in a 70% drop in the death rate from uterine cervical cancer, one of the leading causes of cancer deaths in women (see Glossary item, Precancer, Precancerous condition). By 1947, we had overwhelming epidemiologic evidence that cigarettes caused lung cancer. The first polio vaccine and the invention of oral contraceptives came in 1954. By the mid-1950s, the sterile surgical technique was widely practiced, bringing a precipitous drop in post-surgical and post-partum deaths. The elucidation of the molecular basis of sickle cell anemia came in 1956 [12,13]. The major discoveries of the fundamental chemistry and biology of DNA came in the 1950s.

Perhaps the greatest advances in the common diseases, in the past several decades, have been in the realm of heart disease. The role of statins in the prevention of heart attacks and strokes, improvements in cardiac surgery, and the use of stents to open narrowed arteries are major therapeutic success stories (see Glossary item, Brain attack). Nonetheless, few would argue that the benefits from these interventional measures would be dwarfed by the benefits enjoyed by individuals who adapted healthy eating habits, exercised regularly, attained a trim habitus, and avoided smoking; sensible life choices available prior to 1950.

The National Cancer Institute is the largest of the research institutes at the NIH, receiving about 10% of the total NIH budget. Despite intense effort by generations of medical scientists, the cancer death rate today is about the same as it was in 1970 [14]. Though there has been a drop in the cancer death rate that has extended from the last decade of the twentieth century to the present, that drop was preceded by a rise in the cancer death rate from 1970 to the early 1990s. The two-decade rise followed by a two-decade drop was shaped by both the rise and consequent fall of smoking. Countries that had a drop in smoking prior to the U.S. saw a drop in cancer death rates prior to the U.S. drop. Countries in which smoking is on the increase have increasing rates of cancer death. For the common cancers (lung, colon, prostate, breast, pancreas, esophagus), progress has been impressive, extending survival times after diagnosis; but the overall death rate from the common cancers has not changed appreciably.

The Human Genome Project is a massive bioinformatics project in which multiple laboratories contributed to sequencing the 3 billion base pairs encoding the full, haploid human genome (see Glossary item, Haploid). The project began its work in 1990, a draft human genome was prepared in 2000, and a nearly complete genome was announced in 2003, marking the start of the so-called post-genomics era. One of the purposes of the project was to find the genetic causes of common diseases. Although we have learned much about the genetics of the common diseases, most of what we have learned has only served to teach us that the genetics of common diseases are much more complex than we had anticipated. Common diseases are associated with hundreds of gene variations, and the gene variations that we have found explain only a small portion of the observed heritability of common diseases [15,16]. Early studies using polygenic variants to predict risk of developing common diseases have not been clinically useful [15].

If the rate of scientific accomplishment were dependent upon the number of scientists on the job, you would expect that progress would be accelerating, not decelerating. According to the National Science Foundation, 18,052 science and engineering doctoral degrees were awarded in the U.S. in 1970. By 1997, that number had risen to 26,847, nearly a 50% increase in the number of graduates at the highest level of academic training [17]. The growing work force of scientists failed to advance science at rates achieved in an earlier era, with fewer workers.

While the overall rate of medical progress has slowed over the past half century, research funding has accelerated. In 1953, according to the National Science Foundation, the total U.S. expenditure on research and development was $5.16 billion, expressed in current dollar values. In 1998, that number had risen to $227.173 billion, greater than a 40-fold increase in research spending [17]. There has not been a commensurate 40-fold increase in scientific discoveries.

The U.S. Department of Health and Human Services has published a sobering document entitled "Innovation or Stagnation: Challenge and Opportunity on the Critical Path to New Medical Products" [18]. The authors note that fewer and fewer new medicines and medical devices are reaching the Food and Drug Administration. Significant advances in genomics, proteomics, and nanotechnology have not led to equivalent advances in the treatment of common diseases. The last quarter of the twentieth century has been described as the "era of Brownian motion in health care" [19]. Wurtman and Bettiker, in their review of medical treatments, commented that, "Successes have been surprisingly infrequent during the past three decades. Few effective treatments have been discovered for the diseases that contribute most to mortality and morbidity" [20].

2.3 WHY MEDICAL SCIENTISTS HAVE FAILED TO ERADICATE THE COMMON DISEASES

"One does not discover new lands without consenting to lose sight of the shore for a very long time."

—Andre Gide

Suppose we lived in a society where every adolescent and adult smoked two or three packs of cigarettes every day. Of course, the incidence of lung cancer, chronic obstructive pulmonary disease, emphysema, and other smoking-associated disorders would skyrocket. Still, we would be less likely to associate smoking with common diseases than we would be if only a small proportion of society were smokers. The cornerstone of research into the causal mechanisms of disease involves comparing disease occurrences in a group of individuals who share a particular trait (e.g., smoking), against a group of individuals who lack the trait (e.g., non-smokers). When everyone smokes, there is no basis to make a comparison.

Suppose there were familial clusters of lung cancer (i.e., some families at higher risk than others). You could start checking to see if high-risk families have certain sets of genes that account for lung cancer heritability. Imagine that you start finding hundreds of gene variants that seem to separate the high-risk families from the low-risk families. How do you begin to determine which of those genes contribute to the pathogenesis of lung cancer? Keep in mind that, because high-risk families are small study populations, you may find it impossible to assign any statistical significance to your findings.

In the last decades of the twentieth century, scientists hoped that the common diseases, like the rare diseases, were each caused by a single, disease-specific genetic mutation. Once the mutation was found, it could be targeted with a drug. Most scientists today will admit that the common diseases of humans are much more complex than they had ever imagined.

2.3.1 Rule—We may have reached the limit by which we can understand the common diseases through direct genetic studies.

Brief Rationale—The common diseases of humans are complex, and biological complexity cannot be calculated, predicted or solved, even with supercomputers.

An objective review of the genetics of common diseases yields only bad news. With no exceptions, the common diseases are genetically complex. Attempts at predicting the behavior of common diseases, based on detailed, yet incomplete, knowledge of their complex genetic attributes, have led to failure after failure [21–23].

Not to be discouraged, data analysts believe that with the right algorithm, and the right supercomputer, the complexities of common diseases can be predicted. This belief is based, in no small part, on the assumption that organisms and cells behave much like non-biological devices composed of many parts, each performing some well-defined function, according to well-defined laws of physics, and interacting to produce a predictable and repeatable effect. Physicians have bought into this fantasy. When a sampling of physicians was asked to rank the areas in which they needed additional genetics training, their number one choice was the "genetics of common disease" [24,25].

2.3.2 Rule—Biological systems are much more complex than naturally occurring non-biological systems (i.e., galaxies, mountains, volcanoes) and man-made physical systems (e.g., jet airplanes, computers).

Brief Rationale—The components of biological systems, unlike the components of non-biological systems, have multiple functions, dependencies, and regulatory systems. We cannot predict how any single component of a biological system will react under changing physiologic conditions.

The grim truth is that biological systems are nothing like man-made physical systems. When an engineer builds a radio, she knows that she can assign names to components, and these components can be relied upon to behave in a manner that is characteristic of its type. A capacitor will behave like a capacitor, and a resistor will behave like a resistor. The engineer need not worry that the capacitor will behave like a semiconductor or an integrated circuit. What is true for the radio engineer does not hold true for the biologist [26].

In biological systems, components change their functions depending on circumstances. For example, cancer researchers discovered a protein that plays an important role in the development of cancer. This protein, p53, was once considered to be the primary cellular driver for human malignancy. When p53 mutated, cellular regulation was disrupted, and cells proceeded down a slippery path leading to cancer. In the past few decades, as more information was obtained, cancer researchers have learned that p53 is just one of many proteins that play a role in carcinogenesis, but the role changes depending on the species, tissue type, cellular micro-environment, genetic background of the cell, and many other factors (see Glossary item, Carcinogenesis). Under one set of circumstances, p53 may play a role in DNA repair; under another set of circumstances, p53 may cause cells to arrest the growth cycle [26,27]. It is difficult to predict a biological outcome when pathways change their primary functionality based on cellular context. Various mutations in the TP53 gene have been linked to 11 clinically distinguishable cancer-related disorders, and there is little reason to assume that the same biological role is played in all of these 11 disorders [28].

Likewise, the Pelger–Huet anomaly and hydrops-ectopic calcification-moth-eaten (HEM) are both caused by mutations of a gene, coding for the lamin B receptor. The Pelger–Huet anomaly is a morphologic aberration of neutrophils wherein the normally multi-lobed nuclei become coffee bean-shaped, or bilobed, with abnormally clumped chromatin. The condition is called an anomaly, rather than a disease, because despite the physical abnormalities, the affected white cells seem to function adequately. HEM is a congenital chondrodystrophy that is characterized by hydrops fetalis (i.e., accumulations of fluid in the fetus), and skeletal abnormalities. It would be difficult to imagine any two diseases as unrelated as Pelger–Huet anomaly and HEM. How could these disparate diseases be caused by a mutation involving the same gene? As it happens, the lamin B receptor has two separate functions: preserving the structure of chromatin and serving as a sterol reductase in cholesterol synthesis [29]. These two different and biologically unrelated functions in one gene product account for two different and biologically unrelated diseases.

A gene's role may be influenced by other genes, a phenomenon called epistasis (see Glossary item, Epistasis). Likewise, the role of a gene is influenced by the temporal expression of the gene (e.g., at precise moments of organismal development), and by its sequential activation (e.g., preceding or succeeding sequential steps in multiple pathways). The activity of a protein encoded by a gene can be influenced by subtle variations in amino acid sequence, by three-dimensional structure, by chemical modifications of the protein, by quantity of the protein, by location of the protein molecules in cells, and by the type of cell in which the protein is expressed. Attempts to predict the functional effect of single or multiple gene variations are typically futile [30,31].

The most complex man-made physical systems are laughably simplistic compared to human genetics. The fastest supercomputers cannot cope with networks of systems whose individual objects behave in unpredictable and indescribable ways.

With a few exceptions, the common diseases of humans are products of modern life; hence, they are relatively new diseases. Heart disease, diabetes, obesity, hypertension were not major scourges of ancient man. Neither are they common among modern men who lack modern conveniences, such as fresh food, comfortable shelter, potable water, and hygienic plumbing. It can be assumed that sets of gene variants that predispose us to most of the common diseases are old genes faced with new tasks. It is reasonable to expect that the genes that seem to associate themselves with common diseases may vary in different populations of humans, living under different environments. If this turns out to be the case, such variations may make an impossible job (determining clinical phenotype from genotype) even more impossible.

Infectious diseases, to the contrary, have been around for a very long time. In Chapter 7, we will see that the human genome has evolved to cope with infectious organisms that have specifically evolved to live in our bodies (see Glossary item, Genome).

REFERENCES

1. Total Midyear Population for the World: 1950–2050. United States Census Bureau. Available from: http://www.census.gov/population/international/data/idb/worldpoptotal.php, viewed May 21, 2013.
2. Deaths by Age, Sex and Cause for the Year 2008. World Health Organization. Available from: http://www.who.int/healthinfo/global_burden_disease/estimates_regional/en/index.html, viewed May 19, 2008.
3. U.S. and World Population Clocks. U.S. Census Bureau. Available from: http://www.census.gov/main/www/popclock.html, viewed July 20, 2011.
4. The World Factbook. Central Intelligence Agency, Washington, DC, 2009.
5. Hoyert DL, Heron MP, Murphy SL, Kung H-C. Final data for 2003. Natl Vital Stat Rep 54(13) April 19, 2006.
6. NIH Budgets by Institute and Funding Mechanism, FY 1998–2013. American Association for the Advancement of Science. Available from: http://www.aaas.org/spp/rd/fy2013/health13pTBL.pdf, viewed May 21, 2013.
7. Cancer Research Funding. National Cancer Institute. Available from: http://www.cancer.gov/cancertopics/factsheet/NCI/research-funding, viewed August 7, 2013.
8. Cancer Incidence by Age. Cancer Research UK. Available from: http://www.cancerresearchuk.org/cancer-info/cancerstats/incidence/age/, viewed November 13, 2013.
9. Tsai SP, Lee ES, Hardy RJ. The effect of a reduction in leading causes of death: potential gains in life expectancy. Am J Public Health 68:966–971, 1978.
10. Central Intelligence Agency World Factbook. Rank-order life expectancy at birth. https://www.cia.gov/library/publications/the-world-factbook/rankorder/2102rank.html.
11. Guessous I, Gwinn M, Khoury MJ. Genome-wide association studies in pharmacogenomics: untapped potential for translation. Genome Med 1:46, 2009.
12. Pauling L, Itano HA, Singer SJ, Wells IC. Sickle cell anemia, a molecular disease. Science 110:543–548, 1949.
13. Ingram VM. A specific chemical difference between globins of normal and sickle-cell anemia hemoglobins. Nature 178:792–794, 1956.
14. Berman JJ. Precancer: The Beginning and the End of Cancer. Jones and Bartlett, Sudbury, 2010.
15. Wade N. A decade later, genetic map yields few new cures. The New York Times June 12, 2010.
16. Manolio TA, Collins FS, Cox NJ, Goldstein DB, Hindorff LA, Hunter DJ, et al. Finding the missing heritability of complex diseases. Nature 461:747–753, 2009.
17. National Science Board. Science & Engineering Indicators—2000. National Science Foundation, Arlington, VA, 2000 (NSB-00-1).
18. Innovation or Stagnation: Challenge and Opportunity on the Critical Path to New Medical Products. U.S. Department of Health and Human Services, Food and Drug Administration, 2004.
19. Crossing the Quality Chasm: A New Health System for the 21st Century. Quality of Health Care in America Committee, eds. Institute of Medicine, Washington, DC, 2001.
20. Wurtman RJ, Bettiker RL. The slowing of treatment discovery, 1965–1995. Nat Med 2:5–6, 1996.
21. Cecile A, Janssens JW, vanDuijn CM. Genome-based prediction of common diseases: advances and prospects. Hum Mol Genet 17:166–173, 2008.

22. Ioannidis JP. Is molecular profiling ready for use in clinical decision making? Oncologist 12:301–311, 2007.
23. Venet D, Dumont JE, Detours V. Most random gene expression signatures are significantly associated with breast cancer outcome. PLoS Comput Biol 7:e1002240, 2011.
24. Calefato JM, Nippert I, Harris HJ, Kristoffersson U, Schmidtke J, Ten Kate LP, et al. Assessing educational priorities in genetics for general practitioners and specialists in five countries: factor structure of the Genetic-Educational Priorities (Gen-EP) scale. Genet Med 10:99–106, 2008.
25. Julian-Reynier C, Nippert I, Calefato JM, Harris H, Kristoffersson U, Schmidtke J, et al. Genetics in clinical practice: general practitioners' educational priorities in European countries. Genet Med 10:107–113, 2008.
26. Madar S, Goldstein I, Rotter V. Did experimental biology die? Lessons from 30 years of p53 research. Cancer Res 2009(69):6378–6380, 2009.
27. Zilfou JT, Lowe SW. Tumor suppressive functions of p53. Cold Spring Harb Perspect Biol 1:a001883, 2009.
28. Vogelstein B, Lane D, Levine AJ. Surfing the p53 network. Nature 408:307–310, 2000.
29. Waterham HR, Koster J, Mooyer P, van Noort G, Kelley RI, Wilcox WR, et al. Autosomal recessive HEM/Greenberg skeletal dysplasia is caused by 3-beta-hydroxysterol delta(14)-reductase deficiency due to mutations in the lamin B receptor gene. Am J Hum Genet 72: 1013–1017, 2003.
30. Chi YI. Homeodomain revisited: a lesson from disease-causing mutations. Hum Genet 116: 433–444, 2005.
31. Gerke J, Lorenz K, Ramnarine S, Cohen B. Gene environment interactions at nucleotide resolution. PLoS Genet 6:e1001144, 2010.

Chapter 3

Six Observations to Ponder while Reading this Book

3.1 RARE DISEASES ARE BIOLOGICALLY DIFFERENT FROM COMMON DISEASES

"The study of rare diseases offers a way of implementing the tools and procedures that will later be used in more widespread applications of genomic medicine."

—from Institute of Medicine workshop summary [1]

3.1.1 Six observations that distinguish common diseases from rare diseases

1. Rare diseases typically occur in a young population. Common diseases typically occur in adults, increasing in frequency with age.
2. Rare diseases usually occur with a Mendelian pattern of inheritance. The most common diseases may sometimes cluster in families, but they are, without exception, non-Mendelian.
3. Rare diseases often occur as syndromes, involving several organs or physiologic systems, often in surprising ways; most common diseases are non-syndromic (see Glossary items, Syndrome, Non-syndromic).
4. Environmental factors play a major role in the cause of common diseases; much less so in the inherited rare diseases.
5. The difference in rates of occurrence of the rare diseases compared with the common diseases is profound, often on the order of a thousand-fold, and sometimes on the order of a million-fold.
6. There are many more rare diseases than there are common diseases.

3.2 COMMON DISEASES TYPICALLY OCCUR IN ADULTS; RARE DISEASES ARE OFTEN DISEASES OF CHILDHOOD

"I conclude that for a number of diseases the mutation rate increases with age and at a rate much faster than linear. This suggests that the greatest mutational health hazard in the human population at present is fertile old males."

—James F. Crow [2]

Rare Diseases and Orphan Drugs. http://dx.doi.org/10.1016/B978-0-12-419988-0.00003-1

We know that many of the common diseases are caused by long-term exposures to causal agents. Heart attacks often follow decades of dietary indiscretion; lung cancer and emphysema often follow decades of exposure to cigarette smoke. Humans tend to gain weight throughout most of their lives; obesity is predominantly a disease of adults.

All of the life-threatening diseases of childhood are rare. We can infer that childhood diseases do not develop as adult diseases do; there just isn't the time for it. When a newborn comes into the world with a severe lung disease, or an anatomic abnormality, or a tumor, we can be certain that the lung disease will not be emphysema, the anatomic abnormality will not be obesity, and the tumor will be fundamentally different from the common cancers that occur in adults.

3.2.1 Rule—There is almost no overlap in the types of tumors that occur in children, all of which are rare, and the common tumors that occur later in life.

Brief Rationale—The tumors of adults are different from the tumors of children because these two sets of tumors have different causes and different pathogeneses.

The common tumors that occur in adults arise in tissues exposed to exogenous carcinogenic agents, through air and water, and food (see Glossary item, Carcinogen). The carcinogen exposures that cause the common cancers are prolonged, typically throughout the lives of the affected individuals (e.g., sunlight, cigarettes, and carcinogens in food and water).

The tumors that occur in children are caused, for the most part, by inherited mutations or by errors that occur during development. The tumors of infancy and childhood tend to arise from primitive cells or from the mesoderm (the embryonic layer between the epithelial layers), or the neural tube, or the neural crest. With some exceptions, the tissues giving rise to tumors in children (e.g., brain cells, lymphocytes) are not the tissues giving rise to tumors in adults (e.g., lung cells, gut cells, skin epidermis). Several biological conditions account for the profound differences between adult tumors and childhood tumors, and these will be discussed in Chapter 8.

Are there exceptions to the rule that rare diseases occur in the young, and common diseases occur in older individuals? Of course, some inherited diseases will be triggered by events that seldom occur in childhood. For example, susceptibility to anesthesia-induced malignant hyperthermia can be caused by an inherited mutation in any of several genes, including the ryanodine receptor gene (see Glossary item, Malignant). This potentially fatal condition arises clinically as an idiosyncratic, rapidly progressing fever that occurs during a surgical procedure. Dantrolene sodium is used to treat the acute episode. Affected persons are at risk from the moment of their birth, but if they manage to avoid anesthesia during their early years, their condition will not be evident until adulthood. When an individual develops anesthesia-induced malignant hyperthermia, family members should be counseled that they may also be at risk.

Huntington disease is a rare autosomal dominant inherited neurologic disease. It produces progressive dyskinesia and rigidity, with symptoms usually beginning in adults between the ages of 35 and 44. As an inherited rare disease, why does Huntington target adults? Actually, Huntington disease can affect children. Preclinical cognitive signs can be measured, in some cases well over a decade before neurologic symptoms arise.

The age at which any disease manifests clinically is determined by its pathogenesis, the cellular events that follow an underlying cause, such as a gene mutation, and that lead to a pathologic condition. Pathogenesis can be brief (e.g., cellular respiratory arrest occurring seconds after cyanide ingestion), or prolonged (e.g., mesothelioma developing four decades following asbestos exposure). In the case of Huntington disease, a mutation in the Htt gene sets in motion a series of events that, over time, leads to cell death of neurons, particularly those located in the caudate, the putamen, and the substantia nigra. Understanding the pathogenesis of Huntington disease is one of the great challenges of medicine; research into this disease has yielded important discoveries in genetics and cell biology (see Glossary item, Anticipation). The lesson learned is that, for the rare inherited diseases, the clinical phenotype may occasionally manifest in adults, but the process begins at conception.

So strong is the age difference between the common diseases and the rare diseases that the occurrence of a common disease in a very young person should prompt a search for a rare genetic origin. For example, a heart attack occurring in a 15-year-old girl should prompt the search for an inherited dyslipidemia or clotting disorder. In some cases, a common disease occurring in a young person should prompt the search for a toxin or an environmental hazard (e.g., tobacco chewing in an adolescent baseball player leading to oral cancer; radiation exposure in a child with thyroid cancer).

Physicians should never forget that diagnoses that fit poorly with the age of a patient can signal a misdiagnosis. For example, pathologists should be cautious when rendering a malignant diagnosis on vulvar melanocytic lesions growing on teenaged girls. Most malignant melanomas arise in sun-exposed regions on middle-aged or older individuals. There is little reason to expect to find a malignant melanoma on the vulva of a teenager. As it happens, a rare type of benign nevus (i.e., mole) occurs on the vulva of young women, and such lesions mimic the appearance of malignant melanomas [3]. Needless to say, a misdiagnosis of malignant melanoma of the vulva in a young girl would have terrible consequences.

3.3 RARE DISEASES USUALLY OCCUR WITH A MENDELIAN PATTERN OF INHERITANCE. COMMON DISEASES ARE NON-MENDELIAN

"Most [rare diseases] result from a dysfunction of a single pathway due to a defective gene. Understanding the impact of a single defect may therefore yield insights into the more complex pathways involved in common diseases which are generally multifactorial."

—Segolene Ayme and Virginie Hivert in "Report on rare disease research, its determinants in Europe and the way forward," 2011 [4]

In 1865, Mendel published his laws of inheritance, which were universally ignored until about 1900. In 1915, Thomas Hunt Morgan integrated Mendel's laws into what was then known about the role of chromosomes as carriers of genetic material. In the 1950s, when DNA was found to be the carrier molecule of genetic information, the concept of Mendelian genetics was integrated into molecular biology (see Glossary item, Mendelian inheritance).

Back in 1909, nearly a half century before we understood that our inheritance was encoded in molecules of DNA, scientists knew that inborn errors of metabolism had a pattern of inheritance much like the pattern reported by Mendel for his pea garden [5].

Mendelian inheritance, in one sentence, comprises the now familiar genetic diseases inherited from one or both parents with patterns typical of autosomal dominant, autosomal recessive, or with linkage to the X- or Y-chromosome. A more complete definition is found in the Glossary (see Glossary items, Mendelian inheritance, X-chromosome, Y-chromosome). **Mendelian inheritance is the low-hanging fruit of clinical genetics.** When the inheritance is Mendelian, the cause of the disease is monogenic. We will see in Section 9.2 that monogenic diseases, the simplest form of genetic diseases, carry complexities that Mendel could not have imagined (see Glossary item, Monogenic disease).

Non-Mendelian inheritance is a murky topic (see Glossary item, Non-Mendelian inheritance). Whenever a biological concept is named for what it is not (i.e., not Mendelian), rather than being named for what it is, you can expect to encounter a certain degree of confusion and ignorance. Suffice it to say that a disease has a non-Mendelian pattern if knowledge of disease occurrences in ancestors cannot yield the simple inheritance ratios for offspring that Mendel might have predicted. In general, diseases that exhibit non-Mendelian inheritance occur in family clusters, but predicting which offspring will be affected is impossible.

We shall see in later chapters that several biological processes can account for a non-Mendelian pattern of inheritance, but polygenic inheritance plays a role in most of the examined common diseases. For now, let us concentrate on the inheritance patterns of polygenic diseases.

A polygenic disease is caused by variations in numerous genes that work in concert to produce a disease or to heighten susceptibility to disease. Imagine that a common disease is caused by a set of 10 variant genes that, together, confer susceptibility to an environmental toxin. How would you predict that an offspring will develop the disease? If the gene variation were rare, each variant might have as high as a 50% chance of appearing in the offspring's DNA, but there are 10 genes involved, and the chance of all of them being passed to the offspring would be small. If the variant were common within the population, then inheritance odds would increase, and we would need to take into account homozygosity (i.e., the gene on both chromosomes from one parent being variant), as well as the likelihood that the other parent carried the gene variation (see Glossary items, Homozygosity, Uniparental disomy). If one of the 10 gene variants were necessary to produce disease, while the other nine genes had less effect, then the calculations would change. If there were alternate gene variants that could substitute for, supplement, or modify any of the original 10 gene variants, then the calculations would change again. In point of fact, the inheritance of polygenic diseases defies prediction; it is all too complex.

How many genes are necessary to produce a non-Mendelian pattern of inheritance? Just two may do. Bruning and coworkers developed a digenic model of type 2 diabetes in mice (see Glossary item, Digenic disease). Like the common disease in humans, diabetes arose in offspring in an age-dependent manner, and the pattern of inheritance was non-Mendelian [6].

The rare disease Bardet–Biedl syndrome is characterized by obesity in infancy, retinal dystrophy, polydactyly, and abnormalities of multiple organs. In most cases, Bardet–Biedl syndrome is a monogenic rare disease with an autosomal recessive pattern of inheritance. In a small percentage of cases, Bardet–Biedl syndrome is polygenic, and does not exhibit the usual Mendelian pattern of inheritance. These exception cases are caused by three mutations occurring at two of the loci known to be associated with the syndrome [7].

In a well-controlled experiment, in a simple yeast cell system, Gerke and coworkers tried to predict the outcome for a set of four gene variants known to influence a specific yeast phenotype, in this case, yeast sporulation efficiency [8]. As expected, genotype could not predict phenotype; four genes made the system too complex to predetermine sporulation efficiency in progeny (see Glossary item, Phenotypic heterogeneity).

3.3.1 Rule—No common disease is monogenic.

Brief Rationale—In the past several decades, medical scientists have found thousands of rare diseases, each with a monogenic cause. Scientists have not found a single instance wherein a monogenic cause accounts for all the cases of a common disease.

Every good scientist knows that the absence of a positive finding can never constitute proof of a negative finding. Nonetheless, there is a long, unbroken tradition of searching for, and failing to find, a monogenic cause for any of the common diseases. Accumulated experience would suggest that the common diseases of clinical importance are all polygenic.

Quibblers might argue that glucose-6-phosphate dehydrogenase deficiency (G6PD) is an exception to the rule: a common disease with a monogenic cause. One gene is involved, and the number of people with the deficiency is large, approximately 400 million people worldwide. Most people with G6PD deficiency are totally asymptomatic, and some might say that the deficiency does not rise to the level of a disease; it is more like a trait. Some individuals with G6PD will develop hemolysis after ingesting certain types of drugs, foods, or chemicals (e.g., primaquine, sulfonamides, fava beans, methylene blue, naphthalene, nalidixic acid, aspirin). Others with the same deficiency will be unaffected by the same substances. Why does nature preserve this potentially harmful trait? Having the G6PD trait protects against *Plasmodium falciparum*, the most serious form of human malaria. The trait is most common in geographic areas where malaria is, or has been, endemic. This issue is revisited in Section 11.1.

3.4 RARE DISEASES OFTEN OCCUR AS SYNDROMES, INVOLVING SEVERAL ORGANS OR PHYSIOLOGIC SYSTEMS, OFTEN IN SURPRISING WAYS. COMMON DISEASES ARE TYPICALLY NON-SYNDROMIC (SEE SECTION 10.1)

"I learned very early the difference between knowing the name of something and knowing something."

—Richard Feynman

Imagine the following scenario. An automobile manufacturer orders a quantity of steel to be used in the productions of its cars. When the steel was prepared, a rare mistake was made at the steelyard, and the wrong amount of carbon was added to the molten mix. Consequently, all of the steel used in the production of 20 cars is of poor quality. The cars that rolled off the assembly line looked like any other car, but soon after they hit the roads, various parts of the car began to fail: the engine, suspension struts, axel joints, and chassis are the first to go. In this example, the inherent deficiency (poor steel) is expressed everywhere, but causes diseases in a syndromic fashion (i.e., causing malfunction in several systems or parts of the car, but not others).

The 20 cars all suffer from a rare syndrome caused by one specific defect in a basic building material. The defect was demonstrated when a materials engineer examined the steel of the affected cars using a specialized microscope. The steel from unaffected cars produced the same day, on the same assembly line, had a normal appearance.

When an error is introduced into the constitutive fabric of a system (e.g., automobile) or organism (e.g., human), its affects are likely to occur in several parts or systems, and the effects are likely to occur early. Many rare diseases are caused by a single genetic defect that occurs in every cell of the organism to produce an assortment of malfunctioning parts early in the life of the organism (i.e., a syndrome).

It is easy to see why rare diseases, produced by monogenic errors, result in syndromic disorders. It is much more difficult to understand why some monogenic diseases are non-syndromic (i.e., producing disease in a single organ).

3.4.1 Rule—When a rare disease is non-syndromic, some particular combination of conditions must apply.

Brief Rationale—Additional conditions, beyond the single genetic defect underlying the rare disease, constrain the expression of disease to a specific organ.

An example may clarify how additional conditions, imposed on a genetic defect, result in non-syndromic rare diseases. Xeroderma pigmentosum, described in Section 4.3, is a monogenic inherited disorder in which affected persons cannot efficiently repair DNA damage induced by ultraviolet (UV) light. Since UV light penetrates into the skin, but no further, the clinical phenotype of xeroderma pigmentosum is essentially limited to the skin and the cornea. Unless sunlight is scrupulously avoided, affected individuals will typically develop multiple skin cancers at an early age. The defect in DNA repair is present in every organ in the body. Nonetheless, the disease is non-syndromic, isolated to tissues covered by a squamous epidermis. The disease is non-syndromic because the expression of disease is conditional upon exposure to UV light.

3.4.2 Rule—Single gene disorders tend to be syndromic; polygenic/multifactorial disorders tend to be non-syndromic.

Brief Rationale—Single gene disorders are caused by a gene alteration that is present in every cell in the body; hence, any tissue has a chance of suffering a functional or anatomic abnormality due to the gene alteration. Polygenic disorders are caused by a combination of gene variants that occur in the normal human population (i.e., the variant genes are not defective). The expression of disease follows a collection of events and environmental influences occurring over time. The likelihood that these occur in many different tissues is remote; hence, most polygenic diseases are non-syndromic.

Common diseases have high specificity (because they are caused by the accumulation of damages and adverse events that culminate in one part or system breaking down. Returning to the car analogy, we see that older cars tend to collect damage in the parts that get the most use, or the most exposure, or the least maintenance.

3.4.3 Rule—Eponymic disorders (i.e., diseases with a name of a person) are usually syndromic.

Brief Rationale—It can be too taxing to name a syndromic disease by listing the various organs and abnormalities that comprise the syndrome. It is much easier to apply a person's name to the disease, and be done with it.

For example, it is easier to remember Adams–Oliver syndrome than to remember aplasia cutis congenita with terminal transverse limb defects, possibly including congenital heart defect, and frontonasal cysts. When no eponym comes to mind, a syndrome may take the name of an acronym. For example, LEOPARD syndrome is an acronym with each letter of the acronym composed of the first letter of a clinical component of the syndrome: Lentigines, Electrocardiographic conduction abnormalities, Ocular hypertelorism, Pulmonary stenosis, Abnormal genitalia, Retarded growth, Deafness. Ironically, the pattern of skin mottling caused by the lentigines calls to mind the appearance of a leopard. In this case, the acronym is also a descriptor for the most visually prominent component of the syndrome.

3.4.4 Rule—A high proportion of diseases caused by regulators of transcription are syndromic.

Brief Rationale—Regulators of transcription have many functions, affecting many genes, and may produce changes in more than one organ, at more than one moment in development [9,10].

On a simplistic level, a neuron differs from a gut lining cell because the neuron has high levels of the proteins normally found in neurons and low levels of the kinds of proteins normally found in gut lining cells (see Glossary item, Differentiation). The opposite would be true for the epithelial cells lining the gut (see Glossary item, Epithelial cell). Through regulation of gene expression, transcription factors, in concert with other regulatory processes, create brain cells, kidney cells, liver cells, and the hundreds of distinctive types of cells that populate the tissues of our bodies (see Glossary items, Gene regulation, Transcription factor, Cell type). Because transcription factors play an important role in development, the inherited diseases caused by transcription factor mutations tend to produce syndromes characterized by developmental anomalies in several different tissues. For example, mutation of the transcription factor TBX5 causes Holt–Oram syndrome, characterized by thumb anomalies and atrial septal defect, and some cases having ventricular septal defect and phocomelia. A mutation in the transcription factor WT1 causes WAGR syndrome, characterized by Wilms tumor-aniridia-genitourinary anomalies-mental retardation syndrome [11].

3.4.5 Rule—Common diseases can be conceptualized many different ways, all of which are objectively correct.

Brief Rationale—Because many conditions and factors can produce a common disease, it is impossible to exclude any single mechanism as a valid cause.

Let us return to the automobile analogy. Imagine that you have owned your car for about a decade. One morning, you reach to adjust your rear-view mirror, and in the next moment you find that the entire mirror assembly is broken and has fallen onto the dashboard. In your mind, you review some of the possible causes of your current predicament. You may have clumsily caused the mirror or the mount to break when you reached over to make an adjustment. The mount may have broken from long-term wear; all those morning adjustments may have loosened or fractured parts. The mirror and the mount may have had design flaws. If they had built the mirror correctly, then your adjustments would not have loosened the mount. The mirror and mount may have been built with inferior materials; too much plastic and not enough metal. The mirror may have arrived damaged from the factory due to a machining hiccup. Human error in the factory may have contributed to the problem. Could the quality assurance inspector have overlooked a quality flaw?

The example of the broken rear-view mirror demonstrates that it is nearly impossible to assign a single cause to a commonly occurring flaw that occurs after a long period of normal use. This applies even when the dysfunction is very simple (i.e., a broken rear-view mirror). Common flaws are common precisely because many different factors may contribute to the flaw. In the case of the broken rear-view mirror, it is quite possible that all or none of the listed causes could have applied. For example, the mirror may have been knocked down, it may have been loosened before it fell, it may have been designed poorly, it may have had defective parts, and so on.

Consider lung cancer, an all-too-common disease. Is it caused by smoking, or tobacco addiction, or air pollution, or inherited susceptibility genes, or uncharacterized co-factors, or an insufficiency of dietary anti-oxidants, or inadequate screening for bronchial precancerous conditions, or so on? Common diseases can be conceptualized in many different ways.

3.4.6 Rule—Common diseases have many causes; that is why they are common. Rare diseases have a small number of causes; that is why they are rare.

Brief Rationale—Common diseases have many contributing causes. It is impossible to think that all of these causes will activate the same pathways, in the same sequence, and in the same timeframe, for each instance of disease. It is much more likely that an assortment of pathways leads eventually to a collection of pathologic conditions that share a similar phenotype. In the case of rare diseases, many of which are caused by a specific mutation in a specific gene, the pathways follow the same course, over a similar timeframe, to produce very similar phenotypic outcomes in an age-restricted population (e.g., young children).

Now, imagine that one morning, as you seat yourself in your car, and as you glance at the rear-view mirror, you notice that the mirror surface has turned emerald green. Everything is otherwise as you would expect, but the mirror is the wrong color! This is a rare event. At the moment, you cannot determine why the mirror has turned green, but you are confident that there must be a simple explanation. A rare disease, like a rare event, is conceptually simple and often has a single, specific cause.

3.5 ENVIRONMENTAL FACTORS PLAY A MAJOR ROLE IN THE CAUSE OF COMMON DISEASES; LESS SO IN THE INHERITED RARE DISEASES

"As well as providing new approaches to carrier detection, prenatal diagnosis, and treatment of single gene disorders, these advances [understanding of the basic molecular pathology of single gene disorders] promise to provide important information about the pathophysiology of many common polygenic diseases."

—Sir David Weatherall [12]

Several observations lead us to infer that the common diseases are caused largely by environmental influences. First, the common diseases (e.g., cancer, heart attack, stroke, obesity, diabetes, hypertension) all occur preferentially in adults. Diseases with a pure genetic cause typically manifest in young individuals. The occurrence of common diseases in older individuals would suggest that these diseases are caused by accumulated effects from external influences; not from inborn genetic errors. Second, most of the common diseases are conclusively linked to environmental factors (e.g., smoking and lung cancer, alcohol abuse and cirrhosis, sun exposure and skin cancers, excessive salt intake and hypertension, unsanitary drinking water and diarrheal diseases, obesity and heart disease). Finally, if the common diseases had a purely genetic cause, we might expect natural selection to gradually reduce the populations of individuals that carry the gene; hence, over time, the disease-causing genes would become rare variants, and the common diseases would become uncommon (see Glossary item, Natural selection). We see little evidence of a decline in the common diseases of humans.

Suppose, to the contrary, that the environment had little or no impact on the common diseases. In that case, we would expect that genetic influences would be the major or the exclusive factor determining whether an individual will develop a given common disease. Though most common diseases of humans show significant heritability, the contribution is often modest. For the common diseases occurring in adults, there is considerable discordance among close relatives, even among monozygotic twins [13,14]. Schizophrenia is considered to be a disease with a strong genetic component. Even so, about half of monozygotic twin pairs are discordant for schizophrenia.

A high genetic concordance for disease is, with almost no exceptions, the reserve of the simple monogenic diseases. The more complex and common a disease becomes, the lower the genetic concordance. The lower the genetic concordance, the closer we must look for environmental causes. **So strong is the genetic influence on the monogenic rare diseases that an increase in the occurrences of a rare disease, within a population of unrelated individuals, should prompt a thorough search for a phenocopy disease**. A phenocopy disease is a clinical phenotype, produced by one or more environmental factors, that mimics a genetic disorder. Often the phenocopy disease will be caused by a toxin that affects the same biological pathway that accounts for the clinical phenotype of the genetic disease. Phenocopy diseases will be discussed in Section 9.5.

The astute reader may find the gene/environment dichotomy somewhat lacking as an explanation for the age-dependent nature of the common diseases. If you assume that environmental influences assert themselves in a cumulative way, over long periods of time, eventually producing common, often chronic, conditions in later life, then you must infer that the cells of the body pass the accumulating damage onto succeeding generations of cells until the disease finally emerges. How can damage caused by environmental agents pass itself to succeeding generations of cells without the participation of genes? In some cases, the answer lies in the epigenome, but we are getting ahead of ourselves (see Glossary item, Epigenome). This topic will be discussed in detail in Section10.2.

3.6 THE DIFFERENCE IN RATES OF OCCURRENCE OF THE RARE DISEASES COMPARED WITH THE COMMON DISEASES IS PROFOUND, OFTEN ON THE ORDER OF A THOUSAND-FOLD

"If I didn't believe it, I would never have seen it."

—Anon

You need to wonder, is there some biological factor that keeps the incidence of the monogenic diseases low and the incidence of polygenic diseases high? Darwinian selection keeps the incidence of life-threatening monogenic diseases low; individuals with serious childhood diseases will be less likely to procreate and to pass disease genes onto others.

In the case of the polygenic diseases, there is no natural process of selection that would cull disease genes from the general population. If a disease occurs in late adulthood, as is often the case for polygenic diseases, Darwinian selection may not apply; affected individuals will have an opportunity to procreate. More importantly, though, Darwinian selection cannot operate efficiently on a set of polygenic disease genes. If the variant genes that cause a polygenic disease are common polymorphisms (i.e., naturally occurring gene variants observed in populations), then the variants may serve a useful purpose in concert with other genes, in some subset of cells, or under certain sets of conditions. For example, a gene that down-regulates the number of mitochondria contained in cells may be useful under anoxic conditions, when mitochondrial oxygenic respiration is low, and useful in red cell differentiation, when mitochondria are normally eliminated. If such a gene were somehow removed, its absence may reduce susceptibility to a particular disease while simultaneously introducing a new cellular defect. When dealing with a polygenic disease, selecting against the variant genes may have adverse consequences. As the number of genes involved in a polygenic disease increases, the overall effect of selecting against individual variant genes becomes unpredictable and chaotic.

3.6.1 Rule—Every common disease was, at some point, a rare disease.

Brief Rationale—Every epidemic begins with a solitary case. Common diseases are equivalent to epidemics that settle in to stay.

There is an inherited immunodeficiency of cattle caused by a deficiency of leukocyte adhesion factor. Affected cattle are homozygous for a gene allele that codes for a substitution in a single amino acid in its protein product (see Glossary item, Allele). Heterozygotes (i.e., cattle with an unpaired mutant allele) are common in the U.S., with a carrier rate of about 10%. Every cattle with a mutant allele is a descendant from one bull, whose sperm was used to artificially inseminate cows in the 1950s and 1960s [15]. A disease that was essentially non-existent in 1950 became a common scourge of the dairy industry within a half-century, all due to the founder effect amplified by modern animal husbandry (see Glossary item, Founder effect).

In the past century, we have seen many rare diseases of humans become commonplace. Here are some examples:

- **Heart disease.** Increased availability of cheap fatty and sweet foods, combined with social factors that favor a sedentary lifestyle, raised the heart attack from a collection of rare, hereditary conditions to one of the most common causes of death in industrialized societies.
- **Colon cancer.** Common in the United States, colon cancer has an incidence of 40/100,000. In Africa and some parts of Asia, colon cancer is a rare disease, with an incidence under 5/100,000 [16] (see Glossary item, Rare disease). Speculation abounds to explain why this is so, but the issue of diet looms large. The low-fiber, low-vegetable, high-meat diet preferred in high-incidence societies, contrasted with the high-fiber, high-vegetable, low-meat diet in the low incidence societies, provides a credible, if unproven, explanation.
- **AIDS.** Late in 1981, a Haitian man presented at Jackson Memorial Hospital in Miami with a constellation of infectious diseases, a strange rash, and mouth lesions of an unfamiliar type. At the time, attending physicians were baffled. Eventually, after a desperate review of the newest literature, a diagnosis of GRIDS (gay-related immune disease syndrome) was rendered. Today, GRIDS, now known as AIDS (autoimmune disease syndrome), is a diagnosis that can be rendered, without hesitation or error, by a first-year medical student. In 1981, there were about a dozen well-documented cases in the U.S. In 2011, 1.7 million people died of AIDS worldwide [17].
- **Lung cancer.** Prior to the popularization of cigarette smoking, lung cancer was extremely rare. Today, lung cancer is the leading cause of cancer deaths in every country where smoking is common.

3.6.2 Rule—Some of yesterday's common diseases are today's rare diseases.

Brief Rationale—The fundamental theory underlying all medical research is that we can eliminate diseases that we fully understand.

What is the natural history of disease, in the absence of human intervention? Imagine what might happen if the human population were suddenly attacked by a virus or bacteria that spreads efficiently from human to human with a nearly 100% kill rate. Let us assume that in one month, nearly 7 billion humans are dead. Let us imagine that a few thousand diseased individuals managed to recover. Another few thousand were infected, as judged by high levels of pathogen-specific antibodies, but did not develop the disease. And another few thousand never seemed to develop an infection (i.e., have no antibodies to the infection). You would expect that the genome of each and every survivor will have a story to tell. Variant genes, present in the general population before the arrival of the plague, appeared in just the right combinations to confer survival on a small subset of individuals. It is likely that each individual had a unique set of "survival genes," but it is also likely that there would be some overlap among some of the survivors, particularly those survivors with familial or ethnic relationships. These new sets of plague survival genes become part of the available gene pool in the group of humans that repopulate the planet.

Time passes. The pathogen that caused the plague is still in the environment. Perhaps it is living in a carrier host (see Glossary items, Host, Intermediate host, Secondary host). Perhaps it is lying dormant on the wall of a deep cave. One day, a descendant of one of the original survivors will lack the resistance of his ancestors. That descendant may encounter the pathogen, develop the full-blown disease, and die. By such a sequence of events, a rare disease (occurring in the first human infected by a plague pathogen) becomes a common disease (eradicating the vast majority of humans), and finishes as a rare disease, infecting those descendants of the original survivors who have "lost" resistance.

An astute reader might raise an objection at this point. It would seem that the transition from a common disease to a rare disease is simply numeric, without any fundamental biological differences to distinguish one from the other. Actually, this is not so. We shall revisit this same example in Section 11.6 after discussing the topic of genetic variation within a population and the topic of new mutations within a population; and describing their different roles in common and rare diseases.

Are biological catastrophes plausible? In the annals of horrifying plagues, nothing compares with myxomatosis. Myxomatosis is a fatal disease of rabbits caused by the myxomatosis virus. The disease is characterized by the rapid appearance of skin tumors (myxomas), followed by severe conjunctivitis, systemic symptoms, and fulminant pneumonia. Death usually occurs 2–14 days after infection. In 1952, a French virologist, hoping to reduce the rabbit population on his private estate, inoculated a few rabbits with myxoma virus. The results were much more than he had bargained for. Within 2 years, 90% of the rabbit population of France had succumbed to myxomatosis [18].

European rabbits, introduced to Australia in the nineteenth century, became feral and multiplied. By 1950, the rabbit population of Australia was about 3 billion. Seizing upon the myxoma virus as a solution to rabbit overpopulation, the Australians launched a myxoma virus inoculation program. In less than 10 years, the Australian rabbit population was reduced by 95% [19]. Nearly 3 billion rabbits died, a number very close to the number of humans living on the planet in the mid-1950s. This plague on rabbits was unleashed by a committee of humans who decided, one day, that it would be expeditious to use a lethal rabbit virus as a biological weapon [18].

Not all the rabbits died. Eventually, a population of rabbits emerged that had developed genetic resistance to the myxomatosis virus [20]. Among the new generations of survivors, myxomatosis is a rare disease.

Within the lifetimes of many humans living today, medicine has witnessed the conversion of common diseases into rare diseases. Polio, yellow fever, pertussis, diphtheria, measles, and botulism are all rare diseases in the developed countries. In all these cases, the drop in disease rates resulted from prevention, not treatment. Protection from polio, yellow fever, pertussis, diphtheria, and measles came in the form of effective vaccines in the 1950s.

Trichinosis, caused by the nematode parasite *Trichinella spiralis*, was common in the U.S. until about the mid-twentieth century (see Glossary item, Parasite). The disease was transmitted by undercooked or uncooked infected pigs. With improved cooking and canning methods, the incidence of disease dropped. Today, there are about a dozen cases of trichinosis reported annually in the U.S. Most of these cases arise from eating undercooked game meat. Improved preservation techniques likewise reduced rates of botulism. The rate of gastric cancer dropped precipitously in the mid-twentieth century, presumably due to the widespread use of refrigeration and other measures to preserve food.

Before the introduction of cervical precancer treatment, cervical carcinoma was the leading cause of cancer deaths in women. Today, in those countries that have not deployed precancer treatment, cervical cancer is still the leading cause of cancer deaths in women [21,22]. Elsewhere, deaths from cervical cancer have dropped to about 20% of prior levels, due to screening for precancerous lesions (i.e., the Pap smear, introduced in the 1940s). Over the next few decades, the incidence of cervical cancer is expected to drop even further, thanks to human papillomavirus vaccines.

In the developed countries, largely due to the widespread availability of potable water, mosquito control, and modern sanitation, a variety of once-common infectious diseases have all but disappeared (e.g., cholera, malaria).

In the case of one common disease, it would seem that preventive measures have resulted in its complete eradication. Smallpox was the first disease for which vaccination was successful. As early as 200 B.C.E. in China and 1000 B.C.E. in India, physicians knew that infection with smallpox conferred immunity against subsequent infection. Based on this observation, Chinese and Indian physicians were probably the first to develop a vaccination, administered nasally, of attenuated virus. Arabic doctors developed their own treatment, consisting of transferring material from an infected pox blister to another person via a small cut. Emmanuel Timoni (1670–1718) was a physician practicing in Constantinople. He introduced the Arabic vaccination process to the West in 1717. In 1796, Edward Jenner (1749–1823) developed a new vaccine from a bovine pox virus (vaccinia) that seemed to confer cross-immunity against smallpox (variola). The word "vaccine" derives from Jenner's choice of inoculum (vaccinia). Jenner's paper describing his smallpox vaccine was

FIGURE 3.1 A cross-section of a gross specimen of lung, exhibiting the pathology of tuberculous pneumonia. Cavities and blebs are noted in the upper lobe, indicating past destruction of the lung tissue. White areas of consolidated inflammation are present throughout the specimen. Notice the small white, round grains that encircle the areas of inflammation. These are miliary granulomas, areas of chronic inflammation that arise at sites of mycobacterial infection. See color plate at the back of the book. (*Source: MacCallum WG. A Textbook of Pathology* [25].)

rejected in 1796 by a peer-reviewed journal [23]. History has vindicated Jenner. Smallpox has been completely eradicated. The last known smallpox victim was Janet Parker [24]. In 1978, Janet Parker was a 40-year-old photographer who lived in Birmingham, England. She worked on the floor above Henry Bedson's smallpox research laboratory. Virus particles escaped from Bedson's laboratory and infected Ms. Parker [24].

Successful treatment for tuberculosis, introduced in the 1950s, nearly wiped out the disease (see Figure 3.1). In the early 1970s, physicians were predicting the total eradication of tuberculosis. For a variety of reasons, the medical assault on tuberculosis has failed, and numerous drug-resistant strains have emerged. Tuberculosis, once a common disease, became a rare disease, and is now on the verge of becoming a common disease.

Mental disorders come in and out of fashion. In Freud's heyday, hysteria was a common mental illness affecting women, characterized by excessive emotional fear and panic. The disease was somehow connected to the womb, from which the disease derived its name (from the Greek "hystera," meaning the uterus). Today, hysteria is a colloquialism, not a medical term. Women still exhibit fear and panic, as do men, but their diagnoses are assigned other names appropriate to their individual situations.

It should be acknowledged that some diseases, both common and rare, have existed only in the fertile imagination of hypochondriacs and gullible physicians. In the late nineteenth and early twentieth centuries, doctors attributed childhood asthma and crib death (now known as sudden infant death syndrome) to enlarged thymus glands; they named the condition status thymicolymphaticus. In the 1920s, doctors radiated enlarged thymus glands of children as a preventive measure against crib death. It is estimated that about 20,000–30,000 people died from cancers produced by "therapeutic" radiation for this and other real or imagined disorders [26]. We now know that status thymicolymphaticus is not a disease. Some children are born with larger thymus glands than other children, but no disease syndrome results from this anatomic disparity. Is it unrealistic to think that some of today's diseases (e.g., vaccine-induced autism, and a host of disorders stemming from fluoridated water) may not actually exist, while other so-called diseases may exist only as normal variants of the human condition (e.g., Asperger syndrome, solar elastosis, balding, mild attention deficits)?

In Chapter 7, we shall see how diseases have shaped the human genome and have accounted for the occasional symbiotic relationship between humans and rare diseases.

3.7 THERE ARE MANY MORE RARE DISEASES THAN THERE ARE COMMON DISEASES

"Man needs more to be reminded than instructed."

—Samuel Johnson

People are generally surprised to learn that there are just a few dozen, certainly less than 100, common diseases. The remainder of diseases, about 7000 of them, are rare. We learned in Section 2.1 that diseases, like most naturally occurring things, obey Pareto's rule. A few common items account for the bulk of occurrences. The remainder of occurrences falls in the long tail of a Zipf distribution. In a perverse sense, common diseases are the rarest types of diseases because there are so few of them.

One of the great biological mysteries is "Why are there any common diseases?" Considering the genomic uniqueness of every individual on the planet, and taking into account the variety of environmental exposures that unique individuals experience throughout their lives, it would seem almost inevitable that every instance of every disease must be unique. Why would millions of unique individuals develop the identical disease, expressing a common clinical phenotype?

Imagine for a moment that you and all of your colleagues have become deranged finger painters. You have access to all the colors of the finger painting rainbow, but instead of composing a simple, organized painting, you jab furiously at the easel again and again, at every possible angle, filling the painting with senseless lines and squiggles that criss-cross one another. Eventually, the painting becomes a gray–green jumble, with no coherent pattern. You look at all of the finger paintings created by your colleagues. Each is a gray–green jumble indistinguishable on casual inspection from your own atrocity.

All complex finger paintings look just about alike. This is true even when the painting is composed of strokes of non-repeating lengths, displayed at every allowable angle, with thousands of seemingly random intersections. Whether the patterns are simple or complex, patterns can be classified under a relatively small set of outcomes [27]. The common diseases are complex diseases, but the more complex the disease, the fewer the possible clinical phenotypes that emerge (see Glossary item, Complex disease).

If we looked very closely at one area on all of the various paintings, we would see all sorts of differences; a green line slanting down on one, a red line arcing upward on another. The micro-effect is unique, but the macro-effect is commonplace. This same phenomenon seems to hold with the common diseases. The argument has been made repeatedly that every common disease is really a collection of unique diseases occurring in individuals that seem to share a broadly similar clinical phenotype [28–30].

3.7.1 Rule—In common diseases, different pathways lead to a somewhat constrained set of clinical phenotypes. In rare diseases, single gene mutations activate a specific pathway producing a characteristic phenotype.

Brief Rationale—Common diseases have many contributing causes. It is impossible to think that all of these causes will activate the same pathways, in the same sequence, and in the same timeframe, for each instance of disease. It is much more likely that an assortment of pathways all lead eventually to a similar phenotype. In the case of rare diseases, many of which are caused by a specific mutation in a specific gene, the pathways follow the same course, over a similar timeframe, to produce very similar phenotypic outcomes in an age-restricted population (e.g., young children).

The role of pathways in the development of the clinical phenotypes of rare diseases and common diseases will be discussed in Section 10.4.

REFERENCES

1. Olson S, Beachy SH, Giammaria CF, Berger AC. Integrating Large-Scale Genomic Information into Clinical Practice: Workshop Summary. The National Academies Press, Washington, DC, 2012.
2. Crow JF. The high spontaneous mutation rate: is it a health risk? Proc Natl Acad Sci USA 94:8380–8386, 1997.

3. Clark WH Jr, Hood AF, Tucker MA, Jampel RM. A typical melanocytic nevi of the genital type with a discussion of reciprocal parenchymal-stromal interactions in the biology of neoplasia. Hum Pathol 29(1 Suppl 1):S1–S24, 1998.
4. Ayme S, Hivert V, eds. Report on Rare Disease Research, its Determinants in Europe and the Way Forward. INSERM, May 2011. Available from: http://asso.orpha.net/RDPlatform/upload/file/RDPlatform_final_report.pdf, viewed February 26, 2013.
5. Garrod AE, Harris H. Inborn Errors of Metabolism. Henry Frowde and Hodder and Stoughton, London, 1909.
6. Bruning JC, Winnay J, Bonner-Weir S, Taylor SI, Accili D, Kahn CR. Development of a novel polygenic model of NIDDM in mice heterozygous for IR and IRS-1 null alleles. Cell 88:561–572, 1997.
7. Eichers ER, Lewis RA, Katsanis N, Lupski JR. Triallelic inheritance: a bridge between Mendelian and multifactorial traits. Ann Med 36:262–272, 2004.
8. Gerke J, Lorenz K, Ramnarine S, Cohen B. Gene environment interactions at nucleotide resolution. PLoS Genet 6(9):e1001144, 2010.
9. Adams J. The complexity of gene expression, protein interaction, and cell differentiation. Nat Educ 1:1, 2008.
10. Heintzman ND, Hon GC, Hawkins RD, Kheradpour P, Stark A, Harp LF, et al. Histone modifications at human enhancers reflect global cell-type-specific gene expression. Nature 459:108–112, 2009.
11. Seidman JG, Seidman C. Transcription factor haploinsufficiency: when half a loaf is not enough. J Clin Invest 109:451–455, 2002.
12. Weatherall DJ. Molecular pathology of single gene disorders. J Clin Pathol 40:959–970, 1987.
13. Chatterjee A, Morison IM. Monozygotic twins: genes are not the destiny? Bioinformation 7:369–370, 2011.
14. Wong AHC, Gottesman II, Petronis A. Phenotypic differences in genetically identical organisms: the epigenetic perspective. Human Mol Gen 14:R11–R18, 2005.
15. Kehrli ME, Ackermann MR, Shuster DE, van der Maaten MJ, Schmalstieg FC, Anderson DC, et al. Bovine leukocyte adhesion deficiency: beta(2) integrin deficiency in young Holstein cattle. Am J Path 140:1489–1492, 1992.
16. World Cancer Research Fund and American Institute for Cancer Research. Food, Nutrition, Physical Activity, and the Prevention of Cancer: A Global Perspective. American Institute for Cancer Research, Washington, DC, 2007.
17. Global Health Observatory. HIV/AIDS. World Health Organization. Available from http://www.who.int/gho/hiv/en/, viewed July 27, 2013.
18. Berman JJ. Taxonomic Guide to Infections Diseases: Understanding the Biologic Classes of Pathogenic Organisms. Academic Press, Waltham, 2012.
19. Spiesschaert B, McFadden G, Hermans K, Nauwynck H, Van de Walle GR. The current status and future directions of myxoma virus, a master in immune evasion. Vet Res 42:76, 2011.
20. Ross J, Sanders MF. The development of genetic resistance to myxomatosis in wild rabbits in Britain. J Hyg (Lond) 92:255–261, 1984.
21. Wabinga HR. Pattern of cancer in Mbarara, Uganda. East Afr Med J 79:193–197, 2002.
22. Nze-Nguema F, Sankaranarayanan R, Barthelemy M, et al. Cancer in Gabon, 1984–1993: a pathology registry-based relative frequency study. Bull Cancer 83:693–696, 1996.
23. Altman LK. When peer review produces unsound science. The New York Times, June 11, 2002.
24. LeVay S. When Science Goes Wrong. Twelve Tales from the Dark Side of Discovery. Plume, New York, 2008. pp. 160–180.

25. MacCallum WG. A Textbook of Pathology, 2nd edition. WB Saunders Company, Philadelphia and London, 1921.
26. Jacobs MT, Frush DP, Donnelly LF. The right place at the wrong time: historical perspective of the relation of the thymus gland and pediatric radiology. Radiology 210:11–16, 1999.
27. Wolfram S. A New Kind of Science. Wolfram Media, 2002.
28. Rennard SI, Vestbo J. The many "small COPDs", COPD should be an orphan disease. Chest 134:623–627, 2008.
29. Crow YJ. Lupus: how much "complexity" is really (just) genetic heterogeneity? Arthritis Rheum 63:3661–3664, 2011.
30. Wade N. Many rare mutations may underpin diseases. The New York Times, May 17, 2012.

Part II

Rare Lessons for Common Diseases

Chapter 4

Aging

4.1 NORMAL PATTERNS OF AGING

The sixth age shifts into the lean and slipper'd pantaloon,
With spectacles on nose and pouch on side,
His youthful hose, well saved, a world too wide
For his shrunk shank; and his big manly voice,
Turning again toward childish treble, pipes
And whistles in his sound. Last scene of all,
That ends this strange eventful history,
Is second childishness and mere oblivion,
Sans teeth, sans eyes, sans taste, sans everything.

—the character Jaques in Shakespeare's *As You Like It*

Despite our most energetic efforts, we understand very little about aging.

4.1.1 Rule—We do not have a scientifically meaningful definition for the diseases of aging.

Brief Rationale—We do not know the cellular basis of aging; hence, we cannot determine whether a disease qualifies as a disease of aging on a cellular basis. The majority of the so-called diseases of aging are conditions that make individuals look like old persons, or they are conditions that happen to occur more often in elderly individuals than in young individuals.

One of the few points that experts in the field of aging can agree on is that the aging process is complex; not caused by any single factor.

4.1.2 Rule—Aging is not caused by a single gene.

Brief Rationale—If aging were caused by a single gene, you would expect rare occurrences of loss-of-function mutations of the gene, leading to instances of human immortality. Outside of science fiction, immortal humans do not exist.

Most of us gauge aging by looking for visible features that always seem to be present in older individuals, and that are absent in youth. For the most part, these signs have very little to do with the biological aging process. The most familiar example is wrinkling and sagging. Wrinkling is a condition produced

Rare Diseases and Orphan Drugs. http://dx.doi.org/10.1016/B978-0-12-419988-0.00004-3

by chronic exposure to ultraviolet (UV) light. Over time, UV light denatures the connective tissue in the dermis, producing a condition called senile elastosis or, more accurately, solar elastosis. Most of the skin changes associated with aging, such as cracking, leathery texture, and poor elasticity (i.e., the ability of skin to regain its flat, tight surface after being stretched or pinched), are the result of chronic UV toxicity.

The other obvious change observed in older individuals is skin sagging. In many individuals, this is most pronounced in the folds of skin that grow under the chin and down the neck. Sagging flesh on older individuals is due entirely to two phenomena. The first is skin growth; humans grow their skin throughout life. This skin accumulates to different degrees in different individuals, depending on their genetically determined propensities for skin growth. The other phenomenon is gravity. Without the effects of gravity, our skin would grow evenly over our body contours. We develop pendulous skin at sites with the least skeletal support (e.g., under chin, under breasts, under our arms).

The changes we see in the skin of older individuals are due to the chronic effects of UV light, skin growth throughout life, and gravity that occur over time. They are not fundamental features of biological aging, because they do not occur in the absence of toxic conditions. Is there any evidence to support this claim? One piece of evidence lies in differences in skin damage among races. The heavily pigmented races have much less wrinkling than the less pigmented races, because they are better shielded from UV light. Yet there is no corresponding extension of life expectancy among the less-wrinkled races, suggesting that damaged skin is unrelated to the aging process. Aside from that, any elderly person can do a simple experiment that will doubtless settle the issue for them. Strip off your clothes and inspect the parts of your body that are not exposed to light and that are not hanging from an anatomic prominence. For some, this would be the lower back or the upper thigh. In almost every case, you will be gratified to learn that this region of skin is unwrinkled, youthfully elastotic (i.e., will snap back in place when pinched), and relatively flat. Aside from a bit of softness due to skin growth, there really is not much difference between these protected regions of skin in elderly individuals and in young individuals. At this point, you can put your clothes back on, if you wish.

If wrinkling and sagging are not part of the aging process, then what physiological processes characterize aging?

The animal kingdom sheds some light on the process. Programmed aging, also known as senescence, is the physiological process whereby the phenotypic features of aging are compelled to appear. In the animal kingdom, the Pacific salmon provides a stark example of senescence. These organisms enjoy a maritime existence that can extend for many years, followed by a brief and tumultuous counter-current swim up their rivers of birth. Here, the exhausted salmon spawns and dies. Rapid senescence is characterized by multi-organ deterioration and immunosuppression.

In humans, there is no precipitating event that heralds an abrupt transition to senescence. As a purely working definition, we can characterize aging as a process wherein frailty and cachexia increase, while the normal physiological responses to stressors (e.g., infections, injuries, extremes of heat and cold, and other environmental stimuli) decrease.

Despite all the aging that goes on in this world, our knowledge of the pathogenesis of frailty and cachexia is limited [1]. We can say the obvious; that frailty is associated with a decrease in muscle (i.e., sarcopenia [2]), but we cannot say much about the underlying causes. It would seem that frailty and cachexia are not part of the defining clinical phenotype of any common disease; they are conditions that wait patiently in the background, looming large at the ends of our lives. If a miracle occurred, and every common disease of humans was eradicated, we would still need to contend with cachexia and frailty.

It would seem that our observations of elderly individuals have failed to explain any of the underlying cellular mechanisms that account for the normal human aging process. The greatest insights into the biology of aging have come from two sources: (1) our discovery of rare organisms that seem to have evaded the aging process, and (2) observations of rare, inherited premature aging diseases in humans.

4.2 AGING AND IMMORTALITY

"The dream of every cell, to become two cells!"

—Francois Jacob

It is important to understand that aging, followed by death, is not a constant feature of living organisms. Some organisms are short lived; some organisms live for a very long time; and other organisms are apparently immortal. Here are a few examples of long-lived organisms:

- Rougheye rockfish (*Sebastes aleutianus*), 205 years [3]
- Freshwater pearl mussel, 210–250 years [4]
- Bristlecone pine tree, 5000 years
- Quaking aspen tree, 80,000 years
- Hydrozoans *Turritopsis dohrnii* and *Hydractinia carnea* may be immortal [5–7]
- Some planarian flatworms appear to live indefinitely [8]

Many species of trees can live hundreds or even thousands of years. Methuselah, a Great Basin bristlecone pine residing in Inyo County, California, is reputed to be about 5000 years old, which seems to be about the observed limit for the lifespan of any individual standing tree.

Some trees self-clone within a copse, producing a group of trees all having the same genetic identity, with new clonal growths replacing dying trees. It is not unreasonable to consider the copse itself as a single biological organism, characterized by one genome and by a stable collection of growing and dying cells. Such clonal organisms are virtually immortal. A copse of quaking aspen, living in Fishlake National Forest, Utah, is estimated to be hundreds of thousands of years old [9].

Farmers in ancient and modern times have benefited from the self-cloning nature of trees by developing the agricultural technique known as coppicing. Young trees are cut to near-ground level, and new, clonal trees reshoot from the stump. By repeated cuttings, the trees are maintained as juveniles. Regularly coppiced trees never seem to age or die; they just spread out from the center. Individual coppiced trees have been maintained for centuries.

Not to be outdone by plants, two adult hydrozoan animals, *Turritopsis dohrnii* and *Hydractinia carnea*, have been observed to reverse their development to produce new polyps capable of growing into adulthood [5]. The cycle from polyp to medusa (i.e., the free-swimming, tentacled adult) to polyp to medusa can continue, seemingly forever.

What is the secret of the long-lived organisms? They all share the ability to continuously grow and regenerate. The long-lived immortal plants grow through their entire lives, getting bigger and bigger. Long-lived animals, such as tortoises, also continue to grow. The seemingly immortal hydrozoans revert to an early stage of development and regenerate the adult organism. **In no case is long life based on the ability of cells to persist as a collection of long-lived non-dividing cells**.

When humans think about immortality, they think in terms of staying the way they are, forever. Every human would like to keep their 25-year-old mind and body forever. We want to have new experiences, but we want to enjoy them as static organisms, without growing old. Nature seems to operate under a very different concept of immortality. In nature, immortality is achieved through cellular renewal and the creation of new individual organisms and species. Each new organism is created with a cell contributed by the parental organism. It happens that human birth is a sexual enterprise, requiring a gamete from the father and a gamete from the mother. For other organisms, sexual reproduction via two organisms is one of three additional options. These are: (1) parthenogenesis, wherein an egg cell self-fertilizes; (2) hermaphroditic reproduction, wherein gametes obtained from male and female sexual organs in the same organism fertilize one another; and (3) somatic reproduction, wherein a somatic cell breaks away from the parent organism to produce a new organism, or when a somatic cell grows a new organism as an attachment to the old organism (see Glossary item, Somatic). In all three cases, nature achieves immortality through rebirth, abandoning the old organism in the process. Adult humans are simply husks that hold gametes. By seeking to preserve our adult forms, we are pursuing a most unnatural goal.

In the next section, we will see that most of what we call aging involves degenerative changes in the cells that nature has abandoned: post-mitotic non-renewable cells in adult organisms.

4.3 PREMATURE AGING DISORDERS

"Medicine can only cure curable diseases, and then not always."

—Chinese proverb

We take aging for granted, as though it were an inevitable process that always unfolds at a natural pace. As with any other cellular system, normal aging can be slowed, accelerated, and damaged. As in every natural process, our genes play an important role. Long-lived parents tend to have long-lived children [10], and monozygotic twins tend to have closer lifespan concordances than unrelated individuals [11].

Of course, if we want to understand aging, we will need to extend our studies beyond simple life-expectancy measurements. There are many diseases that lead to the premature death of individuals, that have no relationship whatsoever to the aging process. A disease that predisposes to cancer or stroke or suicide would not be a disease of premature aging; it would just be a disease of premature death.

When creating a list of the rare diseases of aging, it is important to draw a distinction between diseases that produce conditions that tend to occur in advanced-age individuals and diseases that cause premature aging. A disease that produces age-related cosmetic changes in young individuals (e.g., gray hair, balding, wrinkles, and sagging skin), without producing constitutive aging processes (e.g., cachexia, frailty, and diminished stressor responses), would not be a premature aging disease. Leastways, not in this book. Hence, some of the diseases that are traditionally counted as premature aging syndromes (e.g., cutis laxa, branchiooculofacial syndrome, Ehlers–Danlos syndrome, epidermolysis bullosa simplex with mottled pigmentation, Williams–Beuren syndrome) are not included here.

Here are some inherited disorders of aging:

- Bloom syndrome, as in Werner syndrome (see below), is caused by a defect in a gene encoding a member of the RecQ helicase family of genes that play an important role in DNA replication, repair, recombination, and transcription [12]. Signs of premature aging include reduced immune competence and predisposition to developing diabetes, and early menopause in women. The most striking clinical feature of Bloom syndrome is a heightened risk of developing a wide range of cancers, and this cancer predisposition seems to be directly related to helicase-related DNA instability.
- Cockayne syndrome is characterized by a failure to grow, impaired development of the brain, physical signs of premature aging, photosensitivity, and leukodystrophy (degeneration of the white matter of the brain). The underlying genetic cause of the disease is a mutation in either the ERCC6 gene or the ERCC8 gene that codes for a DNA-binding protein involved in DNA excision repair.

- Dyskeratosis congenita is characterized by three striking morphologic features: abnormal skin pigmentation, nail dystrophy, and leukoplakia (i.e., white patches) in the oral mucosa. Clinically, the most deleterious clinical feature of dyskeratosis congenita is progressive bone marrow failure, which occurs in about 90% of cases. Bone marrow failure seems to result from a defect in cellular telomerase, leading to shortened telomeres, and to limitations on the replicative potential of bone marrow stem cells [13].

  An understanding of the relationship between telomere length and continuous replication of stem cells is key to understanding the biology of dyskeratosis congenita (see Glossary items, Pluripotent stem cell, Totipotent stem cell). Chromosomes are built with a long padding sequence of repetitive DNA at the chromosome tips, and this sequence is called the telomere. Animal cells lose a fragment of DNA from the tip of the chromosome with each cell division. This is because one strand of DNA is replicated as sequential fragments, with each fragment requiring a template sequence beyond its end to initiate replication. The last fragment in the DNA strand has no template for itself and is not replicated. By providing DNA padding at the tips of chromosomes, the telomere sequence sacrifices fragments of itself for the sake of preserving the coding sequences of the chromosome. As all good things come to an end, the telomere exhausts itself after about 50 rounds of mitosis (see Glossary items, Mitosis, Mitotic). At this time, the cell ceases further replication and will eventually die (see Glossary item, Telomere).

  Cells that continually renew throughout life, such as bone marrow stem cells, epidermal cells, and hair follicle cells, can restore their telomeres with an enzyme, telomerase. When such cells lose function in genes encoding for components of the telomerase complex, their ability to divide throughout the lifetime of the organism is shortened. Mutations in the telomerase-associated genes are the underlying cause of many cases of dyskeratosis congenita, and account for the progressive bone marrow failure associated with this syndrome [14]. About half of the cases of dyskeratosis congenita are molecularly undefined [15]. Telomerase gene mutations have also been found in some cases of acquired bone marrow failure [16]. This would suggest that telomerase deficiency, attained through the clonal expansion of a somatic mutation in a bone marrow stem cell, can lead to premature bone marrow aging.
- Mutations in any of at least eight genes, all involved in one way or another with recognizing and repairing DNA damage, account for Fanconi anemia. Individuals with Fanconi anemia have a high likelihood of developing bone marrow failure. Bone marrow failure can precede the development of acute myelogenous leukemia and myelodysplastic syndrome, a type of preleukemia (see Glossary items, Myelodysplastic syndrome, Myelodysplasia). Fanconi anemia does not produce general features of premature aging, such as frailty and cachexia. It is included here because bone marrow failure in Fanconi anemia seems to develop as the result of organ-specific aging of bone marrow stem cells [17]. As in the more typical diseases of premature

aging, the inability to continually renew tissue cells leads to a reduction in organ functionality. A reduction in the number of normal bone marrow stem cells provides an opportunity for the clonal expansion of pre-existing abnormal stem cells, leading to hematologic disorders such as leukemia and myelodysplastic syndrome.

- Hutchinson–Gilford progeria syndrome is a prototypical disease of premature aging characterized by wrinkled skin, atherosclerosis, renal failure, reduction of visual acuity, alopecia, scleroderma (i.e., skin tightening), and a high risk of heart attacks and strokes occurring at a young age. The underlying cause of Hutchinson–Gilford progeria syndrome is the production of progerin, a mutant form of lamin A. Lamin A is a nuclear protein that has important roles in maintaining the shape of the nucleus and in organizing DNA and RNA synthesis. Progerin, the mutant form of lamin A, produces striking abnormalities in the shape of the nucleus, featuring blebs, folds, and herniations of the nuclear envelope [18]. Also found are abnormalities in chromatin structure and increased DNA damage [19,20]. Individuals with Hutchinson–Gilford progeria seem to have a dysfunction of stem cells, limiting their ability to renew differentiated cells [19,21].
- Werner syndrome is a progeria syndrome with less severe symptoms than those associated with Hutchinson–Gilford progeria syndrome. It is characterized by scleroderma-like skin changes (i.e., skin tightening) with calcifications, cataracts, premature atherosclerosis, diabetes, and facial aging. Werner syndrome is caused by a defect in the WRN gene encoding RecQ helicases [12]. DNA helicases play a role in DNA replication, repair, recombination, and transcription. With multiple deficiencies in the DNA processing activities, it is not surprising that cells from individuals with Werner syndrome demonstrate chromosomal instability and a reduction in replication cycles (i.e., the total number of times a cell can replicate before becoming post-mitotic) (see Glossary item, Post-mitotic).
- Wolfram syndrome 2 is characterized by early onset diabetes, optic atrophy, and a shortened lifespan. It is caused by a mutation in the CISD2 gene, which encodes a protein associated with the outer mitochondrial membrane (see Glossary item, Mitochondria). Cisd2-null mice develop a progressive mitochondriopathy associated with defective respiration and with mitochondrial breakdown. These mice demonstrate premature aging and early death [22].
- Xeroderma pigmentosum was described in Section 3.4. Affected individuals cannot efficiently repair DNA damage induced by UV light. Skin cancers develop at an early age in sun-exposed skin. The mainstay of treatment is avoidance of daylight. Life expectancy is shortened, and signs of premature aging are seen. Xeroderma pigmentosum is usually listed among the diseases of aging, but most of the changes are confined to the skin and are simply the result of excess skin damage, not of a constitutive process that accelerates the aging process.

Based on the observation that some of the premature aging diseases have defects in DNA repair, it was hypothesized that the longevity of animal species is determined by the species-specific rate of DNA repair. Species that had a high rate of DNA repair were expected to have a long lifespan. Species with low rates of DNA repair would be short lived. Though some data supported this hypothesis, a reanalysis of the data found little evidence to favor earlier conclusions [23].

In 2009, Walker and coworkers reported a case study of a *sui generis* condition observed in a 16-year-old girl who had the appearance and anthropometric traits of an 11-month infant [24]. External and internal organs were infantile, including brain structure. After fetal development and birth, she had failed to mature into early childhood or adolescence. In a sense, her condition is the opposite of the premature aging conditions. The extreme rarity of this condition (i.e., more rare than the very rare monogenic disorders that produce premature aging) suggests that a simple loss of function in a single gene is unlikely to be at fault. This strange and sad case raises many questions about human development and aging, but, at this time, there are no answers.

Though aging is a naturally occurring process, it is also a disease. It is a true disease, like any other disease, because it causes the decline of function in various organs; it leads to frailty and a reduced ability to cope with physiological stressors; and it leads inevitably to death. A disease that causes premature aging is a disease that produces all of the aforementioned features at an early age.

When we examine diseases of premature aging, we find that the underlying mechanisms of these diseases are manifold: chromatin instability (Hutchinson–Gilford progeria); DNA instability (Werner syndrome); accumulation of toxic cellular products (tauopathies and prion diseases); mitochondrial degeneration (Wolfram syndrome); telomere shortening (dyskeratosis congenita). What do all these syndromes of diverse etiology have in common? The answer to this question is the topic of the next, and final, section of this chapter.

4.4 AGING AS A DISEASE OF NON-RENEWABLE CELLS

"Life is a concept."

—Patrick Forterre [25]

There are two types of cells in the body: cells that are capable of dividing, and cells that are not capable of dividing. An understanding of the relationship between dividing and non-dividing cells tells us a great deal about the physiologic process of aging, including which tissues age, which tissues do not age, and how tissues and organs are likely to be affected by the rare diseases of premature aging.

Humans grow rapidly *in utero*. After birth, growth continues through adolescence, tapering off as we enter early adulthood. Ideally, humans maintain about the same height and weight in late adulthood as they had in early adulthood. Though our bodies plateau during adulthood, the tissues are undergoing constant renewal. Vigorous, continual cell renewal is most evident in three tissues: the epidermis of the skin, the mucosal lining cells of the gut, and the blood forming cells of the bone marrow. In each of these three tissues, cellular renewal proceeds according to a simple principle: a stem cell divides to produce another stem cell plus a post-mitotic fully differentiated cell that lives for a while, doing whatever it was intended to do, and then dies. Because the stem cell replaces itself with a new stem cell when it divides, the total number of stem cells stays more or less constant throughout adult life.

Let us examine the process of tissue renewal in skin, gut, and bone marrow. The skin is covered by a thin epidermis that lies atop a continuous connective tissue sheath known as the dermis (see Figures 4.1 and 4.2). The bottom layer of the epidermis, directly adjacent to the underlying dermis, is called the basal layer and contains regenerating cells that divide to produce another regenerating cell and a non-dividing epidermal cell that is incapable of further division. The non-dividing cells are referred to as post-mitotic cells. With the exception of the bottom layer of regenerating cells, the full thickness of the epidermis is post-mitotic. These post-mitotic epidermal cells gradually fill their cytoplasm with keratin and flatten out to produce a protective barrier covering our bodies. Flattened epidermal cells are squamous, from the Latin

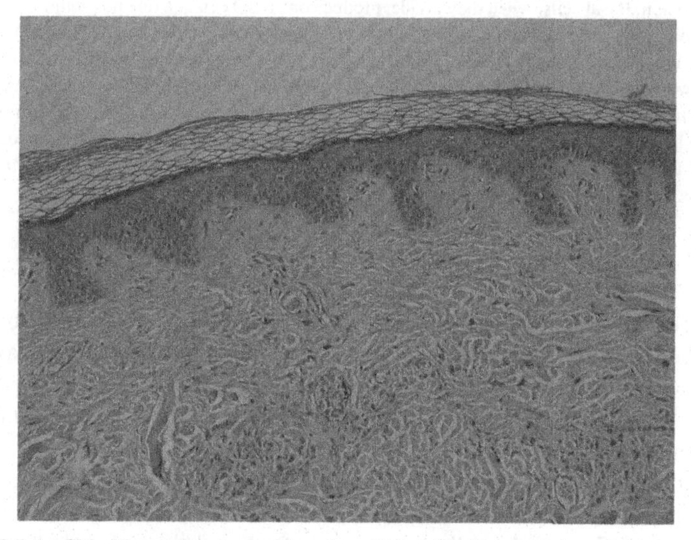

FIGURE 4.1 Skin biopsy showing epidermis overlying the fibrous tissue of the dermis. The epidermis is thinner than the dermis, and contains layers of cells, each layer having characteristic morphologic features. The epidermis has an undulating lower border, with papillae known as rete pegs, jutting down into the dermis. See color plate at the back of the book. *(Source: Wikimedia Commons, acquired as a public domain image.)*

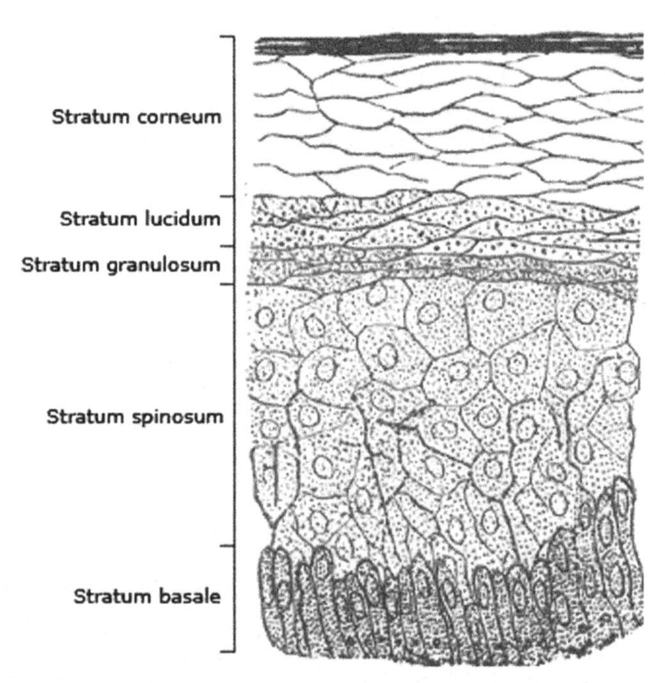

FIGURE 4.2 A graphic representing the various layers of the epidermis. The lowest or basal layer of cells contains the regenerating cells of the epidermis, the only cells of the epidermis capable of cell division. Basal cells, when they divide, produce one basal cell and one post-mitotic cell. The post-mitotic cell is pushed up into the next higher cell layer, and it continues to rise through the epidermis as it is pushed up by successive post-mitotic progeny of the basal layer. As it rises, it flattens out, fills with keratin, and eventually loses its nucleus. At this point, it is little more than a squamous flake, sitting atop the epidermis. The cells at the very top of the epidermis eventually slide off into the air to become floating specks of dust. Common dandruff consists of clumps of squamous cells sloughed from the stratum corneum. See color plate at the back of the book. *(Source: Wikimedia Commons, acquired as a public domain image.)*

root meaning scale. Aside from serving as bricks in a wall, the post-mitotic squamous cells are "dead men walking." Their fate is to rise to the top layer of the epidermis, where they slough off into the environment. The dancing house dust that we see in a beam of light is composed of sloughed post-mitotic squamous cells.

Similarly, the entire gastrointestinal tract is lined by a mucosal surface consisting primarily of non-dividing enterocytes. Under normal conditions, cell division is confined to the cells at the very bottom of the crypts and glands that line the alimentary tract. The post-mitotic enterocytes eventually slough into the gut lumen, and add to the bulk matter of stool.

In the bone marrow, a cascade of stem cells produces the fully differentiated red cells, white cells, and platelets that circulate in our blood. The circulating blood cells are post-mitotic. The red cells have shucked their nuclei and their mitochondria, reducing themselves to little more than bags of hemoglobin. The circulating post-mitotic red blood cells persist in the blood for a few months, after which they are phagocytized by the spleen and by other constituents of the reticuloendothelial system, the physiological equivalent of municipal garbage collectors. Phagocytized red blood cells are replaced by new red blood cells, so that the total number of circulating red blood cells stays fairly constant.

Why is it important to know how the epidermis, the gut, and the bone marrow produce post-mitotic cells from a subpopulation of continuously renewing cells?

4.4.1 Rule—The epidermis, the gut, and the bone marrow do not age.

Brief Rationale—These three tissues are constructed to continuously regenerate. Continuously regenerating tissues, like continuously regenerating animals and plants, do not senesce. It is not unusual to find elderly individuals with no histopathological signs of degeneration in these three tissues.

If we look at the gut, epidermis, and bone marrow of elderly persons, we find that the basal layer of the epidermis, the regenerating cells of the gut crypts and gland, and the progenitor hematopoietic cells are all dividing normally, just as they had in youth. It seems to be a general rule of nature that continuously dividing tissues do not senesce.

There are exceptions, of course. Disease processes that target the renewing cells of skin, gut, or bone marrow will have adverse consequences. For example, Fanconi anemia (see above) produces bone marrow failure, apparently by targeting hematopoietic stem cells [17]. Experiences with rare disorders of self-renewing tissues, such as Fanconi anemia and dyskeratosis congenita, remind us that any cellular system can be disrupted. We are not alone; in nematodes and in mice, conditions that deplete stem cells will produce premature aging [26,27].

4.4.2 Rule—Long-lived or immortal organisms have continual cell growth.

Brief Rationale—Aging is a degenerative process that occurs in cells that have lost the ability to divide. Organisms that maintain a population of cells that grow continuously or that maintain a permanent source of stem cells (i.e., cells that renew themselves and that renew other cells in the organism) can only experience aging in the non-dividing subpopulation.

While the skin, gut, and bone marrow are self-renewing systems, there are a variety of tissues in the body that become post-mitotic early in life, and remain so. These would include cartilaginous cells and oocytes. Other cells such as neurons, muscle cells, and connective tissue cells have a limited ability to divide in adulthood. The long-lived post-mitotic cells are all slowly degenerating throughout life.

4.4.3 Rule—On a cellular basis, aging is a process confined to non-renewable cell populations.

Brief Rationale—Long-lived cells that cannot replace themselves, such as fully differentiated neurons, muscle cells, and cartilage cells, have no biological destiny other than degeneration and death.

As non-dividing cells undergo wear and tear, or suffer damage that cannot be repaired, they will die. The tissues in which these damaged cells reside will function with diminished capacity. For example, osteoarthritis is a chronic disease that occurs from repeated episodes of bone crunching on its cartilage cushion within joints. Osteoarthritis occurs primarily in weight-bearing joints, such as knees and hips. Over a lifetime, the cartilage is frayed and eroded. Injured chondrocytes do not divide, or they divide with insufficient zest to restore a normal cartilaginous cushion. As erosion of the cartilaginous lining continues, an inflammatory reaction develops in the joint. The inflammatory reaction produces pain, swelling, and associated clinical symptoms.

Consider oocytes. All of the oocytes that a woman will produce are present *in utero*, reaching a peak of about 7 million cells at 5 months' gestation. After the peak is reached, about 3 months before birth, the oocytes begin to die; they are not replaced. The number of live oocytes declines until the number falls below a threshold of 1000, triggering menopause [28]. In this instance, as in every other example of human tissues undergoing aging, the process involves cells that cannot regenerate.

Frailty is a universal feature of old age. After the age of about 50, muscle mass gradually declines. The frailty associated with extreme aging is due, in part, to progressive sarcopenia. Muscle cells atrophy (i.e., reduce their size), die, and are not renewed. Frailty occurs because muscle cells were not designed to renew themselves continuously and indefinitely.

It was once thought that the brain cells you were born with are the same cells that you will die with; that brain cells do not divide. It is now known that regeneration (i.e., the growth of new neurons) occurs throughout life. This may be so, but new growth comes from reserve cells, not from fully differentiated neurons. Cell division cannot occur in a cell that becomes very large, like a neuron, and has appendages (i.e., an axon and dendrites) extending to and from other cells, sometimes over great distances (up to several feet in the case of motor neurons innervating foot muscles). Axons are ensheathed by a dependent network of periaxonal cells (i.e., oligodendrocytes in the central nervous system and Schwann cells in the peripheral nervous system). Neurons are transfixed anatomically, and cannot round up to divide. Hence, the fully mature neuron has little or no regenerative opportunity. Consequently, many of the cellular changes that we associate with aging take place in neurons. The dementia that accompanies aging is due to the inability of injured neurons to repair or replace themselves.

The tauopathies are disorders wherein tau protein accumulates within neurons. Tau proteins are involved in the stabilization of microtubules in every cell throughout the body, but they accumulate to the greatest extent in the neurons of the central nervous system. If a fully differentiated neuron cannot clear its tau proteins, it will suffer progressive damage, leading to cell death. Though tau proteins are ubiquitous, the tauopathies always develop as neurodegenerative disorders. Examples of diseases in which tau proteins are found include: Alzheimer's disease, progressive supranuclear palsy, argyrophilic grain disease, corticobasal degeneration, dementia pugilistica, a form of Parkinsonism known as Lytico–Bodig disease or as Parkinson–dementia complex of Guam, a form of Parkinsonism linked to chromosome 17, frontotemporal dementia, frontotemporal lobar degeneration, Hallervorden–Spatz disease, lipofuscinosis, meningioangiomatosis, Pick's disease, a rare tumor of neurons known as ganglioglioma [29], subacute sclerosing panencephalitis, lead encephalopathy, tangle-predominant dementia, and tuberous sclerosis.

The prion diseases are another example of disorders that target non-dividing neurons. The term prion was introduced in 1982 by Stanley Prusiner [30]. Prions are the only infectious agent that contains neither DNA nor RNA. A prion is a misfolded protein that can serve as a template for proteins of the same type to misfold, producing globs of non-functioning protein, causing cells to degenerate. The site of greatest accumulation of prion protein is in brain cells.

Though few scientists would consider prions to be organisms, living or otherwise, they are undoubtedly transmissible infectious agents. The most common mode of transmission of prion disease is through the consumption of brains of infected animals.

The cells of the body that are most vulnerable to prion disease are the neurons of the brain. The reason for the particular sensitivity of neurons to prion disease relates to the limited ability of neurons to replicate (i.e., to replace damaged neurons with new neurons), reconnect (to replace damaged connections between a neuron and other cells), and to remove degenerated cells and debris.

There are five known prion diseases of humans, and all of them produce encephalopathies characterized by decreasing cognitive ability and impaired motor coordination. They are: Kuru, Creutzfeldt–Jakob disease, bovine spongiform encephalopathy (known in humans as new variant Creutzfeldt–Jakob disease), Gerstmann–Straussler–Scheinker syndrome, and fatal familial insomnia. At present, all of the prion diseases are progressive and fatal.

Prions have been observed in fungi, where their accumulation does not seem to produce any deleterious effect, and may even be advantageous to the organism [31].

In Section 4.3, we listed the many causative mechanisms underlying the rare diseases of premature aging. Without exception, every disease of premature aging creates a defect in the normal process of cellular renewal. If we understood how to control and maintain stem cell renewal, a feat that nematodes seem to have mastered, then we might understand how to defeat the aging process.

In Chapter 7, we will be discussing cancer, another disorder of cell renewal. **Whereas aging is a disease of cells that cannot divide, cancer is a disease of cells that cannot stop dividing**.

REFERENCES

1. Xue Q. The frailty syndrome: definition and natural history. Clin Geriatr Med 2011; (27): 1–15, 2011.
2. Sayer AA, Robinson SM, Patel HP, Shavlakadze T, Cooper C, Grounds MD. New horizons in the pathogenesis, diagnosis and management of sarcopenia. Age Ageing 42:145–150, 2013.
3. Cailliet GM, Andrews AH, Burton EJ, Watters DL, Kline DE, Ferry-Graham LA. Age determination and validation studies of marine fishes: do deep-dwellers live longer? Exp Gerontol 36:739–764, 2001.
4. Ziuganov V, San Miguel E, Neves RJ, Longa A, Fernandez C, Amaro R, et al. Life span variation of the freshwater pearlshell: a model species for testing longevity mechanisms in animals. Ambio 29:102–105, 2000.
5. Schmich J, Kraus Y, De Vito D, Graziussi D, Boero F, Piraino S. Induction of reverse development in two marine hydrozoans. Int J Dev Biol 51:45–56, 2007.
6. Martinez DE. Mortality patterns suggest lack of senescence in hydras. Exp Gerontol 33:217–225, 1998.
7. Rich N. Can a jellyfish unlock the secret of immortality? The New York Times, November 28, 2012.
8. Tan T, Rahman R, Jaber-Hijazi F, Felix DA, Chen C, Louis EJ, et al. Telomere maintenance and telomerase activity are differentially regulated in asexual and sexual worms. Proc Natl Acad Sci USA 109:4209–4214, 2012.
9. Mitton JB, Grant MC. Genetic variation and the natural history of quaking aspen. BioScience 46:25–31, 1996.
10. Abbott MH, Murphy EA, Bolling DR, Abbey H. The familial component in longevity. A study of offspring of nonagenarians. II. Preliminary analysis of the completed study. Johns Hopkins Med J 134:1–16, 1974.
11. Jarvik LF, Falek A, Kallmann FJ, Lorge I. Survival trends in a senescent twin population. Am J Hum Genet 12:170–179, 1960.
12. Mohaghegh P, Hickson ID. DNA helicase deficiencies associated with cancer predisposition and premature ageing disorders. Hum Mol Genet 10:741–746, 2001.
13. Calado R, Neal Young N. Telomeres in disease. F1000 Med Rep 4(8) 2012.
14. Vulliamy T, Beswick R, Kirwan M, Marrone A, Digweed M, Walne A, et al. Mutations in the telomerase component NHP2 cause the premature ageing syndrome dyskeratosis congenita. Proc Natl Acad Sci USA 105:8073–8078, 2008.
15. Shtessel L, Ahmed S. Telomere dysfunction in human bone marrow failure syndromes. Nucleus 2:24–29, 2011.
16. Yamaguchi H. Mutations of telomerase complex genes linked to bone marrow failures. J Nippon Med Sch 74:202–209, 2007.
17. Zhang X, Li J, Sejas DP, Pang Q. Hypoxia-reoxygenation induces premature senescence in FA bone marrow hematopoietic cells. Blood 106:75–85, 2005.
18. Mallampalli MP, Huyer G, Bendale P, Gelb MH, Michaelis S. Inhibiting farnesylation reverses the nuclear morphology defect in a HeLa cell model for Hutchinson-Gilford progeria syndrome. PNAS 102:14416–14421, 2005.
19. Scaffidi P, Misteli T. Lamin A-dependent misregulation of adult stem cells associated with accelerated ageing. Nat Cell Biol 10:452–459, 2008.
20. Shumaker DK, Dechat T, Kohlmaier A, Adam SA, Bozovsky MR, Erdos MR, et al. Mutant nuclear lamin A leads to progressive alterations of epigenetic control in premature aging. Proc Natl Acad Sci USA 103:8703–8708, 2006.

21. Liu GH, Barkho BZ, Ruiz S, Diep D, Qu J, Yang SL, et al. Recapitulation of premature ageing with iPSCs from Hutchinson-Gilford progeria syndrome. Nature 472:221–225, 2011.
22. Chen YF, Kao CH, Chen YT, Wang CH, Wu CY, Tsai CY, et al. Cisd2 deficiency drives premature aging and causes mitochondria-mediated defects in mice. Genes Dev 23:1183–1194, 2009.
23. Promislow DE. DNA repair and the evolution of longevity: a critical analysis. J Theor Biol 170:291–300, 1994.
24. Walker RF, Pakula LC, Sutcliffe MJ, Kruk PA, Graakjaer J, Shay JW. A case study of "disorganized development" and its possible relevance to genetic determinants of aging. Mech Ageing Dev 130:350–356, 2009.
25. Forterre P. The two ages of the RNA world, and the transition to the DNA world: a story of viruses and cells. Biochimie 87:793–803, 2005.
26. Ishii N, Fujii M, Hartman PS, Tsuda M, Yasuda K, Senoo-Matsuda N, et al. A mutation in succinate dehydrogenase cytochrome b causes oxidative stress and ageing in nematodes. Nature 394:694–697, 1998.
27. McLaughlin PJ, Bakall B, Choi J, Liu Z, Sasaki T, Davis EC, et al. Lack of fibulin-3 causes early aging and herniation, but not macular degeneration in mice. Hum Mol Genet 16:3059–3070, 2007.
28. Fogli A, Rodriguez D, Eymard-Pierre E, Bouhour F, Labauge P, Meaney BF, et al. Ovarian failure related to eukaryotic initiation factor 2B mutations. Am J Hum Genet 72:1544–1550, 2003.
29. Brat DJ, Gearing M, Goldthwaite PT, Wainer BH, Burger PC. Tau-associated neuropathology in ganglion cell tumours increases with patient age but appears unrelated to ApoE genotype. Neuropathol Appl Neurobiol 27:197–205, 2001.
30. Prusiner SB. Novel proteinaceous infectious particles cause scrapie. Science 216:136–144, 1982.
31. Michelitsch MD, Weissman JS. A census of glutamine/asparagine-rich regions: Implications for their conserved function and the prediction of novel prions. PNAS 97:11910–11915, 2000.

Chapter 5

Diseases of the Heart and Vessels

5.1 HEART ATTACKS

"The chief problem in historical honesty isn't outright lying. It is omission or de-emphasis of important data."

—Howard Zinn

One of the great ironies of pathology (i.e., the study of disease) is that heart attacks, the most common cause of sudden death, are not a primary disease of the heart (see Glossary item, Pathologist). Basically, the heart is a surprised victim of a systemic process that has nothing to do with the heart, at least not directly.

For those who die as a result of their first heart attack, their hearts are typically free of any significant disease, until the moment after an occlusion blocks a major coronary vessel. In most cases, there is no early disease process within the cardiac muscle, the conduction system, or the valves of the heart. The heart does not contribute to the pathologic process that leads to the heart attack. The event that causes the heart attack occurs within a medium caliber vessel, near its origin at the root of the aorta, where the vessel hugs the surface of the heart.

In a sense, the reason that heart attacks are common is the result of poor evolutionary design. If we had two hearts, much like we have a spare kidney, adrenal, and gonad, then we might survive a heart attack. If we had a better vascular system for the heart, with more coronary arteries and more anastomoses (connections between vessels), then heart attacks would be less common and the consequences would be less severe.

A heart attack occurs when a portion of a coronary artery is blocked. Heart muscle distal to the blockage becomes anoxic (i.e., deprived of the oxygen that would have been delivered by the coronary artery under normal circumstances), and dies (see Figures 5.1–5.4). It is the anoxic state of cardiac muscle, and all the pathologic changes that follow, that accounts for the pain and injury that constitutes the heart attack. The location of the blockage influences the location and extent of the cardiac muscle damage, which in turn influences the prognosis (i.e., likelihood of recovery).

Rare Diseases and Orphan Drugs. http://dx.doi.org/10.1016/B978-0-12-419988-0.00005-5

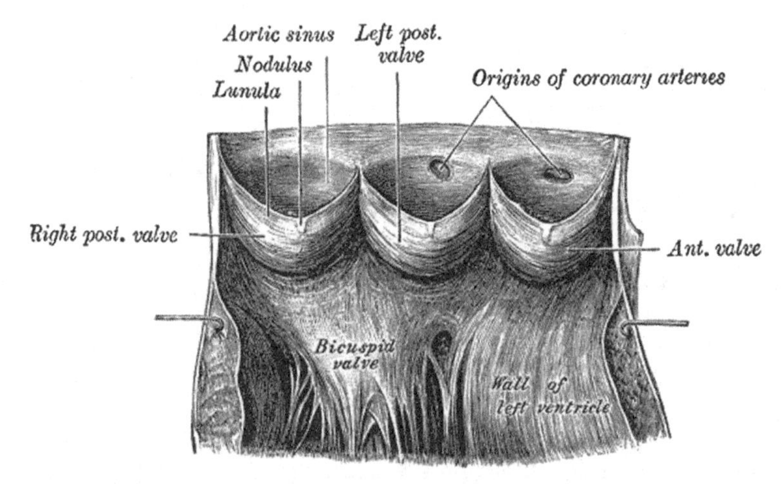

FIGURE 5.1 **The aortic valve of the heart.** Note that the two coronary arteries (i.e., left and right) originate as small ostia (i.e., holes), each leaving from a valve. These two small ostia deliver the arterial blood supply to the heart muscle. (*Source: Wikimedia Commons, acquired as a public domain image.*)

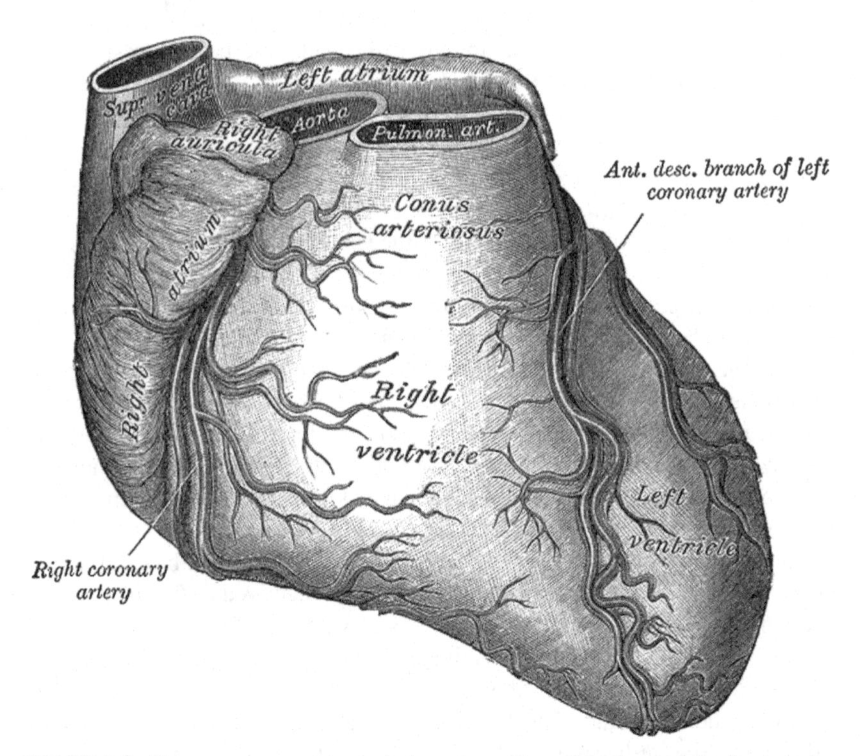

FIGURE 5.2 The coronary arteries travel along the surface of the heart, delivering branches that penetrate into the heart muscle. See color plate at the back of the book. (*Source: Wikimedia Commons, acquired as a public domain image.*)

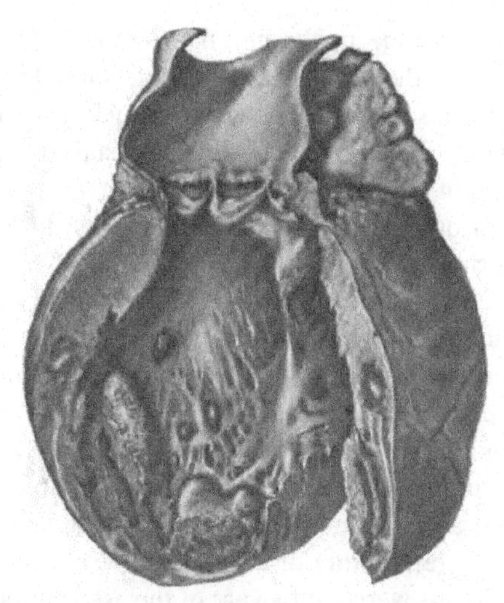

FIGURE 5.3 Myocardial infarction. Leading up from the apex of the heart (i.e., the bottom tip) is extensively scarred, hemorrhagic, and thinned muscle wall, in contrast to the thicker, uniformly dusky color or the heart muscle at the heart's base (i.e., top). The endocardium (i.e., the lining of the heart) near the apical scar has attached thrombus, seen here as bulbous tissue insinuating among trabeculae carneae (i.e., thin strands of muscle lining the endocardium). (*Source: MacCallum WG. A Textbook of Pathology* [1].)

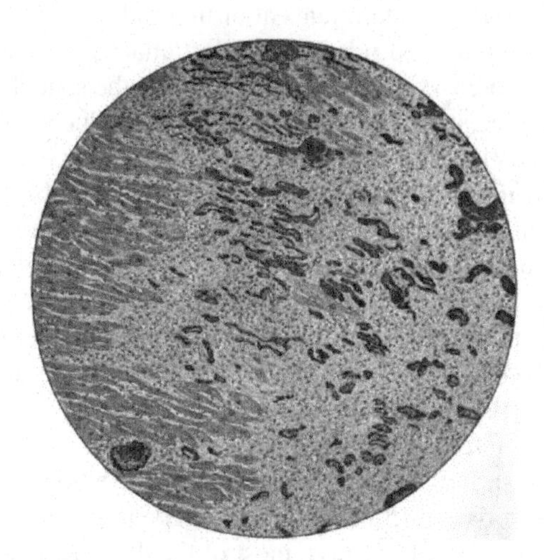

FIGURE 5.4 Histopathology of heart involved by early myocardial infarction. To the left of the image are surviving muscle fibers. To the right, the heart muscle has been replaced by inflammatory cells, vessels, and granulation tissue (i.e., early scar tissue). See color plate at the back of the book. (*Source: MacCallum WG. A Textbook of Pathology* [1].)

What causes the blockage in the coronary artery that leads to the heart attack? Atherosclerosis is the accumulation of fatty or calcified plaques in the wall and lumen of arteries. An atherosclerotic plaque can enlarge to the point that blood flow is blocked, and the heart muscle distal to the blockage dies of hypoxia (i.e., insufficient oxygen).

Thrombi can form on areas of the arterial wall that lack the smooth inner surface typical of normal arteries (see Figures 5.5 and 5.6). When the atherosclerotic plaque serves as a nidus for thrombus formation, blockage can occur when the plaque plus the thrombus act together to narrow the vessel lumen. These two phenomena, atherosclerosis plus thrombus, acting alone or in concert, are common causes of heart attack.

The Framingham Heart Study, which began in 1961, is often given credit for establishing the connection between high cholesterol levels and heart disease. Although the Framingham Heart Study provided important statistical evidence, based on a careful study of a large number of individuals, it is historical fact that physicians were well aware of the association between cholesterol and heart attacks decades prior to 1961 (see Glossary item, Association). Rather early in the twentieth century, the common blood chemistry tests that we use today were established, and these tests were clinically interpreted much like they are interpreted today. An association between high cholesterol levels and arteriosclerosis was recognized by 1921 [2]. The genetic link between cholesterol and heart attacks was understood in 1938, when it was shown that patients with familial hypercholesterolemia had a high risk of developing heart disease [3]. Twenty-five years later, observations of families with familial hypercholesterolemia revealed two distinct forms of the disease. A homozygous form affected infants at birth, produced blood cholesterol levels of about 800 mg/dl, and resulted in heart attacks in children as young as 5 years of age. A heterozygous form, occurring in the same families, produced lower levels of cholesterol, 300–400 mg/dl, and produced heart attacks at the age of 35–60 years [3].

Observations that heart attacks occurred in inherited hypercholesterolemia syndromes inspired a search for cholesterol-lowering drugs. In the 1970s, Akiron Endo found that several species of fungi extrude defensive compounds that inhibit the synthesis of cholesterol in fungal pathogens. Endo studied 6000 fungal compounds, eventually finding mevastatin, the first effective inhibitor of human HMG-CoA reductase, the rate-limiting enzyme in the cholesterol biosynthetic pathway [4]. Other statins followed.

A subfield of internal medicine is devoted to the dyslipidemias. Without going into detail, various types of lipoproteins in the blood transport cholesterol and other lipids to tissues [5]. Dyslipidemias can be monogenic, polygenic, or acquired. Regardless of their cause, most of the dyslipidemias are treated with statins.

Applying a little imagination, it is easy to see that there are many different mechanisms whereby a blockage of the coronary artery may occur. The blockage

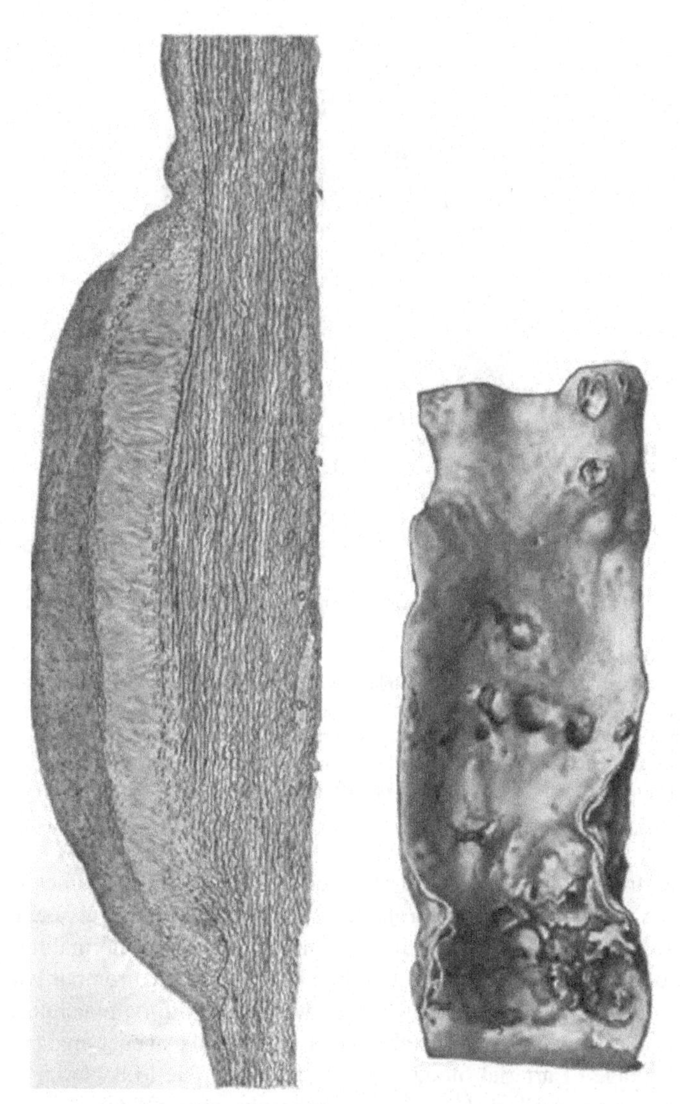

FIGURE 5.5 Histopathologic section of an aortic thrombus (left) and gross specimen of an opened aortic segment displaying its internal surface (right). The histopathology shows a raised atheroma. The surface of the atheroma has a thin fibrinous coating over a fatty streak. The underlying muscular wall of the vessel is, at this time, relatively intact. The gross specimen shows numerous atherosclerotic foci, some of which are protruding atheromas, such as the atheroma shown in the histopathologic section on the left, while others are calcified, irregular, and flattened. These complex, calcified plaques are often associated with degenerative changes in the underlying vessel wall. (*Source: MacCallum WG. A Textbook of Pathology* [1].)

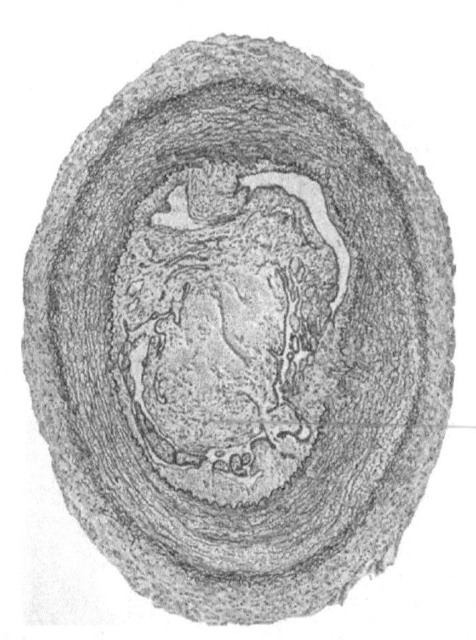

FIGURE 5.6 Thrombus in artery. A fibrinous thrombus fills the lumen of a muscular artery. The artery features a thick media (i.e., muscular wall with circumferentially aligned fibers underlying the wavy intimal lining). Filling the lumen (i.e., the inside of the artery, which would be carrying blood under normal circumstances) is a thrombus. Notice that the thrombus has numerous lined vessels running through it (i.e., the long-thin empty spaces within the thrombus). These vessels within the thrombus arise through a biologic process called re-canalization, in which new vessels grow, and blood flows, at a reduced rate and volume through old thrombi. See color plate at the back of the book. (*Source: MacCallum WG. A Textbook of Pathology* [1].)

may result from a muscle spasm of the arterial wall, causing the lumen to close. Weakening in the wall of the coronary artery might lead to a dissection (i.e., internal tear) that pushes part of the wall into the lumen. Abnormalities of the surfaces of the arterial walls may raise the likelihood that a thrombus will form, leading to a blockage. An aneurysm (i.e., weakness with outpouching) of the muscle wall might cause a large thrombus to form at the aneurysmal site, leading to blockage of arterial blood flow. An infection due to bacteria, virus, or fungus within the wall of the vessel might produce an inflammatory reaction with thrombus and granulation tissue, producing a blockage. An inherited or acquired blood disorder characterized by hypercoagulation (i.e., tendency to form clot) might lead to thrombus formation and consequent blockage. As it turns out, each of these wild scenarios applies in some cases.

A rare cause of heart attacks arises in some cases of Hutchinson–Gilford progeria syndrome, a disorder of premature aging characterized by a dysfunction of stem cells [6] (see Section 4.3) [7]. Individuals with this disease develop arteriosclerotic disease in childhood. Histopathologic findings indicate that the smooth muscle cells in the wall of affected arteries are replaced by fibrous tissue. It has been suggested that arteriosclerosis in Hutchinson–Gilford progeria syndrome begins with senescence of the vascular smooth muscle cells, and this primary cellular defect leads to the arteriosclerotic changes in the vessel, and to the occurrence of heart attacks at a very early age [8].

There are very few diseases that occur as the result of one exclusive pathophysiologic process. Heart attacks are always preceded by one event: the blockage of a coronary artery. Nonetheless, we have seen that there are many different pathological processes that can eventually lead to the blockage of a coronary artery. Each of these processes may be associated with inherited traits that raise the likelihood that a blockage will occur. Likewise, each process that increases the likelihood of a blockage may have an acquired (i.e., non-inherited) cause.

Heart attacks are easy to explain; the pathogeneses of most other diseases are less easily understood. When we think about cancer, we know a lot about the genetic and phenotypic changes that characterize cancer cells, but we have much to learn about the events that occur throughout the long process that leads from a normal cell to a malignant tumor. This topic will be discussed in Chapter 8.

5.2 RARE DESMOSOME-BASED CARDIOMYOPATHIES

"If you want to make an apple pie from scratch, you must first create the universe."

—Carl Sagan

Just like cars, humans have an electrical system. The pistons in a car's engine cannot produce a coordinated combustion cycle unless an electrical spark fires the fuel at just the right time. A human heart cannot beat unless an electrical impulse passes through the ventricles causing them to contract in a coordinated way to produce a heartbeat.

To understand how electrical/mechanical rhythms are carried through the heart, and through the nervous system in general, it is necessary to go back a billion years to a time when the first animals appeared on Earth.

Animals are thought to have evolved from simple, spherical organisms floating in the sea, called gallertoids. The living gallertoid sphere was lined by a single layer of cells enclosing a soft center in which fibrous cells floated in extracellular matrix.

As the gallertoids evolved to extract food from the seabed floor, they flattened out. These early gallertoids had invented a novel way of locking cells together, the desmosome attachment. The desmosome supported the formation of multicellular organisms containing multiple organs.

Three classes within Class Eukaryota (organisms with nucleated cells) are multicellular, and all three contain specialized cells and organs: Class Plantae, Class Fungi, and Class Animalia. Classes Plantae and Fungi build hard, permanent tissues using cells made rigid by a cellulose wrap, in the case of plants, and a chitin wrap, in the case of fungi. Animals, unlike plants and fungi, have non-rigid cell walls (i.e., not hardened by chitin or cellulose). The desmosome, an innovation found exclusively in animals, enabled animals to create sturdy ducts, glands, epidermis, and other component structures of organs, using only baggy, fluid-filled, spherical cells. Here is how it works. The taut epithelial structures composed of polygon-shaped epithelial cells are something of an illusion. Each animal cell that appears to be shaped like a polygon is actually a sphere (see Figure 5.7).

It happens that whenever two soft spheres are pushed together, the conjunction of the cells is a flattened surface (see Figure 5.8). When multiple cells of the same size are pushed together, each cell takes the shape of a regular polygon, and the net effect is a honeycomb-shaped network of cells called an epithelium.

An epithelial structure built from a group of crowded, soft spheres will disassemble into individual, round cells, unless they are somehow fastened together permanently. The invention of the desmosome and the gap junction permitted animals to button up epithelial tissues without the aid of external building materials (such as cellulose or chitin).

Desmosomes and gap junctions, unique to Class Animalia, create a leak-proof continuum of epithelial cells (see Figure 5.9). All animals contain epithelial cells that line the external surface of the animal (i.e., the skin), the gastrointestinal tract, and most of the internal organs (e.g., liver, pancreas, salivary glands).

The first evolutionary achievement that can be credited to the desmosome is the development of the bastula, a stage in embryogenesis that is unique to animals (see Figure 5.10). The cells of the early embryo are held together by desmosomes and gap junctions, and begin to secrete fluid. Because the junctions holding cells together are water-tight, fluids secreted by the cells accumulate in a central cavity. An embryo, with a fluid-filled center, is known as a blastula.

Presumably, the earliest animal, the gallertoids, were lined by non-rigid epithelial cells, zipped together with tight junctions, producing a jelly-like fluid

FIGURE 5.7 Spheres representing single epithelial cells, which are always smooth, round, soft, and non-polyhedral when isolated and suspended in liquid.

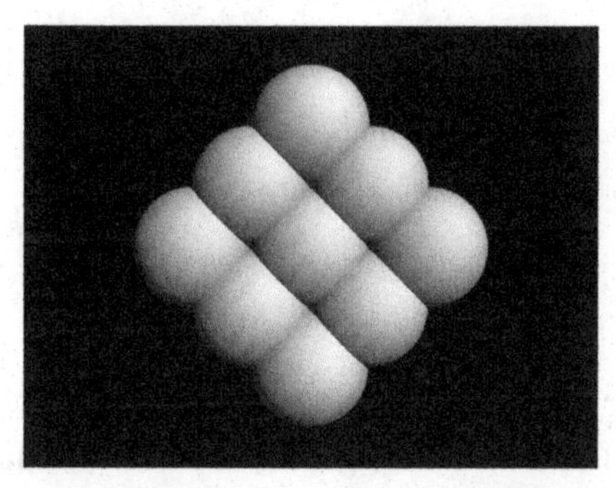

FIGURE 5.8 Touching spheres, representing epithelial cells that are pushed together in a growing tissue. Notice that where spheres touch, flat surfaces are produced. The effect of many spheres pushing together is a polyhedral network, simulating an epithelium.

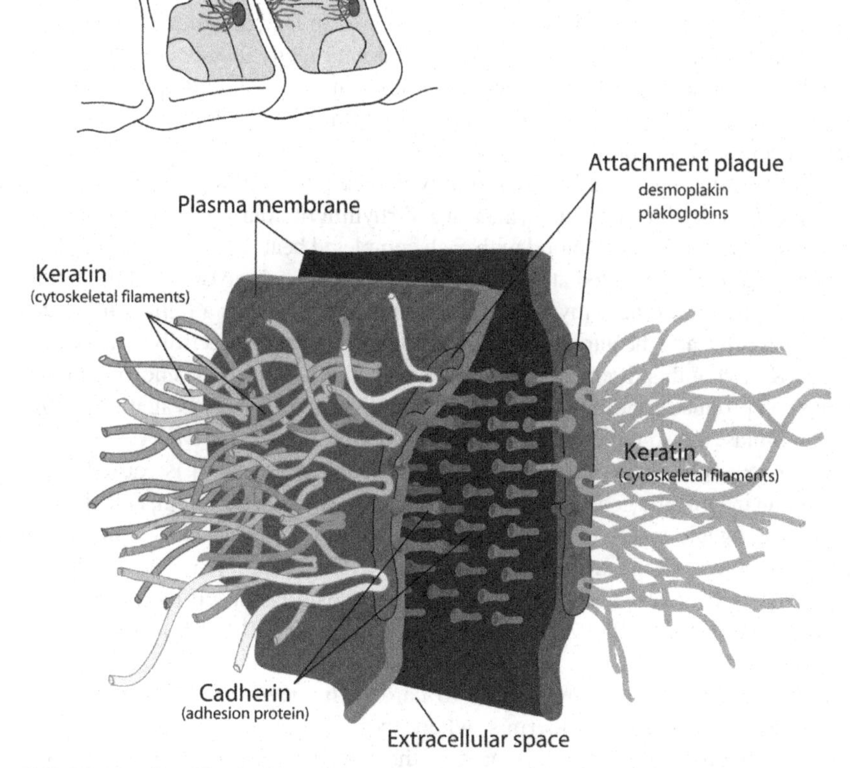

FIGURE 5.9 Graphic of a desmosome. The desmosome is much like a button that holds together the flat-surfaced areas where cells touch one another. The net effect is to produce a permanent epithelial (i.e., polyhedral) network of cells. Some cells are particularly rich in desmosomes, such as keratinocytes, cardiac myocytes, and the cells that ensheath nerve fibers. See color plate at the back of the book. (*Source: Wikipedia, created and released into the public domain by Mariana Ruiz, LadyofHats.*)

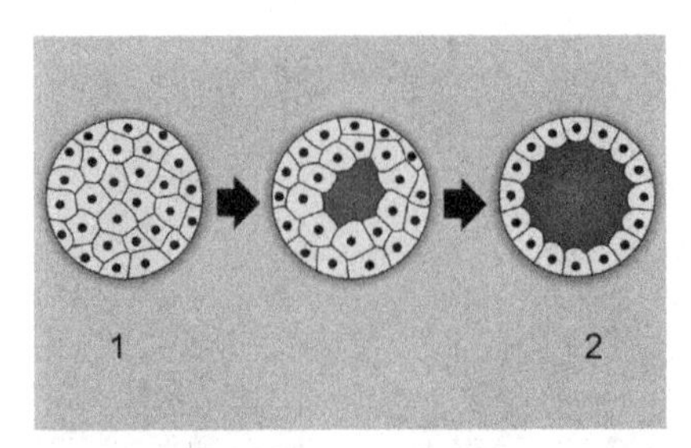

FIGURE 5.10 Graphic of blastulation. The early, solid, embryo secretes fluid into a central viscous, the developing blastula. Blastulation is accomplished with specialized membrane channels that transport ions and water, and with desmosomes, that provide a water-tight boundary between adjacent cells. Among multicellular organisms, the blastulated embryo is found exclusively in members of Class Animalia. See color plate at the back of the book. (*Source: Wikipedia, created and released into the public domain by Pidalka44.*)

filling its center [9]. After the blastula had evolved, additional complex tissues and organs, such as the stomach, came into being. In fact, all animals other than the primitive sponges and placazoans, a relict with many features of the gallertoids, have stomachs. The fundamental biological feature that defines animals, and distinguishes animals from all other organisms, can be traced to soft cells and specialized junctions.

Fascinating, but what has this to do with heart arrhythmias? The heart is a muscle with rhythm. It is the synchronized rhythm of connected myocytes that pumps out about 70 ml of blood with each completed beat. Desmosomes and gap junctions are concentrated at the longitudinal end of each myocyte, at the point of contact with the next myocyte. The desmosome provides a tight continuum between cells, and the gap junctions mediate the passage of electrolytes between cells. Knowing this, it comes as no surprise that mutations in desmosomal proteins can produce conditions wherein cardiac muscle pathology is aggravated by arrhythmias; the so-called arrhythmogenic cardiomyopathies [10].

For example, arrhythmogenic right ventricular dysplasia-8 is caused by a mutation in DSP, the gene encoding desmoplakin (see Glossary item, Dysplasia). Arrhythmogenic right ventricular cardiomyopathy-12 is caused by a mutation in the gene encoding junction plakoglobin (i.e., JUP gene). A form of arrhythmogenic right ventricular cardiomyopathy/dysplasia is caused by a heterozygous mutation in the PKP2 gene, which encodes plakophilin-2, a protein of the cardiac desmosome.

In addition to their importance to cardiac myocytes, desmosomes play an essential role in the epidermis, where they appear in high concentrations on interlocking squamous cells and at the interface between the epidermis and the dermis. Desmosomes keep the epidermis from rubbing off when friction is applied to skin. Desmosomal disorders produce diseases characterized by blistering, acantholysis (e.g., epidermal cells falling apart from one another), and keratoses characterized by thickened and hardened regions. Variations in one gene, desmoplakin, can result in lethal acantholytic epidermolysis bullosa, keratosis palmoplantaris striata-2, and skin fragility–woolly hair syndrome.

The story of desmosomes began with its evolutionary invention a billion years ago, then led to its role in creating the blastula, then the gastrointestinal tract, and finally differentiated organs, including epidermis and heart. It should not be surprising that mutations of desmosomal proteins produce rare syndromes characterized by a combination of cardiac and skin pathology. Naxos syndrome, also known as diffuse non-epidermolytic palmoplantar keratoderma with woolly hair and cardiomyopathy, is caused by a mutation in the plakoglobin gene. Dilated cardiomyopathy with woolly hair and keratoderma is caused by a mutation in the DSP gene, as is a similar syndrome in which the dermatologic features include a pemphigus-like skin disorder [11].

Most of the isolated conduction defects of the heart are channelopathies, and are discussed in the next section.

5.3 SUDDEN DEATH AND RARE DISEASES HIDDEN IN UNEXPLAINED CLINICAL EVENTS

"Life, too, is like that. You live it forward, but understand it backward. It is only when you stop and look to the rear that you see the corpse caught under your wheel."

—Abraham Verghese from his book *Cutting for Stone*

There is a popular conception, fortified by forensic pathologists portrayed on television, that an autopsy always determines the cause of death. Not so; particularly in cases of sudden deaths that occur naturally (i.e., not due to murder or accident). A careful autopsy study of 322 cases of natural sudden death, published in 1988, revealed that heart attacks accounted for 59% of cases [12]. Other morphologically verified heart diseases accounted for an additional 7.5% of cases (e.g., valvulopathies, cardiomyopathies, and anatomic abnormalities). Non-cardiac diseases accounted for 28% of cases (e.g., pulmonary emboli, stroke). Alcohol caused about 3% of sudden deaths. In 3.4% of cases, no cause of death was found [12]. Other studies indicate that if studies are limited to the young (i.e., younger than age 36), up to one-third of cases of sudden natural deaths are unexplained [13].

In the past, when an autopsy failed to reveal a cause of death, the pathologists inferred that the death was "physiologic," signifying that a lethal event occurred that did not produce tissue changes that could be diagnosed by a pathologist. The lethal event most likely to produce death, without evidence of cellular disease, was thought to be ventricular arrhythmia. Of course, every disease has pathology; if you know where to look for it. In cases of "physiologic" sudden deaths, you need to look at genes.

Today, we know the genetic causes of many types of lethal arrhythmias, and we know that all of the arrhythmias that lack a structural explanation (i.e., conditions without anatomic or histologic pathology) are channelopathies [14]. These are diseases caused by defects in the channels that conduct ions across the membranes of specialized cells (e.g., sodium channel, potassium channel, chloride channel, calcium channel). Most of the channelopathies involve the passage of ion waves through the so-called excitable tissues: nervous system, skeletal muscle, or the conduction system of the heart. The channelopathies of excitable tissues include: alternating hemiplegia of childhood, Brugada syndrome, Dravet syndrome (severe myoclonic epilepsy of infancy), episodic ataxia, erythromelalgia, generalized epilepsy with febrile seizures, plus familial hemiplegic migraine, hyperkalemic periodic paralysis, hypokalemic periodic paralysis, long QT syndrome, malignant hyperthermia, myotonia congenita, neuromyotonia, nonsyndromic deafness, paramyotonia congenita, retinitis pigmentosa, short QT syndrome, and Timothy syndrome.

Other channelopathies have a pathogenetic mechanism that does not involve the passage of an electronic wave through tissues. The channelopathies of non-excitable tissues include: Bartter syndrome, congenital hyperinsulinism, cystic fibrosis, malignant hyperthermia, and mucolipidosis type IV.

To complicate matters, channelopathies can result from environmental agents that target channels, or from channel-specific autoimmune diseases [15]. Venoms, including tetrodotoxin, saxitoxin, and ciguatoxin, all target sodium channels. Myasthenia gravis is an acquired autoimmune disease characterized by progressive muscle weakness. The disease phenotype is produced by autoantibodies reacting with nicotinic acetylcholine receptors. Reduced activity of nicotinic acetylcholine receptors upsets the normal flux of sodium ions that is required for muscle contraction.

Here are a few genetic channelopathies known to cause sudden natural death:

- **Long QT syndrome** is a collection of closely related genetic disorders that are all characterized by an electrocardiogram with a prolonged QT segment (see Figure 5.11). A long QT segment is associated with an increased risk of an irregular heartbeat known as torsades de pointes. Torsades de pointes, in turn, may lead to sudden death from ventricular fibrillation. Genetic mutations in potassium channels, sodium channels, or calcium channels can lead to long QT syndrome. Long QT syndrome can also result from a mutation in the gene encoding ankyrin, a protein that anchors ion channels to the plasma membrane. The three major genes causing long QT syndrome are: KCNQ1, KCNH2, and SCN5A [13].
- **Brugada syndrome** is a frequent cause of sudden natural deaths. Brugada syndrome is most common in individuals of Asian descent. The syndrome affects males nearly 10 times as often as females. It can cause the death of individuals of any age, but men in the 30–50 year age group are at highest risk. It is characterized by one of several abnormal electrocardiogram patterns that indicate a risk of progressing to ventricular fibrillation [16].

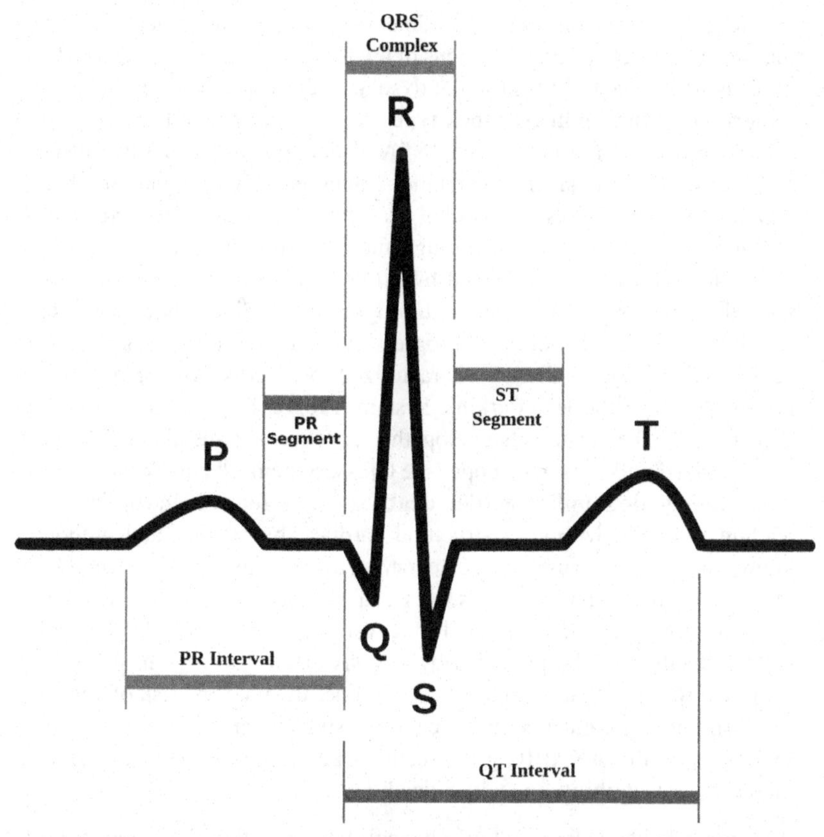

FIGURE 5.11 An electrocardiogram tracing of the electrical conduction through the heart in a single, normal heartbeat. The QRS complex indicates the depolarization of the right and left ventricles. See color plate at the back of the book. (*Source: Wikipedia, created by Agateller (Anthony Atkielski), and released into the public domain.*)

Mutations in sodium channel genes or in calcium channel genes can cause Brugada syndrome. Most cases occur in the sodium channel gene SCN5A, with more than 160 disease-causing SCN5A variants discovered to date.

- **Catecholaminergic polymorphic ventricular tachycardia** is characterized by episodes of syncope (i.e., fainting) or lipothymia (i.e., feeling as though one is about to faint), often occurring at moments of exertion or emotional excitement. Such episodes are caused by ventricular tachycardia, which carries the risk of progressing to ventricular fibrillation and sudden death. As its name suggests, catecholaminergic polymorphic ventricular tachycardia can be elicited by adrenaline, a catecholamine. Prophylaxis usually involves treatment with a beta-blocking agent. Mutations in any one of several genes (RYR2, CASQ2, and TRDN) may cause this disease.

The RYR2 gene encodes the ryanodine receptor 2, a component of cardiac muscle calcium channels. Mutations of the RYR2 gene can also cause a form of arrhythmogenic right ventricular dysplasia, discussed in the prior section.

- **Progressive familial heart block type IA** is a common cardiac conduction defect, caused by a mutation in SCN5A sodium channel gene. It is responsible for more pacemaker implantations than any other conduction defect. Physiologically, progressive familial heart block type IA is characterized by interference with conduction through the His–Purkinje conduction fibers. The clinical phenotype produced by progressive familial heart block type IA is the same as that observed with progressive cardiac conduction defect, also known as Lenegre disease. Progressive cardiac conduction defect is an acquired degenerative disease characterized by sclerosis of the specialized muscle fibers of the His–Purkinje system. Progressive cardiac conduction defect, unlike the genetic channelopathies, produces tissue changes that can be visualized under a microscope (see Glossary item, Phenocopy disease).
- **Non-monogenic familial sudden death** is a term for sudden cardiac arrest leading to death, without a structural cardiac abnormality and without a known single gene cause. This category of disease fits with the generalized observation that monogenic diseases can produce similar clinical phenotypes as polygenic diseases, and vice versa (see Section 12.2). Familial sudden deaths may be preceded by various electrocardiogram patterns or heartbeat rhythms that seem to confer a higher than normal risk of ventricular fibrillation, the most common cause of sudden cardiac arrest. Common variations in the AKAP10 gene seem to be statistically associated with increased susceptibility to sudden death [17].

As a general observation, sodium channels have a particularly important role in the heart, skeletal muscle, and nervous system. The sodium channel genes are designated SCN genes, and rare inherited alleles of various SCN gene mutations can cause multiple disorders, including sudden death. A few of the SCN gene diseases are listed here:

- SCN1A gene
  - febrile seizures, familial, 3A
  - familial hemiplegic migraine-3
- SCN1B gene
  - generalized epilepsy with febrile seizures plus, type 1
  - Brugada syndrome-5
- SCN2A gene
  - benign familial neonatal-infantile seizures-3
  - early infantile epileptic encephalopathy-11
- SCN4A gene
  - hypokalemic periodic paralysis type 2
  - form of congenital myasthenic syndrome

- SCN5A gene
  - Brugada syndrome-1 (with over 100 gene variants)
  - LQT3 (long QT segment-3)
  - sick sinus syndrome, congenital form [16]
  - familial atrial fibrillation, type 10 [16]
  - dilated cardiomyopathy, some cases [16]

It should be noted that some channelopathies produce lethal arrhythmias without producing unexplained deaths. These syndromes involve multiple tissues, facilitating an "explained" diagnosis in most or all cases. For example, Timothy syndrome, caused by a calcium channel gene defect, may cause life-threatening arrhythmias in infants. It is characterized by webbed fingers and toes, congenital heart disease, immune deficiency, intermittent hypoglycemia, cognitive abnormalities, and autism [18].

There are also inherited arrhythmias that are not channelopathies. These are associated with histologic alterations of conduction tissue that expert pathologists can evaluate. Examples of inherited arrhythmias that might be detected at autopsy are Wolff–Parkinson–White syndrome [19] and non-congenital sick sinus syndrome [20].

If an autopsy, performed on an individual who died of natural sudden death, yields no cause of death, then screening for channel gene mutations is prudent. Tester and colleagues searched for channel gene mutations in 173 consecutive sudden death autopsies for which no cause of death was determined by histopathologic examination [21]. Gene evaluation was limited to the long QT syndrome genes (KCNQ1, KCNH2, SCN5A, KCNE1, and KCNE2) and to the catecholaminergic polymorphic ventricular tachycardia type 1 gene (RYR2). Mutations of the target genes were found in 45 autopsies, accounting for 26% of the previously unexplained deaths.

In cases of natural sudden death, it is important to find the precise cause of death. Lethal conditions that were never diagnosed during an individual's life may pose an enduring threat to family members. The autopsy may represent the last opportunity by which family members can understand the cause of death of their family member, and by which they can assess their own genetic burdens [22]. Sudden unexpected death is a tragedy that family members and hospital staff are seldom prepared to deal with on an emotional or an intellectual level [23]. Research into the rare causes of arrhythmias will help pathologists clarify the causes of sudden unexplained deaths, and will ultimately reduce such deaths in family members.

5.4 HYPERTENSION AND OBESITY: QUANTITATIVE TRAITS WITH CARDIOVASCULAR CO-MORBIDITIES

"Only theory can tell us what to measure and how to interpret it."

—Albert Einstein

Though hypertension influences the development of a great many serious co-morbidities (e.g., renal failure, stroke, heart failure), those of us inclined to dwell on technicalities will insist that hypertension is not a disease; it is a physiologic measurement. Hypertension occurs when our blood pressure rises above a certain quantitative threshold, but there is no specific pathologic finding that characterizes hypertension; nor is there a specific clinical phenotype that tells us that an individual with hypertension is ill. It is best to think of hypertension as a quantitative trait that signals a problem somewhere within the human system.

Imagine for a moment that you are an engineer assigned to monitor a massive steam engine, to ensure that it is performing at an acceptable level. If a problem arises, your job is to call in the correct specialist. Every hour, you look at the pressure gauge located on one of the steam pipes leaving the engine. The gauge has a needle that sweeps through a range of pressures, and is marked by red zones indicating pressure readings that are too high or too low. The needle happens to point midway between the red zones, indicating a normal pressure. The gauge has a second meter for temperature; it too reads normal. You notice that the pressure gauge needle is jiggling slightly at a very fast pace. The fast jiggle tells you that all the pistons in the steam engine are moving, and with about the same pressure kick. If some of the pistons in the engine were failing, the jiggle of the needle would be erratic. A seasoned engineer could determine whether pistons were down or misfiring, just by looking at the jiggle in the gauge.

One day, as you make your customary inspections, you notice that the pressure gauge's needle has risen into the red zone. You know that if the pressure continues to rise, pipes will break and delicate parts in the engine will fail. You open a valve that vents steam into the air, and the pressure returns to normal. You immediately call an engine specialist, who tells you to maintain a normal system pressure by continuing to vent steam. He indicates that he will arrive in a few moments. When he arrives, you anxiously ask him what could have gone wrong. He answers, grimly, that it could be almost anything and that he will need to run a complete diagnostic review of all the subsystems.

Readers who have worked in medical wards undoubtedly recognize that the preceding steamy story was a parable for the medical procedure known as "taking vitals." At regular intervals, the vital signs of every patient in a hospital ward are measured and recorded: respirations, temperature, and blood pressure. If these three measurements fall within normal limits, it is a safe bet that the patient is medically stable. When an abnormal measurement is taken, a physician must be called to work up the problem in the hopes of finding a correctable cause. Fever is not a disease, rapid respiration is not a disease, and high blood pressure is not a disease; they are quantitative indicators that something is amiss.

What is the quantitative measure of hypertension? Definitions vary, but an often-used cut-off is a systolic blood pressure exceeding 140 mmHg, or a diastolic pressure exceeding 90 mmHg (see Glossary item, Blood pressure). It is estimated that 25% of adults and over 1 billion people worldwide are hypertensive [24,25]. Because high blood pressure is a quantitative trait and not a disease, the majority of the occurrences of hypertension cannot have a monogenic cause. Theory, strengthened by empiric observations, informs us that quantitative traits have multi-factorial causes, and that inherited quantitative traits have non-Mendelian inheritance. The non-Mendelian origin of inherited quantitative traits has been recognized since the early studies of R.A. Fisher in 1919, a topic that will be discussed in more detail in Section 11.2 [26–28]. Research scientists could have saved themselves a great deal of effort over the past few decades searching for a specific genetic cause for commonly occurring cases of hypertension had they simply recognized that hypertension is a quantitative trait and not a disease.

We typically find that hypertension co-occurs with rare diseases such as fibromuscular dysplasia, hyperaldosteronism, and various channelopathies; and common diseases such as metabolic syndrome, stroke, and left ventricular hyperplasia (see Glossary item, Metabolic syndrome). We shall see here, as we shall see again and again throughout this book, that we learn a great deal about common conditions by studying rare co-morbidities, because rare conditions typically have a simple pathogenesis that can be observed and analyzed.

Fibromuscular dysplasia is a rare condition of arteries wherein pathological growth of the artery's muscular wall produces a functional narrowing of the artery at the dysplastic site. Fibromuscular dysplasia occurs most often in young-to-middle aged women, but cases have occurred at every age and in either gender. Its cause is unknown.

When fibromuscular dysplasia occurs in a renal artery, the blood flow to the kidney distal to the point of narrowing is reduced, thus producing an orchestrated physiological response of the renin–angiotensin–aldosterone system that produces hypertension.

Here is how the renin–angiotensin–aldosterone system works. Specialized cells located at the root of the glomeruli (i.e., the juxtaglomerular cells) release renin into the general circulation when the blood pressure drops. Renin is involved in a pathway that produces a powerful vasoconstrictor (i.e., angiotensin II) in the lungs. This same vasoconstrictor stimulates the adrenal cortex to secrete aldosterone (part of the mineralocorticoid system), which causes the kidney to increase its absorption of sodium and water, thus increasing the volume of fluid in the body. Increased blood volume produces an increase in blood pressure. In summary, when fibromuscular dysplasia reduces the blood flow to the kidney, the kidney responds as if there were a system-wide drop in blood pressure, setting into motion two connected pathways that increase blood pressure. Because the hypertensive response does not "turn off" the localized hypotensive effect of fibromuscular dysplasia, the renin–angiotensin–aldosterone response stays "on" permanently, contributing to ever-worsening hypertension.

Observations of hypertension caused by fibromuscular dysplasia of the renal artery would suggest that variants of any components of the renin, angiotensin, or aldosterone system could contribute to quantitative alterations in blood pressure. As it happens, most of the rare monogenic and Mendelian forms of hypertension are associated with proteins involved, in one way or another, with the transport of electrolytes in the renal tubules, resulting in increased retention of sodium, increased volume of body fluid, and the enlistment of the mineralocorticoid system [29–32,24] (see Figure 5.12).

Observations on rare causes of hypertension dovetail with medically proven methods for treating and preventing hypertension. Standard therapies for treating hypertension include drugs that target the angiotensin pathway (i.e., angiotensin converting enzyme inhibitors, angiotensin receptor blockers, renin inhibitors, and diuretics). The mainstay of prevention is dietary salt reduction [33].

Genome wide association studies have yielded several dozen genes associated with commonly occurring hypertension [34,25] (see Glossary item, Genome wide association study). These associated genes seem to account for a very small portion of the occurrences of hypertension in the general population [34]. The function of the majority of the associated genes is unknown at present.

There are numerous genetic and environmental causes of hypertension, targeting a wide variety of cellular pathways and anatomic sites. As examples, the different causes of hypertension may include: overactivity of the renin–angiotensin system; defects at various sites of the renal tubule, arterial wall pathology; and increased salt consumption. Regardless of the underlying cause of hypertension,

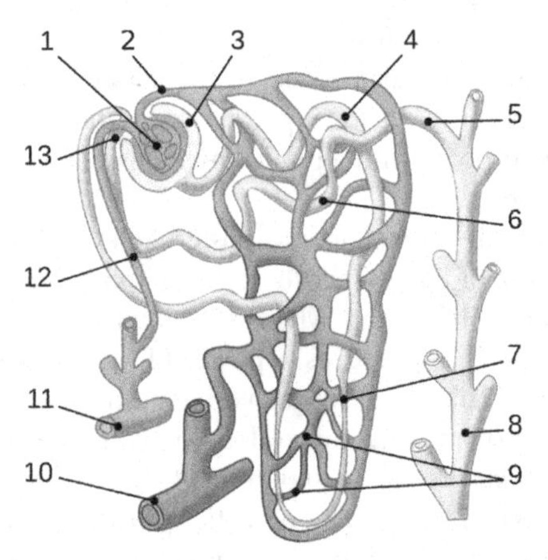

FIGURE 5.12 Illustration of a single nephron, demonstrating specific anatomic components targeted by inherited forms of hypertension: (1) glomerulus; (2) efferent arteriole; (3) Bowman capsule; (4) proximal convoluted tubule; (5) collecting duct, target of Liddle syndrome; (6) distal convoluted tubule, target of Gitelman syndrome; (7) loop of Henle, wherein the thick ascending limb is targeted in Bartter syndrome; (8) Bellini duct; (9) capillaries; (10) arcuate vein; (11) arcuate artery; (12) afferent arteriole; (13) juxtaglomerular apparatus, effector of the renin–angiotensin–aldosterone system, an active pathway employed in the pathogenesis of various causes of hypertension, including hypertension associated with renal artery dysplasia. See color plate at the back of the book. (*Source: Wikipedia, and released into the public domain.*)

all inherited and acquired forms of the disease produce hypertension through the same, final pathway: increased net salt balance, leading to increased intravascular volume, leading to augmented cardiac output, leading to elevated blood pressure [24]. Because all causes of hypertension produce an increase in net salt balance, almost all individuals with hypertension will respond to treatment with diuretics such as hydrochlorothiazide or furosemide, which reduce the reabsorption of sodium in the kidneys. A common, final mechanism accounting for all causes of hypertension is an example of disease convergence. Disease convergence is an extremely important concept, as it provides an opportunity to treat many different diseases with a single medication, if they converge to the same pathway. The topic of disease convergence will be discussed in greater detail in Section 10.1.

Obesity, like hypertension, is a quantitative trait with co-morbidities. There is nothing pathological about the adipose tissue in an obese individual; it is the quantity of the adipose tissue that raises health concerns.

Consider the following scenario. A 300-pound, 30-year-old smoker visits his private physician, complaining of difficulty breathing. After a quick history and physical examination and some simple blood work the physician finds that the patient has hypertension, hyperglycemia, a complex dyslipidemia, osteoarthritis of knees and ankles, chronic bronchitis, sleep apnea, back pain, GERD (gastroesophageal reflux disorder), halitosis, chronic fatigue, depression, and frequent absenteeism from work.

The doctor knows that all of these problems are the direct result of the patient's morbid obesity. If the patient were put on a strict diet, with exercise, all of his medical issues would likely vanish, even the halitosis, which is secondary to GERD.

Obesity, like hypertension, is an example of a disease that has co-morbidities [35]. Also, like hypertension, obesity can be perceived as a quantitative trait, measurable on a bathroom scale, rather than as a disease, definable by histopathologic criteria (see Glossary item, Histopathology). Like other quantitative traits, obesity is associated with many different pre-existing gene variants found in the general population [36]. Though there are several rare monogenic diseases in which obesity is part of the clinical phenotype, there seem to be no monogenic diseases for which obesity is the sole phenotypic condition. Like hypertension, the significance of obesity is its propensity to create co-morbid diseases.

In the U.S., physicians have not enjoyed much success with the only 100% effective treatment for obesity: weight reduction via consistent caloric restriction and nutritional counseling. More typically, physicians treat obese patients on a symptom-by-symptom basis. Obesity-related type 2 diabetes must be controlled; otherwise the patient will develop blindness, foot ulcerations, amputations of feet and legs, painful neuropathy, etc. Long-term treatment with oral hypoglycemic agents is prescribed; eventually followed by insulin injections. The obese patient may learn that his hypertension is out of control, and that he must begin treatment with one or more expensive medications. His treatment for hypertension will be a life-long process. His dyslipidemia will also require life-long treatment. His sleep apnea will need to be evaluated, and he will probably need to sleep with an assistive breathing device fitted to the mouth and face. GERD will require endoscopy, esophageal biopsies, and medications. Chronic bronchitis will require a consultation from a pulmonary medicine specialist. An oxygen tank and antibiotics may be recommended. An orthopedic consultation will be called to evaluate the patient's knees and ankle arthritis. The orthopedic surgeon will probably recommend arthroscopic procedures. Back pain will probably require various pain killers, muscle relaxants, corticosteroid treatments, and possibly surgery. Fatigue and depression will lead to psychiatric examinations, perhaps treated with various tranquilizers, anti-anxiety and anti-depressant medications. The many daily, long-term medications may interact with one another in an unpredictable and adverse manner, producing strange incapacitating symptoms that no doctor will fully understand.

Considering the negative health effects of obesity, one would think that every overweight individual would take appropriate measures to reduce weight. Not so; the global incidence of obesity is increasing. Projections call for 2.16 billion overweight and 1.12 billion obese individuals by 2030 [37]. **Because obesity is common, and because it raises the risk of many serious conditions, including heart disease and cancer, it has become the most devastating disease of developed countries. In theory, obesity is the simplest disease to diagnose (i.e., stand on a scale) and treat (i.e., eat less).**

Our inability to eradicate obesity relates to the startling dichotomy between the simplicity of energy balance and the complexity of human behavior. As anyone who has battled obesity will tell you, it is difficult to control one's appetite. Where exactly does the problem lie? Obesity runs in families. Analyses of twin and adoption data indicate that genetics accounts for about 40–70% of the variance in weight observed in populations [38,39]. When scientists search for susceptibility genes associated with obesity, they find hundreds of gene candidates [36,40]. All these gene associations, combined, seem to account for only a small percentage of the differences in weight among individuals [41].

5.4.1 Rule—Common diseases that have phenotypic overlap with a rare disease will often have genotypic overlap as well.

Brief Rationale—Polymorphisms are common in the population. If a rare disease gene is known to cause a particular phenotype, it is reasonable to expect that functional variations of the rare disease gene will contribute to the clinical phenotype expressed in a polygenic disease.

Some of the genes found to be associated with obesity in the general population are known to cause obesity-associated rare diseases when they occur as germline monogenic disorders (see Glossary item, Germline). For example, separate mutations in a gene on chromosome 20p12 cause both McKusick–

Kaufman syndrome and Bardet–Biedl syndrome-6 [42]. Both of these rare diseases are associated with obesity. Gene polymorphisms of the gene causing McKusick–Kaufman syndrome and Bardet–Biedl syndrome-6 are associated with metabolic syndrome, a common condition characterized by obesity, high triglycerides, hypertension, and hyperglycemia, occurring in up to 25% of the U.S. adults (see Glossary item, Polymorphism) [43,35].

The pharmaceutical industry is on perpetual alert for a magic bullet against obesity. To date, every magic bullet has missed the mark, but hope springs eternal. Leptin is a protein that influences body weight by reducing food intake and increasing energy expenditure. Inherited leptin deficiency is a rare cause of morbid obesity occurring in childhood [44]. A Phase III clinical trial testing the effectiveness and safety of leptin as an anti-obesity drug is under way and scheduled for completion in 2015 [45] (see Glossary item, Clinical trial).

REFERENCES

1. MacCallum WG. A Textbook of Pathology, 2nd edition. WB Saunders Company, Philadelphia and London, 1921.
2. Myers VC. Practical Chemical Analysis of Blood. C.V. Mosby Company, 1921.
3. Goldstein JL, Brown MS. Cholesterol: a century of research. HHMI Bull 16:1–4, September 20, 2012.
4. Tobert JA. Lovastatin and beyond: the history of the HMG-CoA reductase inhibitors. Nat Rev Drug Discov 2:517–526, 2003.
5. Biggerstaff KD, Wooten JS. Understanding lipoproteins as transporters of cholesterol and other lipids. Adv Physiol Educ 28:105–106, 2004.
6. Scaffidi P, Misteli T. Lamin A-dependent misregulation of adult stem cells associated with accelerated ageing. Nat Cell Biol 10:452–459, 2008.
7. Liu GH, Barkho BZ, Ruiz S, Diep D, Qu J, Yang SL, et al. Recapitulation of premature ageing with iPSCs from Hutchinson-Gilford progeria syndrome. Nature 472:221–225, 2011.
8. Capell BC, Collins FS, Nabel EG. Mechanisms of cardiovascular disease in accelerated aging syndromes. Circ Res 101:13–26, 2007.
9. Tyler S. Epithelium—the primary building block for metazoan complexity. Integr Comp Biol 43:55–63, 2003.
10. Delmar M, McKenna WJ. The cardiac desmosome and arrhythmogenic cardiomyopathies from gene to disease. Circ Res 107:700–714, 2010.
11. Alcalai R, Metzger S, Rosenheck S, Meiner V, Chajek-Shaul T. A recessive mutation in desmoplakin causes arrhythmogenic right ventricular dysplasia, skin disorder, and woolly hair. J Am Coll Cardiol 42:319–327, 2003.
12. Thomas AC, Knapman PA, Krikler DM, Davies MJ. Community study of the causes of "natural" sudden death. BMJ 297:1453–1456, 1988.
13. Winkel B, Larsen M, Olesen M, Tfelt-Hansen J, Banner J, et al. The prevalence of mutations in KCNQ1, KCNH2. J Cardiovasc Electrophysiol 23:1092–1098, 2012.
14. Kass RS. The channelopathies: novel insights into molecular and genetic mechanisms of human disease. J Clin Invest 115:1986–1989, 2005.
15. Kleopa KA. Autoimmune channelopathies of the nervous system. Curr Neuropharmacol 9:458–467, 2011.

16. Veerakul G, Nademanee K. Brugada syndrome: two decades of progress. Circ J 76:2713–2722, 2012.
17. Kammerer S, Burns-Hamuro LL, Ma Y, Hamon SC, Canaves JM, Shi MM, et al. Amino acid variant in the kinase binding domain of dual-specific A kinase-anchoring protein 2: a disease susceptibility polymorphism. Proc Natl Acad Sci USA 100:4066–4071, 2003.
18. Splawski I, Timothy KW, Sharpe LM, Decher N, Kumar P, Bloise R, et al. Ca(V)1.2 calcium channel dysfunction causes a multisystem disorder including arrhythmia and autism. Cell 119:19–31, 2004.
19. Cole JS, Wills RE, Winterscheid LC, Reichenbach DD, Blackmon JR. The Wolff-Parkinson-White syndrome: problems in evaluation and surgical therapy. Circulation 42:111–121, 1970.
20. Evans R, Shaw DB. Pathological studies in sinoatrial disorder (sick sinus syndrome). Br Heart J 39:778–786, 1977.
21. Tester DJ, Medeiros-Domingo A, Will ML, Haglund CM, Ackerman MJ. Cardiac channel molecular autopsy: insights from 173 consecutive cases of autopsy-negative sudden unexplained death referred for postmortem genetic testing. Mayo Clin Proc 87:524–539, 2012.
22. Behr ER, Dalageorgou C, Christiansen M, Syrris P, Hughes S, Tome Esteban MT, et al. Sudden arrhythmic death syndrome: familial evaluation identifies inheritable heart disease in the majority of families. Eur Heart J 29:1670–1680, 2008.
23. Huisma FF, Potts JE, Gibbs KA, Sanatani S. Assessing the knowledge of sudden unexpected death in the young among Canadian medical students and recent graduates: a cross-sectional study. BMJ Open 2:e001798, 2012.
24. Lifton RP, Gharavi AG, Geller DS. Molecular mechanisms of human hypertension. Cell 104:545–556, 2001.
25. International Consortium for Blood Pressure Genome-Wide Association Studies. Genetic variants in novel pathways influence blood pressure and cardiovascular disease risk. Nature 478:103–109, 2011.
26. Fisher RA. The correlation between relatives on the supposition of Mendelian inheritance. Trans R Soc Edinb 52:399–433, 1918.
27. Ward LD, Kellis M. Interpreting noncoding genetic variation in complex traits and human disease. Nature Biotechnol 30:1095–1106, 2012.
28. Visscher PM, McEvoy B, Yang J. From Galton to GWAS: quantitative genetics of human height. Genet Res 92:371–379, 2010.
29. Lifton RP. Molecular genetics of human blood pressure variation. Science 272:676–680, 1996.
30. Wilson FH, Kahle KT, Sabath E, Lalioti MD, Rapson AK, Hoover RS, et al. Molecular pathogenesis of inherited hypertension with hyperkalemia: the Na-Cl cotransporter is inhibited by wild-type but not mutant WNK4. Proc Natl Acad Sci USA 100:680–684, 2003.
31. Bahr V, Oelkers W, Diederich S. Monogenic hypertension. J. Med Klin (Munich) 98:208–217, 2003.
32. Warnock DG. Liddle syndrome: genetics and mechanisms of Na+ channel defects. Am J Med Sci 322:302–307, 2001.
33. Hideaki Nakagawa H, Katsuyuki Miura K. Salt reduction in a population for the prevention of hypertension. Environ Health Prev Med 9:123–129, 2004.
34. Cowley AW Jr, Nadeau JH, Baccarelli A, Berecek K, Fornage M, Gibbons GH, et al. Report of the national heart, lung, and blood institute working group on epigenetics and hypertension. Hypertension 59:899–905, 2012.
35. Ford ES, Giles WH, Dietz WH. Prevalence of the metabolic syndrome among US adults: findings from the third national health and nutrition examination survey. JAMA 287:356–359, 2002.

36. Yang W, Kelly T, He J. Genetic epidemiology of obesity. Epidemiol Rev 29:49–61, 2007.
37. Kelly T, Yang W, Chen CS, Reynolds K, He J. Global burden of obesity in 2005 and projections to 2030. Int J Obes (Lond) 32:1431–1437, 2008.
38. Maes HH, Neale MC, Eaves LJ. Genetic and environmental factors in relative body weight and human obesity. Behav Genet 27:325–351, 1997.
39. Strachan T, Read AP. Human Molecular Genetics, 2nd edition. Wiley-Liss, New York, 1999. Available from: http://www.ncbi.nlm.nih.gov/books/NBK7564/, viewed July 25, 2013.
40. Jia G, Fu Y, Zhao X, Dai Q, Zheng G, Yang Y, et al. N6-methyladenosine in nuclear RNA is a major substrate of the obesity-associated FTO. Nat Chem Biol 7:885–887, 2011.
41. Loos RJF. Recent progress in the genetics of common obesity. Br J Clin Pharmacol 68:811–829, 2009.
42. Katsanis N, Beales PL, Woods MO, Lewis RA, Green JS, Parfrey PS, et al. Mutations in MKKS cause obesity, retinal dystrophy and renal malformations associated with Bardet-Biedl syndrome. Nature Genet 26:67–70, 2000.
43. Hotta K, Nakamura T, Takasaki J, Takahashi H, Takahashi A, Nakata Y, et al. Screening of 336 single-nucleotide polymorphisms in 85 obesity-related genes revealed McKusick-Kaufman syndrome gene variants are associated with metabolic syndrome. J Hum Genet 54:230–235, 2009.
44. Montague CT, Farooqi IS, Whitehead JP, Soos MA, Rau H, Wareham NJ, et al. Congenital leptin deficiency is associated with severe early-onset obesity in humans. Nature 387:903–908, 1997.
45. Low-Dose Leptin and the Formerly-Obese. National Institute of Diabetes and Digestive and Kidney Diseases (NIDDK) Clinical Trial Identifier: NCT00073242. Available from: http://www.clinicaltrials.gov/show/NCT00073242, viewed July 25, 2013.

Chapter 6

Infectious Diseases and Immune Deficiencies

6.1 THE BURDEN OF INFECTIOUS DISEASES IN HUMANS

"We are just 'a volume of diseases bound together'."

—John Donne

How many of the world's 56.4 million annual deaths can be attributed to infectious diseases? According to the World Health Organization, in 1996, about 33% of human deaths were attributable to infections [1]. Of course, the numbers vary depending on how you count causes of death. Infection is often the final blow capping any chronic and debilitating disease.

To gain a perspective on the toll engendered by infectious diseases, it is useful to consider the damage inflicted by just a few of the organisms that infect humans. Malaria infects up to 500 million people, killing about 2 million people each year [1]. About 2 billion people are infected with *Mycobacterium tuberculosis*. Most of these infected individuals will not develop active disease, but about 3 million people die of tuberculosis annually [1]. Each year, about 4 million children die from lung infections, and about 3 million children die from infectious diarrheal diseases [1]. The grouped rotaviruses are one of many causes of diarrheal disease (Group III viruses). In 2004, rotaviruses were responsible for about half a million deaths, mostly in developing countries [2].

Worldwide, about 350 million people are chronic carriers of hepatitis B, and about 100 million people are chronic carriers of hepatitis C. In aggregate, about one-quarter (25 million) of these chronic carriers will eventually die from ensuing liver diseases [1]. Infectious organisms can kill individuals through mechanisms other than through the direct pathologic effects of growth, invasion, and inflammation. For example, infections have been implicated in vascular disease. The organisms that contribute to coronary artery disease and stroke include *Chlamydia pneumoniae* and *Cytomegalovirus* [3].

Infections caused by a wide variety of organisms can result in cancer. About 7.2 million deaths occur each year from cancer worldwide. About one-fifth of these cancer deaths are caused by infectious organisms [4]. Hepatitis B alone accounts for about 700,000 cancer deaths each year from hepatocellular carcinoma [5].

Rare Diseases and Orphan Drugs. http://dx.doi.org/10.1016/B978-0-12-419988-0.00006-7

Organisms contributing to cancer deaths include bacteria (*Helicobacter pylori*), animal parasites (schistosomes and liver flukes), and viruses such as herpesviruses, papillomaviruses, hepadnaviruses, flaviviruses, retroviruses, and polyomaviruses (see Glossary item, Retrovirus).

Though fungal and plant organisms do not seem to cause cancer through human infection, they produce a multitude of biologically active secondary metabolites (i.e., synthesized molecules that are not directly involved in the growth of the organism), some of which are potent carcinogens. For example, aflatoxin produced by the fungus *Aspergillus flavus* is possibly the most powerful carcinogen ever studied [6]. Cycasin, a toxin found in the seeds of various trees of Order Cycadales, is a powerful neurotoxin and liver carcinogen [7].

In summary, infectious diseases are the number one killer of humans worldwide, and they contribute to vascular disease and cancer, the two leading causes of death in the most developed countries.

Given all the suffering caused by infectious organisms, you might begin to wonder whether the majority of terrestrial life-forms are devoted to the annihilation of the human species. Not to worry. Only a tiny fraction of the life-forms on Earth are infectious to humans. The exact fraction is hard to estimate because nobody knows the total number of terrestrial species. Most taxonomists agree that the number is in the millions, but estimates range from a few million up to several hundred million (see Glossary item, Taxonomy). For the sake of discussion, let us accept that there are 50 million species of organisms on Earth (a gross underestimate by some accounts). There have been about 1400 pathogenic organisms reported in the medical literature. This means that if you should stumble randomly upon a member of one of a species of life on Earth, the probability that it is an infectious pathogen is about 0.000028.

Of the approximately 1400 infectious organisms that have been recorded somewhere in the medical literature, the vast majority of these are "case report" items; instances of diseases that have, to the best of anyone's knowledge, occurred once or a handful of times. They are important to epidemiologists because today's object of medical curiosity may emerge as tomorrow's global epidemic. Very few of these rare causes of human disease gain entry to a clinical microbiology textbook. Textbooks, even the most comprehensive, cover about 300 organisms (excluding viruses) that are considered clinically important.

It is difficult to rank the common infectious diseases in humans. Some organisms have a very high rate of infection, but produce a relatively low rate of clinical disease and death. Other organisms have relatively low levels of infection, but have a very high virulence, resulting in many deaths.

What follows is a listing of the most common infections that occur in humans, beginning with the organisms that can be found within the majority of humans (e.g., greater than 3.5 billion), and ending with infections involving more than 1 million individuals. Some infections that deserve to be included here (e.g., yellow fever) are omitted for lack of finding a trusted historical data source.

6.1.1 Infections occurring in the majority of humans (i.e., 3.5 to 7 billion cases)

- Demodex is a tiny mite that lives in facial skin. Demodex mites can be found in the majority of humans.
- The BK polyomavirus rarely causes disease in infected patients, and the majority of humans carry the latent virus.
- The JC polyomavirus persistently infects the majority of humans, but it is not associated with disease in otherwise healthy individuals.

6.1.2 Infections occurring in 1 to 3.5 billion humans

- About 2 billion people (of the world's 7 billion population) have been infected with *Mycobacterium tuberculosis*.
- About one-third of the human population has been infected (i.e., about 2.3 billion people) by the only species that produces human toxoplasmosis: *Toxoplasma gondii*.
- *Ascaris lumbricoides*, the cause of ascariasis, infects about 1.5 billion people worldwide, making it the most common helminth (worm) infection of humans [8].

6.1.3 Infections involving 500 million to 1 billion humans

- Various estimates would suggest that worldwide, more than half a billion people are infected with one or another subtypes of *Chlamydia trachomatis*. This would include the various *Chlamydia* organisms and serotypes that account for trachoma and chlamydial urethritis. According to the World Health Organization, there are about 37 million blind persons worldwide. Trachoma, caused by *Chlamydia trachomatis*, is the number one infectious cause of blindness and accounts for about 4% of these cases. The second most common infectious cause of blindness worldwide is *Onchocerca volvulus*, accounting for about 1% of cases [9].
- About 200 million people are infected by schistosomes (i.e., have some form of schistosomiasis).
- Hookworms infect about 600 million people worldwide. Two species are responsible for nearly all cases of hookworm disease in humans: *Ancylostoma duodenale* and *Necator americanus*.
- Scabies is an exceedingly common, global disease, with about 300 million new cases occurring annually.

6.1.4 Infections involving 100 million to 500 million humans

- Hepatitis B infects more than 200 million people worldwide, causing 2 million deaths each year.

- Bubonic plague is credited with killing one-third of the population of Europe in the mid-1300s. Altogether, bubonic plague is estimated to have caused about 200 million deaths. In modern times, plague is rare, but not extinct. Each year, several thousand cases of plague occur worldwide, resulting in several hundred deaths. Virtually all of the contemporary cases occur in Africa.
- Genus *Plasmodium* is responsible for human and animal malaria. About 300–500 million people are infected with malaria worldwide, causing 2 million deaths each year [1,10].
- About 150 million people are infected by the filarial nematodes (genera *Brugia, Loa, Onchocerca, Mansonella,* and *Wuchereria*) [11]. *Wuchereria bancrofti* and *Brugia malayi* together infect about 120 million individuals [11]. Most cases occur in Africa and Asia.
- Smallpox is reputed to have killed about 300 million people in the twentieth century, prior to the widespread availability of an effective vaccine. Smallpox, now extinct, has been referred to as the greatest killer in human history.
- Worldwide, about 100 million cases of acute diarrhea are caused by rotavirus. In 2004, rotavirus infections accounted for about a half million deaths in young children from severe diarrhea [2].

6.1.5 Infections involving 10 million to 100 million humans

- The 1917–1918 influenza pandemic caused somewhere between 50 million and 100 million deaths. Seasonal influenza kills between a quarter million and a half million people worldwide each year. In the U.S., seasonal influenza accounts for about 40,000 deaths annually.
- It is estimated that about 50 million people are infected by *Entamoeba histolytica,* with about 70,000 deaths per year worldwide.
- *Paragonimus westermani,* along with dozens of less frequent species within genus *Paragonimus,* causes the condition known as paragonimiasis. About 22 million people are infected worldwide, with most cases occurring in Southeast Asia, Africa, and South America.
- More than 50 million dengue virus infections occur each year, causing about 25,000 deaths worldwide. Most infections are asymptomatic or cause only mild disease. A minority of cases are severe.
- Between 1918 and 1922, epidemic, louse-borne, typhus (*Rickettsia prowazekii*) infected 30 million people in Eastern Europe and Russia, accounting for about 3 million deaths [12].
- The most common sexually transmitted disease is trichomoniasis (Class Metamonada), with about 8 million new cases each year in North America [13]. The second most common sexually transmitted disease is chlamydia (Class Chlamydiae), with about 4 million new cases each year in North America [13]. Approximately 1.5 million new cases of gonorrhea occur annually in North America, where gonorrhea is the third most common sexually transmitted disease [13]. According to the U.S. Centers for Disease Control

and Prevention, there were about 46,000 new cases of syphilis and 48,000 new cases of HIV reported in the U.S. in 2011 [14].

- Leprosy, also known as Hansen disease, is caused by *Mycobacterium leprae* and *Mycobacterium lepromatosis*. From the 1960s to the 1980s, the number of leprosy cases worldwide was a steady 10–12 million [15]. The introduction of effective multidrug protocols has resulted in many cured cases and in lowered infection rates. Consequently, the number of cases of leprosy has dropped to about 5.5 million worldwide in the 1990s [15]. In 2005, there were about 300,000 new cases reported worldwide [16].
- *Leishmania* species cause leishmaniasis, a disease that infects about 12 million people worldwide. Each year, about 60,000 people die from the visceral form of the disease.
- *Trypanosoma cruzi* is the cause of Chagas disease, also known as American trypanosomiasis. Chagas disease affects about 8 million people [17]. *Trypanosoma brucei* is the cause of African trypanosomiasis (sleeping sickness). The reported numbers of cases may be somewhat unreliable, but it has been estimated that infection with *Trypanosoma brucei* accounts for about 50,000 deaths each year.
- Fasciolopsiasis is caused by *Fasciolopsis buski*, a large (up to 7.5 cm length) fluke that lives in the intestines of the primary host, pigs and humans (see Glossary item, Primary host). The number of humans infected is about 10 million.
- *Clonorchis sinensis*, the Chinese liver fluke (also known as the oriental liver fluke), infects about 30 million people.
- In the year 2000, measles caused approximately 40 million illnesses and about 750,000 deaths worldwide [18].

What kinds of organisms cause infections? All types. It seems to be a condition of terrestrial life that organisms live within one another. Humans can be infected by any of the classical kingdoms of living organisms: Bacteria, Fungi, Animalia, Plantae, and members of the kingdom formerly known as Protoctista, containing the protozoan parasites. Humans can also be infected by non-living organisms (i.e., virus, prion). However, not every class of organism within these kingdoms contains human pathogens. In my book *Taxonomic Guide to Infectious Diseases: Understanding the Biologic Classes of Pathogenic Organisms*, most of the known infections of humans are described, with their culpable organisms assigned to their proper phylogenetic classes [19].

6.2 BIOLOGICAL TAXONOMY: WHERE RARE INFECTIOUS DISEASES MINGLE WITH THE COMMON INFECTIOUS DISEASES

"Bacteria will no longer be conceptualized mainly in terms of their morphologies and biochemistries; their relationships to other bacteria will be central to the concept as well."

—Carl R. Woese [20]

It is impossible to compare the rare and the common infectious diseases without understanding the meaning and purpose of biological taxonomy. Taxonomy is the scientific field that determines how different organisms are related to one another. Plesionyms (i.e., near-synonyms) for "taxonomy" include "classification" and "systematics." The word "taxonomy" is also used by naturalists to describe the product of a taxonomic construction: the hierarchy of ancestral organisms and their descendants, and the list of names assigned to the classes and species of organisms.

To the uninitiated, there is little difference between the life of a taxonomist and the life of a stamp collector. Nothing could be further from the truth. Taxonomy is the grand unifying theory of the biological sciences. The past few decades have revitalized the field of taxonomy by calling upon scientists to apply new techniques (e.g., computational phylogenetics, bioinformatics, and molecular biology) to clarify which species belong to a genus, which genera belong to the same class, and how all the different classes are related.

Taxonomy is indifferent to the frequency of occurrence of an organism in nature or in disease. Organisms that infect a billion humans may occur within the same class of organisms as another species that infects one individual or zero individuals. Two organisms in the same taxonomic class will likely share diagnostic, clinical, and therapeutic traits. When we find a biological trait that informs us that what we thought was a single species is actually two species, we can optimize our methods for preventing, diagnosing, and treating each species.

Taxonomy is serious business. When unrelated organisms are mistakenly assigned to the same class, and when related organisms are separated into unrelated classes, the value of the classification is lost, perhaps forever. When we incorrectly assign rare organisms and common organisms the same name, we cannot develop effective new drugs that are effective against both. Without an accurate classification of living organisms, it would be impossible to make significant progress in the diagnosis, prevention, or treatment of infectious diseases.

The colloquial use of the term "worm" exemplifies a worst case scenario for healthcare workers plainly mystified by the rigors of taxonomy. Laypersons might believe that "worms" constitute a taxonomic class of organisms, and that any member of the "worm" class of organisms can be treated with a standard vermifuge (i.e., a drug that treats worm infections). Tsk, tsk! The term "worm" has no taxonomic meaning; soft, squiggly organisms colloquially known as "worms" are scattered throughout animal taxonomy, with no close relationship to one another. A small squirming organism referred to as a "worm" may be a so-called true worm (i.e., a helminth in Class Platyhelminthes, flatworms, or Class Nematoda, roundworms), or it might be an insect larva, or it may be one of many other classes of organisms. Class Acanthocephala includes the thorny-headed worms. Class Annelida (earthworms) descends from Class Lophotrochozoa, which includes mollusks. Class Nematoda (roundworms) and Class Annelida (earthworms) are more closely related to spiders and clams, respectively, than either one is related to Class Platyhelminthes (flatworms). Many so-called worms are actually the larval forms of animals whose adult stage bears no resemblance to worms. An example is *Linguatula serrata*, the agent causing tongue worm disease. The tongue worm is the larval stage of a crustacean. Likewise, the screw-worm (*Cochliomyia hominivorax*) is actually a type of fly (see Figure 6.1). It is called a screw-worm because the disease is manifested by worm-like larvae growing in skin. The ineptly named "ring-worm" infections are not caused by worms or by any animals; they are fungal infections of the skin. Millipedes, a type of arthropod, resemble annelid worms; their tiny legs notwithstanding. The word "worm" may even refer to a marsupial joey, Class Mammalia, which is typically a smooth hairless slug-shaped organism the size of a jellybean. When a word, such as "worm," cannot be applied to a set of objects that are mutually related, what meaning does it convey? When you misclassify a "worm," you may have a great deal of difficulty finding a suitable treatment.

Infectious diseases may closely mimic one another, leading the unwary physician to mistakenly apply a common diagnosis to an uncommon infection. For example, immune-compromised individuals, especially AIDS patients, may form an exaggerated endothelial growth reaction (i.e., focal proliferation of small vessels) in the skin called bacillary angiomatosis. Bacillary angiomatosis can be caused by *Bartonella henselae*, the same organism that causes cat-scratch fever, and is transmitted by cat scratch, cat bite, and possibly ticks and fleas.

FIGURE 6.1 Larval form of the screw-worm, a poorly chosen and misleading common name for an arthropod whose scientific name is *Cochliomyia hominivorax*. See color plate at the back of the book. (*Source: Wikipedia, produced for the public domain by the Agricultural Research Service, the research agency of the United States Department of Agriculture.*)

Bacillary angiomatosis is also caused by *Bartonella quintana*, transmitted by lice. The gross presentation (i.e., what the lesions look like to the naked eye) and the histologic features (i.e., what the biopsied lesions look like on microscopic examination) of bacillary angiomatosis mimic Kaposi sarcoma, a cancer of vascular origin also occurring in immune-compromised and AIDS patients, but with higher frequency. Kaposi sarcoma is caused by herpesvirus-8. Both bacillary angiomatosis and Kaposi sarcoma occur in AIDS patients, but their prognosis and treatment are very different from one another (see Glossary item, Prognosis). The correct diagnosis requires an astute pathologist who understands and anticipates the rare and the common causes of vascular proliferative lesions in immune-compromised patients.

Neorickettsia sennetsu, formerly known as *Ehrlichia sennetsu*, is the cause of Sennetsu ehrlichiosis. This rare disease is said to mimic a mild case of the common disease infectious mononucleosis. Both diseases produce a mononucleosis (i.e., increased circulating monocytes) and generalized systemic symptoms. Unique among the ehrlichioses, *Neorickettsia sennetsu* is transmitted by trematodes, harbored within fish, and eaten undercooked or uncooked by humans. This disease is a diagnostic and a taxonomic challenge. It can easily be misdiagnosed as common viral mononucleosis, caused by Epstein–Barr virus. It can also be misclassified as a rickettsial disease. *Neorickettsia*, despite its name, is not a type of *Rickettsia* (i.e., not a member of Class Rickettsiaceae). *Neorickettsia* is a member of Class Anaplasmataceae; hence, the disease it produces is an ehrlichiosis, and will have the biological attributes of the other organisms in its class.

The heterokonts (i.e., Class Heterokontophyta) comprises a taxonomically unstable (i.e., subject to many revisions) class of eukaryotic single-celled organisms. The heterokonts, also known as stramenopiles, contain two strikingly different subclasses of organisms: a pigmented group containing various types of algae and diatoms, and a colorless group containing organisms that have morphologic features similar to fungi. Oomycota, a "colorless" class of heterokonts, contains the organisms that produce late blight of potato (*Phytophthora infestans*), and sudden oak death (*Phytophthora ramorum*). There is only one heterokont genus that is infectious in humans: *Blastocystis*. Human infection seems to follow acquisition of the cyst form of the organism through a fecal–oral route. Incidence is highest where humans are exposed to animal feces, implying that animal–human transmission is common. In the U.S. the prevalence rate of infection of *Blastocystis hominis* is 23%, with the highest rates found in the western states [21]. The length of infection varies from weeks to years. Many *Blastocystis* infections do not manifest clinically, and an asymptomatic carrier state is common. Clinical *Blastocystis* disease occurs only rarely, but when it occurs, it may mimic irritable bowel syndrome, a very common disease. Given that the infection rate of *Blastocystis hominis* is high, about 23%, is it worthwhile testing patients with irritable bowel syndrome for *Blastocystis hominis* [22]? If detected, should the patient be treated with metronidazole, an anti-eukaryotic agent known to kill *Blastocystis* [23]? The clinical dilemmas posed by *Blastocystis hominin is* exemplify the tentative relationships between highly prevalent organisms that rarely cause disease and highly prevalent diseases that are rarely caused by organisms.

Pneumocystis is the only pathogenic genus in Class Taphrinomycotina, a type of fungus. Until recently, *Pneumocystis* was presumed to be a member of Class Protoctista, a former kingdom of one-celled eukaryotic organisms (see Glossary item, Protozoa). Early papers concocted a detailed protozoan life cycle for *Pneumocystis*, complete with morphologically distinct developmental stages that included cyst, trophozoite, sporozoite, and intracystic bodies [24]. Owing to molecular analyses, we now know that *Pneumocystis* is a fungus. Though *Pneumocystis* cannot be cultured, we have learned a great deal about the life cycle of *Pneumocystis* by studying a sister organism, *Schizosaccharomyces pombe*, in Class Taphrinomycotina. *S. pombe* can be cultured. By observing *S. pombe*, microbiologists can infer that the yeast form of *Pneumocystis* creates an enclosed cyst, which eventually ruptures, releasing spores. The different forms of *Pneumocystis* comprise the various morphologic forms of the fungus upon which histologic diagnosis is rendered. *Pneumocystis jerovicii* (formerly *Pneumocystis carinii*) is ubiquitous in the environment, but produces pneumonia only in immune-deficient individuals. It serves as an example of a common organism that produces a rare disease that we can understand and treat based on inferences drawn from observations on a taxonomically related model organism.

Genus *Plasmodium* is responsible for human and animal malaria. There are several hundred species of *Plasmodium* that infect animals, but only a half dozen species are known to infect humans [10]. Newly emerging species, causing human disease, may arise from animal reservoirs. For example, *Plasmodium knowlesi* causes malaria in macaque monkeys. It has emerged as a rare cause of human malaria in Southeast Asia, where it has grown in incidence to the point that it currently accounts for about two-thirds of malarial cases in this region. Malaria is commonly diagnosed with an antigen test developed against the common forms of human plasmodia (e.g., *Plasmodium falciparum*). Patients with *P. knowlesi* may have a negative reaction to the standard *Plasmodium* antigen test [25]. If this were to occur, a *P. knowlesi* infection may go undiagnosed, and the patient might not be provided with needed anti-malarial medication. Careful examination of blood will usually indicate the presence of parasites in cases of *P. knowlesi* malaria, and a specific diagnosis can be confirmed with advanced molecular tests. Here is an example of a rare infection, emerging as an endemic infection, whose diagnosis can be missed with the standard testing methods for the common forms of disease.

Until the past decade, members of Class Microsporidia were considered to be protozoa. With molecular techniques, the members of Class Microsporidia have been shown to contain ribosomal RNA sequences typical of fungi [26,27]. Aside from their phylogenetic relationship to fungi, these organisms are strikingly dissimilar in morphology and lifestyle from other fungi, being obligate intracellular parasites that have adapted themselves to parasitic lives in a wide range of eukaryotic organisms (see Glossary item, Obligate intracellular organism). Unlike virtually all other members of Class Fungi, the members of Class Microsporidia lack mitochondria [28], lack a hyphal form, and do not produce multicellular tissue structures. With all these non-fungal properties, taxonomists never entertained the notion that the microsporidia were fungi. Twentieth century taxonomists overlooked an important clue; the microsporidia synthesize chitin, a structural feature found only in opisthokonts, such as fungi. A wide variety of animals are reservoirs for the various species of microsporidia: mammals, birds, insects. The spores are passed in the stools and infect humans through direct contact, water contamination, or through respiration of airborne spores. Preliminary evidence suggests that microsporidial infections are common in humans [29]. Most, but not all, cases of symptomatic microsporidiosis occur in immune-compromised individuals, particularly in patients who have AIDS. Though microsporidiosis is considered a rapidly emerging disease, we lack important and fundamental epidemiologic information. How prevalent is the organism in the immune-competent population? How prevalent is the organism in the population of immune-deficient but asymptomatic patients? How often is a microsporidium the causative agent of diarrheal diseases among different age groups? In which geographic regions does microsporidiosis occur? What are the most important animal reservoirs for human microsporidiosis?

6.2.1 Rule—Our knowledge of the relationships between rare infectious diseases and common infectious diseases is dependent upon our access to an accurate and comprehensive taxonomy of living organisms.

Brief Rationale—There are over 1400 known pathogenic organisms, and it would be impossible to develop individual methods to prevent, diagnose, and treat each of these organisms. Taxonomies drive down the complexity of infectious diseases, and permit us to find the biological and clinical relationships among rare infections and common infections.

Every known disease-causing organism falls into one of 40 well-defined classes of organisms, and each class fits within a simple ancestral lineage. When we learn the taxonomic classes, and their relationships to every other class within the taxonomy, we can start to understand how to treat common infectious diseases based on our understanding of rare infectious diseases, and vice versa. Taxonomic research has become one of the most active and fruitful areas of biomedical research. **Virtually every important advance in clinical microbiology relates in one way or another to taxonomy. Because most pathogenic species are rare, and we depend on a variety of pathogenic species to find the general properties of classes of organisms, we can infer that advances in the common infections will continue to depend, in no small part, on taxonomic research into rare infections**.

6.3 BIOLOGICAL PROPERTIES OF THE RARE INFECTIOUS DISEASES

"The purpose of narrative is to present us with complexity and ambiguity."

—Scott Turow

Of the 1400 named organisms that have been shown to cause infectious diseases in humans, about 300 are sufficiently common to be listed in popular textbooks of infectious diseases. The remainder are obscure and known only to clinical microbiologists and infectious disease specialists. Do these rare infections share a common set of biological properties?

6.3.1 Rule—Common infectious diseases are spread, directly or indirectly, from one infected human to another human.

Brief Rationale—If an organism has succeeded to thwart human defense systems, and if it can spread from person to person, then it is likely to infect lots of persons.

6.3.2 Rule—Rare infectious diseases are seldom transmissible from human to human.

Brief Rationale—Infections that fail to move from person to person cannot effectively spread through a population.

As a general observation, if an infectious disease is rare, then it is not spread directly from human to human. There are exceptions. Some rare diseases are transmissible from human to human. For the most part, these are geographically restricted diseases or diseases that impose restrictive conditions on their propagation. For example, Ebola virus has human–human transmission, but has caused only small outbreaks in Africa. A few single case outbreaks have been reported in laboratories (e.g., Petri dish to human transmission). Ebola has a rapid clinical course and a very high fatality rate. It is generally believed that infectious agents that have no chronic phase, and that rapidly kill the transmitting host, cannot easily infect a large population.

As we will see in Section 6.5, virtually all fungal infections are spread from spores living in air, water, or soil, and are not spread from human to human. There are a few common fungal diseases, but most fungal diseases are rare diseases, and the list of rare fungal infections grows continuously.

Common infections are most often passed from human to human, but there are many instances for which common infections are not directly transmitted by humans. This would include most of the infections caused by parasites with complex life cycles. About one-third of the human population has been infected by *Toxoplasma gondii*, the cause of toxoplasmosis. The most common primary host of *Toxoplasma* is the cat. Humans are the intermediate host. Hence, the mode of infection is often cat to human, rarely human to human. Most *T. gondii* infections are latent. Active disease occurs only rarely, through activation of latent *T. gondii* in immune-compromised individuals. Rare cases of transplacental infections (i.e., human to human) are known to occur.

Malaria is another common disease that is not generally transmitted directly from human to human. Mosquitoes are the vector; drawing blood from an animal reservoir that includes humans. Rare instances of direct human-to-human transmission, via therapeutic blood transfusions, have been reported [30]. Schistosomiasis and the trypanosomiases are additional examples of common human diseases that are not transmitted by humans.

Some of the most common infectious diseases are greatly facilitated by human behavior. The fecal-borne infectious diseases could be eliminated entirely if sanitary conditions were generally available (e.g., *Ascaris lumbricoides* infecting 1.5 billion individuals). Eating habits, particularly those involving eating uncooked animals, account for a huge number of common parasitic infections (e.g., *Clonorchis sinensis*, the Chinese liver fluke, infecting over 30 million individuals). Among the least tractable of endemic infections are the sexually transmitted diseases.

6.3.3 Rule—The list of rare infectious diseases is growing rapidly; the list of common diseases is more or less static.

Brief Rationale—Improvements in the taxonomic designations of infectious organisms, the availability of highly advanced reference laboratories capable of accurately identifying infectious organisms, increases in the number of immune-compromised patients susceptible to infections by organisms that are not otherwise pathogenic, the increased usage of indwelling therapeutic devices, the emergence of new pathogens, and the ease with which infections can be transported from place to place throughout the world, have all contributed to the increase in newly encountered rare infectious diseases.

Many of the "new" rare diseases simply reflect advances in taxonomy (e.g., new names for old organisms, newly designated subclasses and strains of organisms). When the name of an organism changes, so must the name of its associated disease. Consider "*Allescheria boydii*." People infected with this fungal organism were said to suffer from the disease known as allescheriasis. When the organism's name was changed to *Petriellidium boydii*, the disease name was changed to petriellidosis. When the fungal name was changed, once more, to *Pseudallescheria boydii*, the disease name was changed to pseudallescheriosis [31]. Changes in the standard names of a fungus, appearing in the International Code of Botanical Nomenclature, should trigger concurrent changes in the standard nomenclatures of medicine, such as the World Health Organization's International Classification of Disease, and the National Library of Medicine's Medical Subject Headings, and a variety of specialized disease nomenclatures. Some of these nomenclatures update infrequently. When disease nomenclatures lag behind official fungal taxonomy, a profusion of disease names for infection by a single organism will persist in the literature [19].

In the past, the rational basis for splitting a group of organisms into differently named species required, at the very least, heritable functional or morphologic differences among the members of the group. Gene sequencing has changed the rules for assigning new species. For example, various organisms with subtle differences from *Bacteroides fragilis* have been elevated to the level of species based on DNA homology studies. These include *Bacteroides distasonis*, *Bacteroides ovatus*, *Bacteroides thetaiotaomicron*, and *Bacteroides vulgatus* [32,19].

A large portion of the rare infectious diseases are caused by organisms that, under most circumstances, are unable to produce disease. Some of these infections are opportunistic and occur in a highly select subset of vulnerable individuals (e.g., immunosuppressed, diabetic, renal-compromised, neonatal, elderly, malnourished). For the most part, diseases that arise in immune-compromised individuals are either commensals (i.e., typically non-pathogenic organisms that live in our bodies), or subclinical environmental pathogens (i.e., organisms encountered in the environment that do not cause overt disease in the majority of healthy individuals; see Glossary items, Commensal, Opportunistic infection).

Sources of new, rare infections are invasive instruments and catheters, particularly those that dwell inside the body for prolonged periods. Such indwelling devices would include bladder catheters, ventilator tubes and pulmonary assistive devices, shunts, venous and arterial lines, and indwelling drains and tubes. These devices provide a path of entry to a wide variety of organisms that would otherwise be halted by normal anatomic barriers (e.g., skin, mucous membranes). Of the different organisms that invade via indwelling devices, most are bacteria. Fungal disease has occurred in adults who receive intravenous parental nutrition. The fungi grow in the lipid-rich alimentation fluids, and gain entrance via feeding catheters [33].

The bacterial organisms that gain entrance to normally sterile portions of the body, through indwelling devices, come from diverse taxonomic classes (Pseudomonadales, Bacillales, Bacteroidetes, Fusobacteria, and Legionallales). Despite their taxonomic diversity, these organisms seem to share two properties that enable them to pass over and through mechanical devices. These two properties are the ability to secrete biofilms over surfaces, and the ability to live on and glide through biofilms. Biofilms are invisible, slimy coatings composed of polysaccharides and cellular debris. Biofilms provide a sanctuary that is resistant to the antibacterial sprays and solutions commonly used in hospitals. A small portion of bacterial species have the ability to glide through biofilms; thus, they can follow a catheter into the body. For example, *Staphylococcus epidermidis* is a commensal organism that lives on human skin (see Glossary item, Commensal). Chronically ill patients with indwelling catheters are prone to serious urinary tract infections caused by gliding *Staphylococcus epidermidis*. Other organisms that cause catheter-associated hospital infections were previously obscure (e.g., *Leclercia adecarboxylata* [34]). The list of such organisms is growing.

Also growing is the list of infections that have developed resistance to most types of antibiotic treatments. For the most part, these are not rare diseases; they are common diseases that happen to be resistant to antibiotics. Examples are resistant strains of *Staphylococcus aureus*, *Acinetobacter baumannii*, and *Klebsiella pneumoniae*.

They say that travel broadens the mind. It has certainly opened the world to new exposures to rare and exotic diseases. Although the entire land mass of the planet, outside of Antarctica, has been inhabited by humans for at least 12,000 years, much of this area was seldom visited by non-indigenous people before the past century. With the advent of highways that criss-cross continents, and long-distance transportation modalities, diseases that were previously confined to one location have emerged as worldwide threats. The jet plane is considered by some epidemiologists to be the planet's most effective vector for spreading infectious diseases.

6.4 RARE DISEASES OF UNKNOWN ETIOLOGY

"But though the professed aim of all scientific work is to unravel the secrets of nature, it has another effect, not less valuable, on the mind of the worker. It leaves him in possession of methods which nothing but scientific work could have led him to invent, and it places him in a position from which many regions of nature, besides that which he has been studying, appear under a new aspect."

—James Clerk Maxwell in an essay entitled "Molecules" published in 1873

There are a number of rare diseases of unknown etiology. Some of these diseases may be caused by infectious agents.

6.4.1 Rule—A large portion of human diseases of unknown etiology will eventually be shown to have an infectious etiology.

Brief Rationale—It is difficult to satisfy Koch's postulates for every type of infectious disease (see Glossary item, Koch's postulates). Nonetheless, if efforts to find a non-infectious cause of a disease fail, and if the temporal and geographic pattern of disease occurrences resembles the typical pattern of an infectious epidemic, then an infectious etiology is likely.

Here is an incomplete list of the rare or uncommon diseases whose etiologies are unknown:

Acrocyanosis
Balanitis xerotica obliterans
Behçet disease
Benign fasciculation syndrome
Brainerd diarrhea
Cardiac syndrome X
Chronic fatigue syndrome
Chronic prostatitis/chronic pelvic pain syndrome

Cluster headache
Complex regional pain syndrome
Copenhagen disease
Cronkhite–Canada syndrome
Cyclic vomiting syndrome
Dancing mania
Dancing plague of 1518
Danubian endemic familial nephropathy
Eosinophilic granulomatosis with polyangiitis (Churg–Strauss syndrome)
Electromagnetic hypersensitivity
Encephalitis lethargica
Exploding head syndrome
Fibromyalgia
Fields' disease
Functional colonic disease
Giant cell (temporal) arteritis
Gluten-sensitive idiopathic neuropathies
Gorham vanishing bone disease
Granuloma annulare
Granulomatosis with polyangiitis (Wegener's syndrome)
Gulf War syndrome
Hallermann–Streiff syndrome
Heavy legs
Henoch–Schönlein purpura
Interstitial cystitis
Irritable bowel syndrome
Kawasaki disease
Lichen sclerosus
Lytico–Bodig disease
Microscopic polyangiitis
Morgellons disease
Mortimer's disease
Myofascial pain syndrome
New daily persistent headache
Nodding disease
Peruvian meteorite illness of 2007
Picardy sweat
Pigmented villonodular synovitis
Pityriasis rosea
Polyarteritis nodosa
Posterior cortical atrophy
Prurigo nodularis
SAPHO syndrome
Sarcoidosis

Sick building syndrome
Sjögren's syndrome
Spontaneous cerebrospinal fluid leak
Stiff person syndrome
Sudden unexpected death syndrome
Sweating sickness
Synovial osteochondromatosis
Takayasu's arteritis
Torticollis
Trichodynia
Trigger finger
Tropical sprue

Based on past experience, we can infer that some of the diseases of unknown etiology will have an infectious etiology. Whipple disease, previously a disease of unknown etiology, is characterized by organ infiltrations by foamy macrophages (i.e., scavenger cells that "eat" bacteria and debris). The organ most often compromised is the small intestine, where infiltration of infected macrophages in the lamina propria (i.e., a strip of connective tissue subjacent to the epithelial lining of the small intestine) causes malabsorption. Whipple disease is rare. It occurs most often in farmers and gardeners who work with soil.

Whipple disease was first described in 1907 [35], but its cause was unknown until 1992, when researchers isolated and amplified, from Whipple disease tissues, a 16s ribosomal RNA sequence that could only have a bacterial origin [36]. Based on molecular features of the ribosomal RNA molecule, the researchers assigned it to Class Cellulomonadacea, and named the species *Tropheryma whipplei*, after the man who first described the disease, George Hoyt Whipple.

Particularly noteworthy in the case of Whipple disease is that Koch's postulates were not satisfied. Koch's postulates are a set of observations and experimental requirements proposed by Heinrich Hermann Robert Koch in the late 1800s, intended to prove that a particular organism causes a particular infectious disease. For the experimentalist, the most important of the Koch's postulates require the extraction of the organism from a lesion (i.e., from diseased, infected tissue), the isolation and culture of the organism in the laboratory, and the consistent reproduction of the lesion in an animal injected with the organism. In the case of Whipple disease, the bacterial cause was determined without benefit of isolation or culture. The consistent extraction from Whipple disease tissue of a particular molecule, characteristic of a particular species of bacteria, was deemed sufficient to establish the infectious origin of the disease.

If it were possible to isolate and culture *T. whipplei*, it is highly unlikely that the disease could be experimentally transmitted to animals or humans; another opportunity to satisfy Koch's postulates would fail. As a general rule, bacteria in the human body are eaten by macrophages, wherein they are degraded. In the case of *Tropheryma whipplei*, only a small population of susceptible individuals lacks the ability to destroy *T. whipplei* organisms. In susceptible individuals, the organisms multiply within macrophages. When organisms are released from dying macrophages, additional macrophages arrive to feed, but this only results in the local accumulation of macrophages bloated by bacteria. Whipple disease is a good example of a disease caused by an organism but dependent on a genetic predisposition, expressed as a defect in innate immunity, specifically a reduction of macrophages expressing CD11b (also known as macrophage-1 antigen) [37] (see Glossary item, Innate immunity).

Aside from our inability to culture and extract the *T. whipplei* organism, Whipple disease cannot be consistently reproduced in humans because it can only infect and grow in a small portion of the human population. In short, *T. Whipplei* fails to satisfy Koch's postulates. As we learn more and more about the complexity of disease causation, formerly useful paradigms such as Koch's postulates seem inadequate. When we encounter rare diseases of infectious cause, we might expect to find that the pathogenesis of disease (i.e., the biological steps that lead to a clinical phenotype) may require several independent causal events to occur in sequence. In the case of Whipple disease, the infected individual must be exposed to a soil organism, limiting the disease to farmers and gardeners. The organism, residing in the soil, must be ingested, perhaps by the inhalation of dust. The organism must evade degradation by gut macrophages, limiting disease to individuals with a specific type of defect in cell-mediated immunity, and the individual must have disease that is sufficiently active to produce clinical symptoms.

It has been proposed that Koch's postulates be updated to accommodate modern molecular techniques, and to adjust for the complex ways that organisms interact with humans. The very meaning of biological causation has changed as we learn more and more about disease. We now know that there are many instances wherein the infectious agent cannot account for all of the cellular processes that culminate in disease [38]. The general subject of biological causation will be discussed in Section 9.1.

Diseases of unknown etiology do not fit well into the categories of either rare or common diseases. Most, but not all, of these diseases are exceedingly uncommon. The general public is aware of just a few of these mysterious entities: irritable bowel syndrome, sarcoidosis, Gulf War syndrome, and tropical sprue. Like the common diseases, the diseases of unknown etiology tend to occur in adults and have no obvious patterns of inheritance. It seems like a safe bet to say that a portion of the diseases of unknown etiology will be caused by infectious organisms.

6.5 FUNGI AS A MODEL INFECTIOUS ORGANISM CAUSING RARE DISEASES

"Order and simplification are the first steps toward the mastery of a subject."

—Thomas Mann

It is impossible to cover all of the rare infectious diseases, but it is instructive to concentrate on one class of infectious organisms, the fungi, as a general model.

1. **Most of the fungi that are pathogenic in humans grow in, on, or near soil.** Fungi propagate by ejecting asexual or sexual spores into the air, where they are widely dispersed. Humans become infected when they inhale, ingest, or come into skin contact with fungi in the air, soil, and water. Animal vectors are not required. Animals can, however, serve as reservoirs for fungi (e.g., Microsporidia). If a fungus is growing in your environment, the overwhelming likelihood is that you are constantly exposed to numerous potentially infective fungal elements.
2. **Most pathogenic fungi are globally ubiquitous.** Some are restricted to the tropics; a few are found only in temperate zones. Every human being, with the possible exception of polar inhabitants, is exposed to fungi every day, via air, water, or food. To provide some idea of the ubiquitous nature of fungi, it is estimated that, on average, humans inhale about 40 conidia (fungal spores from Class Ascomycota) each hour. Most of these organisms are non-pathogenic under normal circumstances.
3. **Very few, if any, fungal diseases are contagious from person to person.** The few exceptions, such as the tineas, probably infect most persons through an intermediate medium (e.g., wet bathroom floors, moldy towels, shared sandals), and not by direct skin-to-skin infection.
4. **Fungal colonization, which does not result in disease, is quite common.** *Pneumocystis*, *Aspergillus*, *Histoplasma*, and *Coccidioides* are genera that are found in the lungs of many humans, and are life threatening in a small fraction of the infected population. *Malassezia* species, the causes of tinea versicolor and pityrosporum folliculitis, are normal skin flora. Active infections, when they occur, arise from endogenous skin organisms, not through contagion with infected humans. When colonized individuals become immunosuppressed, endogenized fungi may emerge as serious pathogens.
5. **Though there are over a million fungal species, and hundreds of potential fungal pathogens, the vast majority of human fungal diseases can be accounted for by a few dozen genera, falling into four classes: Class Zygomycota, Class Basidiomycota, Class Ascomycota, and Class Microsporidia.**
6. **Most fungal diseases do not occur in immune-competent individuals.** Of the hundreds of fungal infections that can occur in humans, only a dozen or so produce serious disease in healthy persons (e.g., *Cryptococcus gattii*, *Coccidioides immitis*, *Paracoccidioides brasiliensis*, *Blastomyces dermatitidis*, *Histoplasma capsulatum*). *Candida albicans*, *Lacazia loboi*, and *Sporothrix schenckii* infect immune-competent individuals, but typically cause localized disease. Tinea infections occur in healthy individuals and are caused by *Malassezia* species, *Fusarium* species, *Hortaea werneckii*, and the classic dermatophytes (i.e., Genus *Epidermophyton*, Genus *Microsporum*, and Genus *Trichophyton*).

7. **The more severe the immune deficiency, the more aggressive is the fungal infection.** In general, for any given fungal infection, the disease worsens as immune defenses continue to decline. For example, a superficial, localized, and stable fungal infection of the skin, in an individual who is mildly immune depressed, may progress to a systemic, invasive infection if the immune deficiency worsens.
8. **The number of known fungi that are human pathogens is increasing each year.** As the number of immune-compromised patients increases, due to transplants, AIDS, cancer treatment, long-term steroid use, and with the proliferation of medical devices that provide potential entry points for fungi, the number of newly recognized fungal pathogens will increase. It is estimated that there are about 20 new fungal diseases reported each year [31]. If the number of diseases caused by other types of organisms (i.e., bacteria, protists, animals, viruses, and prions) remains steady, then it will not be long before the number of different fungal diseases exceeds the number of different diseases produced by all other organisms combined.

The increase in newly recognized fungal pathogens is partly credited to technical advances. It is now possible to identify heretofore undiagnosed cases of pathogenic species [39]. In the past, when clinical mycology laboratories had fewer available tests, it was common to lump fungal pathogens under a commonly encountered species or genus. For example, *Aspergillus fumigatus* is a common cause of severe pulmonary infections in immune-compromised patients. With advanced typing techniques, an additional 34 species of *Aspergillus* have been isolated from clinical specimens [31].

In the absence of advanced fungal typing techniques, it can be difficult to correctly assign a fungal species name to a clinical specimen. Pathogenic fungi grow within human tissues vegetatively, as an expanding colony of hyphae or yeasts. The vegetative growth phase observed in tissues lacks the characteristic morphologic traits observed in sexual or asexual fungal reproduction. The pathologist who observes fungal infections in human tissues reaches a diagnosis on clinical presentation and on the somewhat non-specific morphologic features of the fungus in biopsied tissue (i.e., length and thickness of hyphae, presence or absence of septations, angularity of branches, etc.). Adding to the general confusion, fungal specimens grown in culture may have a different morphology from those of the same fungus growing in human tissue. This situation is very different from that of bacterial infections, which have the same morphology in tissues as they have in the culture dish. Consequently, a rare type of fungal infection can be misdiagnosed as a common fungal infection, unless an adequate tissue specimen is delivered to a well-equipped microbiology laboratory.

Sometimes, one clinical disease can be produced by any number of different fungal organisms. Mycetoma, also known as Madura foot and as maduromycosis, occurs most often in India, Africa, and South America. It presents as a slowly growing, fungating mass arising in the subcutaneous tissues, usually on the foot. As the mass grows, draining sinuses discharge fluid and hard grains (white, white–yellow or black grains). Mycetomas may become superinfected, making it very difficult to determine the primary pathogen that caused the disease. More than 30 different species of fungi, and several bacteria, have been grown from these lesions. It has been claimed that black grain mycetomas are caused by *Leptosphaeria senegalensis*, *Madurella grisea*, *Madurella mycetomatis*, or *Pyrenochaeta romeroi*. White grain mycetomas are reputedly caused by *Acremonium* species, *Aspergillus nidulans*, *Neotestudina rosatii*, or *Pseudallescheria boydii*. White–yellow grain mycetomas are said to be caused by *Actinomadura madurae*, *Nocardia asteroides*, and *Nocardia brasiliensis*. Brown–red grain mycetomas are said to be caused by *Actinomadura pelletieri* or *Streptomyces somaliensis*. Taken at face value, these claims would indicate that many different organisms, both bacterial and fungal, can produce a disease of remarkably specific, even unique, clinical features. Suffice it to say that clinical science has much to learn about mycetoma.

It is worthy to note that many fungal organisms are unknown; we simply do not know the full list of potential fungal pathogens that live on Earth. Furthermore, many of the known fungal organisms are unnamed. Fungi are classified based on the morphologic features of sexual growth in culture. If a fungal organism cannot be cultured, or if it does not display sexual reproduction in culture, then it cannot be classified with certainty. A special pseudoclass of fungi, deuteromycetes (spelled with a lower case "d," signifying its questionable validity as a true biologic class) has been created to hold these indeterminate organisms until definitive classes can be assigned. At present, there are several thousand such fungi sitting in a taxonomic limbo [31].

A fungal infection with a single organism may produce many different clinical presentations. One or more of the following biological scenarios may unfold when a human is exposed to a fungus. These scenarios are listed in order of increasing clinical consequence:

1. The fungus grows in the external environment, usually in soil or on plants, never interacting in any way with humans.
2. Spores and asexual reproductive forms are emitted into the air. In warm and tropical locations, fungal elements are the predominant particulate matter found in air samples. Humans are exposed constantly to a wide variety of fungi just by breathing (spores and conidia), by ingestion (fungi grow on the plants we eat), and by direct skin contact with fungal colonies in soil and airborne organisms.
3. After exposure, fungi may leave without colonizing (e.g., you inhale them, and then you exhale them, and they're gone).
4. After exposure, fungi may transiently colonize a mucosal surface, such as the oral cavity, the nose, the gastrointestinal tract, the respiratory tract, or the skin. Once on a mucosal surface, an acute allergic response may occur (e.g., sneezing). After a time, the colony fails to thrive due to an inhospitable environment (e.g., insufficient food, poor ionic milieu, effective host immune response).

5. After exposure, fungi permanently colonize the mucosal surface with no clinical effect. *Candida* species commonly colonize the mouth and the vagina. *Aspergillus* species may colonize the respiratory surfaces (e.g., bronchi). In many cases, we simply carry fungal colonies as commensals (organisms that live within us, without causing disease).
6. Colonies persist, but the host reacts with an acute or chronic immune response. Chronic allergic aspergillosis of the bronchi is a good example. The patient may have a chronic cough. Microscopic examination of bronchial mucosa may reveal some inflammation, the presence of eosinophils, and the occasional hypha. Sometimes the host response is granulomatous, producing small nodules lining the bronchi, containing histiocytes and lymphocytes. A truce between the fungal colony and the host response is sometimes attained, in which the fungus colonies never leave, the inflammation never regresses, but the fungus does not invade into the underlying mucosa.
7. Fungi invade through the mucosa into the submucosa and underlying tissue. These locally invasive infections often manifest as a fungal ball, consisting of varying amounts of inflammatory tissue, necrosis, and fungal elements.
8. Fungal elements invade into lymphatics, traveling with the lymph fluid, and producing regional invasive fungal disease along the route of lymphatic drainage. The prototypical example of this process is found in infections with *Sporothrix schenckii*, which typically gains entrance to the skin, from the soil, through abrasions. Infection yields multiple skin papules, emanating from the point of primary infection (usually the hand or the foot), and following the line of lymphatic drainage.
9. Fungal elements invade the walls of blood vessels.
10. Fungal elements grow in the blood.
11. Fungal elements spread throughout the body to produce invasive fungal infections in multiple organs.

A single species of fungus may manifest itself by any and all of these biologic options, depending primarily on host factors.

REFERENCES

1. The state of world health. Chapter 1 in World Health Report 1996. World Health Organization. Available from: http://www.who.int/whr/1996/en/index.html 1996.
2. Weekly Epidemiological Record. World Health Organization 32:285–296, 2007.
3. Muhlestein JB, Anderson JL. Chronic infection and coronary artery disease. Cardiol Clin 21:333–362, 2003.
4. zur Hausen H. Infections Causing Human Cancer. John Wiley and Sons, Hoboken, 2006.
5. DNA Transforming Viruses. MicrobiologyBytes, October 19, 2004.
6. Wales JH, Sinnhuber RO, Hendricks JD, Nixon JE, Eisele TA. Aflatoxin B1 induction of hepatocellular carcinoma in the embryos of rainbow trout (Salmo gairdneri). J Natl Cancer Inst 60:1133–1139, 1978.

7. Spencer P, Fry RC, Kisby GE. Unraveling 50-year-old clues linking neurodegeneration and cancer to cycad toxins: are microRNAs common mediators? Front Genet 3:192, 2012.
8. Crompton DW. How much human helminthiasis is there in the world? J Parasitol 85:397–403, 1999.
9. Resnikoff S, Pascolini D, Etyaale D, Kocur I, Pararajasegaram R, Pokharel GP, et al. Global data on visual impairment in the year 2002. Bull WHO 82:844–851, 2004.
10. Lemon SM, Sparling PF, Hamburg MA, Relman DA, Choffnes ER, Mack A. Vector-Borne Diseases: Understanding the Environmental, Human Health, and Ecological Connections, Workshop Summary, Institute of Medicine (US) Forum on Microbial Threats. Washington (DC): National Academies Press (US), 2008.
11. Foster J, Ganatra M, Kamal I, Ware J, Makarova K, Ivanova N, et al. The Wolbachia genome of Brugia malayi: endosymbiont evolution within a human pathogenic nematode. PLoS Biol 3:e121, 2005.
12. Cowan G. Rickettsial diseases: the typhus group of fevers: a review. Postgrad Med J 76:269–272, 2000.
13. Global Prevalence and Incidence of Selected Curable Sexually Transmitted Infections: Overview and Estimates. World Health Organization, Geneva, 2001.
14. Sexually Transmitted Diseases. U.S. Centers for Disease Control and Prevention. Available from: http://www.cdc.gov/std/syphilis/stdfact-syphilis.htm, viewed October 24, 2013.
15. Noordeen SK, Lopez Bravo L, Sundaresan TK. Estimated number of leprosy cases in the world. Bull WHO 70:7–10, 1992.
16. World Health Organization. Global leprosy situation. Weekly Epidemiological Record 81: 309–316, 2006.
17. Rassi A Jr, Rassi A, Marin-Neto JA. Chagas disease. Lancet 375:1388–1402, 2010.
18. Stein CE, Birmingham M, Kurian M, Duclos P, Strebel P. The global burden of measles in the year 2000: a model that uses country-specific indicators. J Infect Dis 187:S8, 2003.
19. Berman JJ. Taxonomic Guide to Infections Diseases: Understanding the Biologic Classes of Pathogenic Organisms. Academic Press, Waltham, 2012.
20. Woese CR. Bacterial evolution. Microbiol Rev 51:221–271, 1987.
21. Amin OM. Seasonal prevalence of intestinal parasites in the United States during 2000. Am J Trop Med Hyg 66:799–803, 2002.
22. Sekar U, Shanthi M. Blastocystis: consensus of treatment and controversies. Trop Parasitol 3:35–39, 2013.
23. Samuelson J. Why metronidazole is active against both bacteria and parasites. Antimicrob Agents Chemother 3:1533–1541, 1999.
24. Walker J, Conner G, Ho J, Hunt C, Pickering L. Giemsa staining for cysts and trophozoites of *Pneumocystis carinii*. J Clin Pathol 42:432–434, 1989.
25. Fan L, Lee SY, Koay E, Harkensee C. Plasmodium knowlesi infection: a diagnostic challenge. BMJ Case Rep 2013 Apr 22, 2013.
26. Fischer WM, Palmer JD. Evidence from small-subunit ribosomal RNA sequences for a fungal origin of Microsporidia. Mol Phylogenet Evol 36:606–622, 2005.
27. Keeling PJ, Luker MA, Palmer JD. Evidence from beta-tubulin phylogeny that microsporidia evolved from within the fungi. Mol Biol Evol 17:23–31, 2000.
28. Burri L, Williams B, Bursac D, Lithgow T, Keeling P. Microsporidian mitosomes retain elements of the general mitochondrial targeting system. PNAS 103:15916–15920, 2006.
29. Sak B, Kvac M, Kucerova Z, Kvetonova D, Sakova K. Latent microsporidial infection in immunocompetent individuals: a longitudinal study. PLoS Negl Trop Dis 5:e1162, 2011.
30. Kitamura H, Okudela K. Bronchioloalveolar neoplasia. Int J Clin Exp Pathol 4:97–99, 2011.

31. Guarro J, Gene J, Stchigel AM. Developments in fungal taxonomy. Clin Microbiol Rev 12:454–500, 1999.
32. Baron EJ, Allen SD. Should clinical laboratories adopt new taxonomic changes? If so, when? Clin Infect Dis 16(Suppl 4):S449–S450, 1993.
33. Inamadar AC, Palit A. The genus Malassezia and human disease. Indian J Dermatol Venereol Leprol 69:265–270, 2003.
34. de Mauri A, Chiarinotti D, Andreoni S, Molinari GL, Conti N, de Leo M. Leclercia Adecarboxylata and catheter-related bacteremia: review of the literature and outcome of catheters and patients. J Med Microbiol 62:1620–1623, 2013.
35. Whipple GH. A hitherto undescribed disease characterized anatomically by deposits of fat and fatty acids in the intestinal and mesenteric lymphatic tissues. Bull Johns Hopkins Hosp 18:382–393, 1907.
36. Relman DA, Schmidt TM, MacDermott RP, Falkow S. Identification of the uncultured bacillus of Whipple's disease. N Engl J Med 327:293–301, 1992.
37. Marth T, Roux M, von Herbay A, Meuer SC, Feurle GE. Persistent reduction of complement receptor 3 alpha-chain expressing mononuclear blood cells and transient inhibitory serum factors in Whipple's disease. Clin Immunol Immunopathol 72:217–226, 1994.
38. Inglis TJ. Principia aetiologica: taking causality beyond Koch's postulates. J Med Microbiol 56:1419–1422, 2007.
39. Pounder JI, Simmon KE, Barton CA, Hohmann SL, Brandt ME, Petti CA. Discovering potential pathogens among fungi identified as nonsporulating molds. J Clin Microbiol 45:568–571, 2007.

Chapter 7

Diseases of Immunity

7.1 IMMUNE STATUS AND THE CLINICAL EXPRESSION OF INFECTIOUS DISEASES

"In 1736 I lost one of my sons, a fine boy of four years old, by the small-pox, taken in the common way. I long regretted bitterly, and still regret that I had not given it to him by inoculation. This I mention for the sake of parents who omit that operation, on the supposition that they should never forgive themselves if a child died under it; my example showing that the regret may be the same either way, and that, therefore, the safer should be chosen."

—Benjamin Franklin

In Chapter 6, we learned that a given fungal pathogen may produce many different clinical phenotypes. Let us examine how the clinical manifestation of a fungal disease is modified by an individual's immune status.

Species of *Fusarium* can cause corneal keratitis and onychomycosis (fungal nail infection) in otherwise healthy individuals. In immune-compromised patients with very low white blood cell counts, various *Fusarium* species can produce life-threatening disseminated infections. As a general rule, any superficial tinea disease can convert from superficial infections to invasive infections, so-called tinea profunda, in immune-compromised individuals [1].

7.1.1 Rule—Any infection that occurs in a healthy individual can manifest as a more serious infection if the individual becomes immune-compromised.

Brief Rationale—The immune system keeps infections in check. When the immune status is compromised, the clinical expression of an infection worsens.

Every human is infected with a wide assortment of pathogenic bacteria, fungi, viruses, single-celled eukaryotic organisms, and even small animals (e.g., *Demodex folliculorum*). The human body contains many more cells from non-human organisms than human cells. Though most of these organisms are commensals (i.e., living within humans without causing disease under any circumstances), there are many organisms living within us that are opportunistic pathogens (i.e., capable of causing disease, if not controlled by bodily defenses). These opportunistic pathogens are said to be endogenized, indicating that they have adapted to life inside us. We can infer that all these pathogens are held in

Rare Diseases and Orphan Drugs. http://dx.doi.org/10.1016/B978-0-12-419988-0.00007-9

check by our immune systems; when our immune systems are diminished, these pathogens emerge as active infections (see Glossary item, Immune system). It is easy to find examples that demonstrate the point.

Candida species are normal inhabitants of humans. *Candida* species are found on the skin, respiratory tract, gut, and female genital tract of healthy individuals. An ecological balance exists between *Candida* species and various bacterial commensals. When this balance is disrupted by the use of antibiotics, overgrowth of *Candida* species may occur. In addition, as with virtually all of the pathogenic fungi, overt diseases may occur in immune-deficient individuals. Patients undergoing intense chemotherapy are at particular risk for life-threatening *Candida* infections. So-called invasive candidiasis occurs when organisms penetrate through the mucosa into deeper tissue layers. The transition to invasiveness is often accompanied by a change in morphology from the yeast form to elongated cells (pseudohyphae) and hyphae. The most serious stage of candidiasis involves growth in blood (candidemia) and dissemination to distant organs. *Candida albicans* is the best known pathogenic species, but there are many more known pathogenic types, including *Candida dubliniensis*, *Candida glabrata*, *Candida parapsilosis*, *Candida rugosa*, and *Candida tropicalis*.

Legionella species live within amoeba in the environment. Infection occurs after inhalation of the bacteria, and epidemics have been linked to contaminated sources of aerosolized water from water holding systems. The name of the disease and of the organism derives from the first diagnosed epidemic, occurring in members of an American Legion who attended a bicentennial convention in Philadelphia in July 1976. Direct person-to-person spread has not been established. Disease most often occurs in immune-compromised individuals and in the elderly. Infection is usually pulmonary and can be fatal.

When an individual's immunity is diminished, infectious diseases often emerge at sites where pathogenic bacteria are normally held in abeyance.

7.1.2 Rule—The most common site of presentation of infectious disease in individuals who are immune-compromised is the mouth.

Brief Rationale—The mouth is the dirtiest place in the body, with the greatest variety of potentially pathogenic commensals, many of which live exclusively in periodontal tissues. When an immune-deficient state provides opportunistic pathogens with an occasion to grow and invade, the mouth is often the first site of attack.

About 1000 different species of organisms have been isolated from human mouths. For a given individual, 100–200 of these organisms live in the mouth at any given time [2]. Some of these organisms become pathogenic in immune-competent individuals, simply by mechanical insertion deep into the gingival tissues or into the blood. *Eikenella corrodens* is a normal inhabitant of the mouth, but can produce a bacteremia if mechanically forced into the bloodstream (e.g., chewing bite). *Eikenella corrodens* can cause cellulitis, as well as endocarditis. Another oral inhabitant, *Prevotella dentalis*, can also produce a bite bacteremia.

Many of the opportunistic infections associated with AIDS can arise in the mouth: candidiasis, warts (due to human papillomavirus), herpes simplex virus infection, and hairy leukoplakia (due to Epstein–Barr virus). In point of fact, any systemic infections associated with AIDS may occur in the mouth (e.g., *Cytomegalovirus*, herpes zoster virus reactivation, histoplasmosis, *Cryptococcus neoformans*, and the gamut of opportunistic fungi).

AIDS patients are prone to mouth lesions produced by a variety of cancers of viral origin, including: squamous cell carcinoma of HPV (human papillomavirus) origin, Kaposi sarcoma of herpesvirus 8 origin, primary effusion lymphoma of herpesvirus 8 origin, and nasopharyngeal carcinoma associated with Epstein–Barr virus.

Readers may have noticed that the types of cancers arising in AIDS patients all have a viral etiology. This observation has a biological basis and is generalizable.

7.1.3 Rule—The types of cancers that are known to arise soon after immunosuppression (i.e., weeks or months) are all caused by oncogenic viruses.

Brief Rationale—Viruses are capable of inducing tumors rapidly when not kept in check by immune systems. No other cause of cancer produces tumors in adult humans in a short timeframe, with no apparent latency period.

Viral carcinogenesis will be described in Chapter 8. Suffice it to say that carcinogenic viruses dwell within us. In periods of immunosuppression, these viruses rapidly proliferate, and tumors can arise quickly. If normal immune status is quickly restored, some types of virus-induced tumors (e.g., Kaposi sarcoma) will regress (i.e., involute and disappear). Regression is an extremely rare occurrence in advanced cancers of non-viral origin (see Glossary items, Spontaneous regression, Precancer regression) [3].

Because eukaryotes have evolved along with the organisms that live within them, animal species have had ample opportunity to develop defenses against infectious organisms. Likewise, our infectious organisms have developed defenses against these defenses. As examples of human genetic adaptation to infectious organisms, consider the rare inherited autosomal recessive diseases cystic fibrosis and sickle cell anemia. Carriers of one cystic fibrosis gene have a survival advantage against cholera infection (see Glossary item, Carrier). Carriers of one sickle cell gene have a survival advantage against malaria. The cystic fibrosis and sickle cell anemia carrier states are among the most prevalent of the inherited gene mutations. In each case, an altered gene is preserved in populations because the carrier state confers a survival advantage against a pathogenic organism. Nature has made a grim and merciless trade-off; measuring the increased survival of frequently encountered disease carriers against the suffering and deaths incurred by rare individuals with homozygous disease.

While nature has selected humans for their resistance to infectious organisms, nature has likewise selected infectious organisms for their resistance to

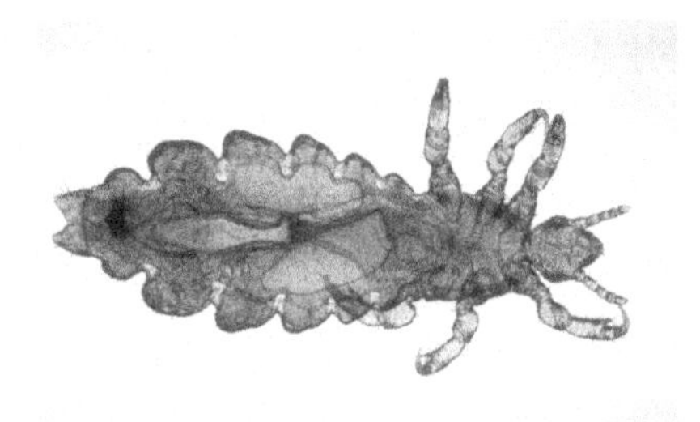

FIGURE 7.1 *Pediculus capitus*, or head louse, an insect specialized to live on the hair of human heads. See color plate at the back of the book. (*Source: Wikipedia, created and released to the public domain by Dr. Dennis D. Juranek of the U.S. Centers for Disease Control and Prevention.*)

humans. *Pediculus humanus capitis* is the organism that causes head lice (see Figure 7.1). This organism only infects humans, and it only lives in the hair that grows on heads. *Pediculus humanus pubis* is the organism that causes crabs. It infects only humans, and it only lives in pubic hair.

All of the thousands of organisms that live within us are genetically selected to survive and prosper in the human environment. Complex immune systems keep these organisms in check, but breakdowns in immunity will result in infections from endogenous and exogenous life-forms.

Humans have three evolved immune systems: innate, intrinsic, and adaptive. The innate immune system is an ancient and somewhat non-specific mechanism deployed by plants, fungi, insects, and most multicellular organisms, including humans [4]. This system recruits immune cells to sites of infection using a protein complex expressed by white blood cells known as the inflammasome (see Glossary item, Inflammasome). Some inflammasome proteins are caspase 1 and 5, PYCARD, and NALP. The inflammasome promotes a variety of chemical mediators known as cytokines. Innate immunity includes the complement system, which acts to clear dead cells. It also includes the macrophage system, which engulfs and removes foreign materials.

The intrinsic immune system is a cell-based (i.e., not humoral) anti-viral mechanism that is always "on" (i.e., not activated by the presence of its target, as seen in adaptive immunity and innate immunity) [5]. Intrinsic immunity has been studied for its role in controlling retrovirus infections (e.g., HIV infection). It is known that intrinsic immunity is not restricted to retroviruses, but its role in blocking infection by other classes of virus is still largely unknown and unstudied. Intrinsic immunity is a newly discovered immune response system about which we have much to learn. One question: Is the intrinsic immune response available to all cell types, or is it strictly restricted to specific responder cells, as is the case with innate immunity and adaptive immunity?

The adaptive immune system adapts to the specific chemical properties of foreign antigens, such as those that appear on viruses and other infectious agents. Adaptive immunity is a system wherein somatic T cells and B cells are produced, each with a unique and characteristic immunoglobulin (in the case of B cells) or T-cell receptor (in the case of T cells). Through a complex presentation and selection system, a foreign antigen elicits the replication of a B cell whose unique immunoglobulin molecule (i.e., so-called antibodies) matches the antigen. Secretion of matching antibodies leads to the production of antigen–antibody complexes that may deactivate and clear circulating antigens, or may lead to the destruction of the organism that carries the antigen (e.g., virus or bacteria). To produce the many unique B and T cells, each with a uniquely rearranged segment of DNA that encodes specific immunoglobulins or T-cell receptors, recombination and hypermutation take place within a specific gene region. This process yields on the order of a billion unique somatic genes, starting with one germinal genome. This amazing show of genetic heterogeneity requires the participation of recombination activating genes (i.e., RAGs). The acquisition of an immunologically active recombination activating gene is presumed to be the key evolutionary event that led to the development of the adaptive immune system that is present in all jawed vertebrates (i.e., gnathostomes). In addition, a specialized method of processing immunoglobulin heavy chain mRNA transcript accounts for the high levels of secretion of immunoglobulin proteins by plasma cells [6]. As one might expect, inherited mutations in RAG genes cause immune deficiency syndromes [7,8].

We learn about the immune system by studying rare diseases. Here are just a few examples of inherited immune deficiency diseases:

- Ataxia telangiectasia is caused by a defect in the ATM gene. The ATM gene regulates a variety of cellular responses to stress, and orchestrates the complex repair of double-stranded breaks in DNA [9]. Ataxia telangiectasia is a syndrome involving several organs and physiologic systems. As its name suggests, it causes ataxia (i.e., imbalanced gait) due to degenerative changes in the cerebellum. It also produces focal areas of small vessel dilation (i.e., telangiectasias). The majority of affected individuals have immune deficits, usually manifested as low levels of immunoglobulins, poor antibody response to vaccines, and a reduction in circulating lymphocytes. Like patients with selective IgA deficiency (see below), they are prone to ear, sinus, and lung infections.
- Chronic granulomatous disease is a genetically heterogeneous group of childhood immune-deficiencies that are all caused by an inability of neutrophils and macrophages to produce a so-called "respiratory burst" (see Glossary items, Genetic heterogeneity, Intra-tissue genetic heterogeneity). The respiratory burst involves production of reactive oxygen species that are toxic to ingested organisms (e.g., bacteria and fungi). Without the respiratory burst, ingested organisms persist, and white blood cells are persistently attracted to the site of infection, eventually producing a granuloma (i.e., a collection of macrophages and fibrocytes with some acute and chronic

inflammatory cells). Catalase-positive organisms are most likely to cause infections in individuals with chronic granulomatous diseases, as these organisms will break down hydrogen peroxide, a respiratory burst molecule, thus exacerbating the deficiency. Studies of chronic granulomatous disease have elucidated some of the important components of the "respiratory burst" pathway and have taught us how ingested organisms are killed inside neutrophils and macrophages [10].

- DiGeorge syndrome is caused by a small deletion of chromosome 22. As one might expect, the loss of a stretch of genes encompassing 3 million base pairs results in developmental anomalies in several organs. The immune deficit is a consequence of thymic hypoplasia. Observations on children with DiGeorge syndrome have added enormously to our understanding of the crucial role of the neonatal thymus in the growth and maturation of T cells.
- Haim–Munk syndrome, Papillon–Lefevre syndrome, prepubertal periodontitis, and aggressive periodontitis are all characterized by early periodontitis, and all are caused by cathepsin C gene mutations [11]. Cathepsin C is a lysosomal proteinase with high activity in inflammatory cells. It is highly expressed in gingival epithelium, where a reduction in its normal activity seems to encourage the emergence of opportunistic infections [11].
- Hyper IgM syndrome is an immunodeficiency disease that results from an inability of B cells to produce classes of immunoglobulins other than IgM. IgM is the default immunoglobulin produced by B cells. A vigorous B-cell response to antigens requires a switch from IgM synthesis over to the other immunoglobulins (e.g., IgG, IgA, IgE). The switching system is complex, and mutations of different genes expressed in T or B cells can cause the hyper IgM syndrome (e.g., CD40LG gene, AICDA gene, CD40 gene, UNG gene). The genes involved in the switching system may also be involved in other pathways used by T or B cells to induce a combined immunodeficiency syndrome in which hyper IgM is just one component [12].
- Hyperimmunoglobulin E syndrome, also known as hyper IgE recurrent infection syndrome, is characterized by immunodeficiency, recurrent infections often involving the skin and the lungs, and a variety of skeletal and connective tissue abnormalities [13]. Most prominent on physical examination of children with the disease is a double row of teeth due to failure of the primary teeth to extrude when the permanent teeth grow in. Because the skin infections may occur as boils, another synonym for this disease is Job's syndrome (a trope to the biblical story of Job and his affliction with boils). The genetic condition is somewhat heterogeneous, with most cases caused by mutations in the STAT3 (signal transducer and activator of transcription 3) gene, while others are caused by Tyk2 mutations as well as currently undetermined mutations in other genes. Another immunodeficiency syndrome, familial chronic mucocutaneous candidiasis-7, is caused by heterozygous mutation in the STAT1 gene. In any case, individuals who have hyperimmunoglobulin E syndrome are caught in a pleiotropic storm produced by alterations in STAT3

pathways that yield a wide variety of developmental and immunologic manifestations (see Glossary items, Pleiotropic, Pleiotypia). In their review paper on the hyper IgE syndromes, Alexandra Freeman and Steven Holland concluded, "Understanding how STAT3 deficiency leads to the many facets of this disease will hopefully help us understand diseases that are more common, such as idiopathic scoliosis, atopic dermatitis, staphylococcal skin abscesses, and the coronary artery aneurysms of Kawasaki disease" [13].

- Immunodeficiency–centromeric instability–facial anomalies syndrome types 1 and 2 are caused by mutation in the DNMT3B gene and in the ZBTB24 gene, respectively. These two diseases are clinically identical and are characterized by an immunoglobulin deficiency resulting in recurrent infections, facial abnormalities, and mental retardation.
- Recurrent invasive pneumococcal infections in a child or adolescent may indicate a deficiency of the innate immune system. Inherited susceptibility to invasive pneumococcal infections can be caused by mutations in genes encoding innate system protein (e.g., IRAK4 gene, NEMO gene) [14]. Conversely, an advantageous polymorphism in the innate system's TIRAP gene (Toll-interleukin 1 receptor protein) confers protection against infections from invasive pneumococcal disease, bacteremia, malaria and tuberculosis [15].
- Selective immunoglobulin A deficiency is the most common defect of humoral immunity, and is characterized by a deficiency of immunoglobulin A (IgA). IgA is an immunoglobulin primarily synthesized by mucosa-lining lymphoid tissue, such as the lymphoid tissue lining the nasal, oral, bronchial, and gastrointestinal mucosal surfaces. It is unusual among the immunoglobulins because it is secreted onto an epithelial surface, rather than being absorbed directly into the bloodstream. As you might expect, individuals with IgA deficiency are susceptible to recurring sino-pulmonary and gastrointestinal infections.
- Severe combined immunodeficiency is a collection of rare congenital disorders characterized by deficiencies of T and B cells, and all are caused, ultimately, by a defect in hematopoietic stem cells. Each type of SCID is caused by a type-specific gene defect (e.g., the gene that encodes IL-7 receptor, a gene defect that results in gamma chain deficiency, a recombination activating gene deficiency, and the gene encoding adenine deaminase). The SCID diseases illustrate how different genetic defects can lead through different pathways to a common clinical phenotype.
- WHIM syndrome (warts, hypogammaglobulinemia, infections, and myelokathexis) is a combined immunodeficiency disease caused by an alteration in the chemokine receptor gene CXCR4 (see Glossary item, Chemokine) [16]. Warts result from a lowered immune repression of papillomaviruses. Likewise, the other phenotypic components of the disease arise from the aberrant chemokine receptor, including myelokathexis, a congenital cytopenia of white cells.

- Wiskott–Aldrich syndrome is characterized by eczema, a low platelet count and small platelets, and a combined immunodeficiency, involving B and T cells. Most affected individuals develop an autoimmune disease, and there is an increased risk of developing lymphoma or leukemia. The disease is caused by an aberration in the WASp gene, encoding a pleiotropic protein expressed in high levels in hematopoietic cells. The expressed protein plays a role in actin polymerization. A normal actin cytoskeleton is necessary for a proper immune synapse, wherein a target antigen-carrying cell and an effector lymphocyte meet, initiating a process in which an antigen-specific response is attained.
- X-linked agammaglobulinemia, also known as Bruton-type agammaglobulinemia, was first described in 1952, and was the first known immunodeficiency disease [17]. The disease occurs in sons of female carriers, and produces severe infections beginning in early childhood. It is caused by an inherited deficiency in a tyrosine kinase enzyme required for normal B-cell maturation. B cells are not produced; hence, immunoglobulins are not produced, resulting in a deficiency of antibody-mediated adaptive immunity. Knowledge of the pathogenesis of X-linked agammaglobulinemia has advanced our understanding of the role of the B cell in immunity.
- Common variable immunodeficiency (CVID) is somewhat of a misnomer. CVID represents a collection of about 150 rare genetic immune disorders. The "common" phenotypic feature of CVID is that all of the included disorders are characterized by hypoglobulinemia (i.e., low levels of IgA, IgG, or one of the other immunoglobulin types). The inclusion of deficiencies of variable types of immunoglobulin molecule accounts for the "variable" in its name. Another name for this condition is "acquired hypogammaglobulinemia," also a poor choice of terminology. The term "acquired" is applied to the disorder because the majority of patients seem to develop their disease between the ages of 20 and 40, indicating that something happened in the first few decades of life that led to the acquisition of the condition. This is not the case. The disease, even in its so-called acquired form, manifests as gradually diminishing IgG levels beginning in early childhood, becoming clinically evident in early adulthood [18]. CVID, even in the aggregate of all its included disorders, is a rare condition, with a prevalence of about 1 in 50,000. Though rare, CVID is the most commonly diagnosed immune disorder of humans.

Though immunodeficiency diseases are rare, infectious diseases are common. Why do we humans develop infectious diseases when we have three separate immune systems to protect us? As animals have developed elaborate defense systems, the organisms that cause disease have developed even more effective attack systems. Small chinks in our immune defenses can open the gates to an infectious disease. Starvation, concurrent disease, extreme youth, extreme age, and any cause of cytopenia (i.e., reduction in circulating blood cells) can all lower our resistance to disease. In addition, it would seem that complex sets of genes predispose otherwise healthy individuals to specific types of infections.

7.1.4 Rule—Immune deficits are usually polygenic or have an environmental cause.

Brief Rationale—We are constantly being reminded of Darwin's cruel game; monogenic causes of immunodeficiency are rare because they reduce fitness. Infectious diseases are extremely common, and every infectious disease marks a defeat in the human body's battle against invasive organisms. When we study families with increased susceptibility to certain types of infection, we seldom observe Mendelian patterns of inheritance; instead, we observe non-Mendelian patterns indicative of polygenic inheritance [19]. Though there are dozens of monogenic immunodeficiency syndromes, they account for a very small fraction of the instances of immune deficiency in the general population.

Sets of variant genes can produce some increase in the risk of infections. Among these variant genes, CISH (cytokine-inducible SRC homology 2 domain) polymorphisms seem to play a significant role. CISH protein variants can increase susceptibility to bacteremia, malaria, and tuberculosis by about 18% [20].

7.2 AUTOIMMUNE DISORDERS

"The continued study of rare variants in autoimmune disease will inform future investigations and treatments directed at rare and common autoimmune diseases alike."

—Mickie H. Cheng and Mark S. Anderson [21]

The adaptive immune system is designed to produce antibodies that can bind specific foreign (i.e., non-self) antigens. As discussed in the prior section, antibody–antigen complexes can neutralize (i.e., inactivate) circulating proteins produced by infectious agents. Furthermore, antibody–antigen reactions occurring on the surface of an infectious organism may elicit an inflammatory response capable of killing the organism. Though adaptive immunity plays an important role in protecting us from infection, errors of the immune system can lead to pathologic conditions in which an adaptive immune response is launched against "self" antigens. As a group, such conditions are called autoimmune diseases.

Autoimmune conditions are common, with over 8.5 million individuals in the U.S. suffering from one or another of these diseases [22]. The common autoimmune diseases are the various autoimmune thyroid diseases (e.g., Graves' disease), type 1 diabetes mellitis, pernicious anemia, rheumatoid arthritis, and vitiligo, together accounting for about 93% of individuals with autoimmune disease in the U.S. One in 31 Americans has an autoimmune disease.

Autoimmune thyroid disease is the most common autoimmune disease, affecting 2–4% of women and up to 1% of men [23]. The incidence of this disease increases with age. More than 10% of the population over the age of 75 has at least biochemical signs of hypothyroidism, and most of these individuals have an underlying autoimmune cause of their abnormal thyroid tests [23].

A list of some of the less common autoimmune diseases would include: multiple sclerosis, rheumatoid arthritis, systemic lupus erythematosus, myasthenia gravis, primary biliary cirrhosis, scleroderma, various types of glomerulonephritis, Goodpasture syndrome, acquired cutis laxa, idiopathic thrombocytopenia purpura, and relapsing polychondritis [22,24].

Acquired autoimmune diseases may mimic genetic diseases. Examples include: autoimmune forms of cutis laxa [25]; autoimmune von Willebrand disease; a form of congenital heart block in offspring of mothers with systemic lupus erythematosus; and several rare nervous system disorders caused by acquired autoimmune channelopathies (see Glossary item, Channelopathy) [26].

Aside from the recognized autoimmune diseases, there are a host of inflammatory and chronic conditions whose causes are unknown (e.g., polymyositis and dermatomyositis, chronic fatigue syndrome). Some of these diseases may eventually be added to the list of human autoimmune diseases.

7.2.1 Rule—The autoimmune diseases as a group are pathogenetically related to one another.

Brief Rationale—All autoimmune diseases involve some of the same components of a complex pathway that leads to the development of antibodies.

As it happens, more than one autoimmune disease can occur in a single individual. Furthermore, families predisposed to autoimmune disease typically have more than one type of autoimmune diseases represented among the lineage of affected members. This finding strongly suggests that different autoimmune diseases are mechanistically related [27].

7.2.2 Rule—The common autoimmune diseases involve one or two organs; seldom more.

Brief Rationale—The adaptive immune system is designed to produce antigen-specific antibodies. Assuming that the defective pathway is limited to the adaptive immune system, the likelihood that a disorder will yield an antibody that cross-reacts with many different tissues is small.

In point of fact, most antibodies are astonishingly specific. For example, type 1 diabetes targets beta cells of the islets of the pancreas. Goodpasture disease involves the production of an anti-basement membrane antigen found in lung and the kidney; and the disease is isolated to these two organs.

7.2.3 Rule—The common autoimmune diseases have a polygenic origin.

Brief Rationale—Though the common autoimmune diseases (i.e., autoimmune thyroid diseases, type 1 diabetes mellitis, pernicious anemia, rheumatoid arthritis,

and vitiligo) tend to run in families, they seldom display a simple Mendelian inheritance pattern. Inheritance that is non-Mendelian usually has a polygenic origin.

Single gene mutations do not cause the common autoimmune diseases. There are various examples of rare, monogenic autoimmune diseases [21]. As always, the monogenic forms of disease are always highly instructive, allowing us to delineate individual steps in the pathogenesis of disease.

An example of a rare, monogenic autoimmune disorder is C1q deficiency. C1q is a protein involved in the normal fixation of antigen–antibody complexes to complement. A deficiency of C1q leads to the production of multiple autoantibodies and reduces cytotoxicity targeted against infectious organisms. Hence, C1q deficiency, as a rare disease, does not follow the rule for common autoimmune diseases; wherein autoimmunity is confined to one or two organs. C1q deficiency results in a syndrome much like systemic lupus erythematosus, along with recurrent and chronic infections.

Autoimmune lymphoproliferative syndrome is another example of an inherited monogenic autoimmune condition characterized by lymphocytosis and any of various types of autoimmune diseases, including autoimmune-mediated cytopenias of blood cells: hemolytic anemia, neutropenia, thrombocytopenia (see Glossary item, Cytopenia). The term autoimmune lymphoproliferative syndrome covers a group of related disorders caused by mutations of FAS, FASLG, CASPASE 8, CASPASE 10, and several RAS genes. These genes play a role in apoptosis, a normal pathway that induces cell death (see Glossary item, Apoptosis). Mutations in these genes result in the persistence of once-active lymphocytes that have outlived their usefulness. The persistent lymphocytes elevate the number of lymphoid cells in circulation and in lymph nodes, and raise the likelihood of an inappropriate autoimmune response.

7.2.4 Rule—Autoimmune disorders that result from a dysfunction of the innate immune system are rare, and tend to produce systemic disease involving multiple tissues [21].

Brief Rationale—The distinction between "self" and "non-self" proteins is a function of the innate immune system [28]. When the immune system cannot ignore "self" antigens, the effects tend to be systemic.

For example, autoimmune polyendocrinopathy–candidiasis–ectodermal dystrophy is a monogenic disease caused by mutations in the autoimmune regulator (AIRE) gene. The AIRE gene has a role in the development of normal immune tolerance, the process whereby the body determines the proteins that are "self" and thus privileged not to elicit an immune response [29]. AIRE permits thymic cells to express a wide variety of proteins that would otherwise be restricted to specific organs (e.g., cardiac-specific proteins, kidney-specific proteins). The expression of diverse proteins within the thymus at a particular stage of fetal development somehow teaches the immune system to quell an adaptive immune reaction to these proteins later in life. Dysfunction of AIRE gene leads to an inappropriate adaptive immunity response against multiple tissues.

7.2.5 Rule—There is an environmental component to the autoimmune diseases.

Brief Rationale—Autoimmune diseases, like any other diseases with a non-genetic component, are not encountered in neonates, and the overall incidence of the autoimmune diseases increases with age.

As we grow older, we encounter more and more infections and environmental substances that can elicit an antibody response. Every antibody response increases the chance of selecting a clone of immunocytes that produce an antibody capable of cross-reacting with a normal constituent of our cells. For example, autoimmune disease can be acquired as a paraneoplastic phenomenon (i.e., a condition caused by a cancer). Cancer cells elicit a wide variety of antibodies. When a cancer elicits antibodies against desmoplakins, a component of normal desmosomes, the immune response can produce an immunologic phenocopy of pemphigus [30]. As discussed previously in Section 5.2, desmosomes are found in high concentration in the epidermis, where they hold the various layers of keratinocytes together, and where they bind the epidermis to the underlying dermis. Processes that disrupt desmosomal integrity tend to produce blistering diseases such as pemphigus.

Autoimmune conditions can result from bacterial disease. Following infection with *Streptococcus pyogenes*, an immune response cross-reacting between bacterial antigens and normal host proteins (e.g., heart muscle proteins, glomerular basement membranes) may lead to rheumatic fever or to glomerulonephritis.

7.2.6 Rule—Physiologic systems influence the development of autoimmune diseases.

Brief Rationale—Autoimmune diseases can occur in men or women, but most occur preferentially in women. The preferential occurrence of virtually every autoimmune disease in individuals of a particular gender suggests that some intrinsic physiologic condition contributes to disease susceptibility.

Goodpasture's disease is an exception to the rule, occurring more often in men. For all the autoimmune diseases taken together, women are 2.7 times more likely to develop disease than men [22]. This would suggest that the physiological state of the organism, not simply the summation of genetic and environmental conditions, influences the development of autoimmune disease.

REFERENCES

1. Kobayashi M, Ishida E, Yasuda H, Yamamoto O, Tokura Y. Tinea profunda cysticum caused by *Trichophyton rubrum*. J Am Acad Dermatol 54(2 Suppl):S11–S13, 2006.
2. Zimmer C. How microbes defend and define us. The New York Times July 12, 2010.

3. Berman JJ. Neoplasms: Principles of Development and Diversity. Jones & Bartlett, Sudbury, 2009.
4. Vilmos P, Kurucz E. Insect immunity: evolutionary roots of the mammalian innate immune system. Immunol Lett 62:59–66, 1998.
5. Yan N, Chen ZJ. Intrinsic antiviral immunity. Nat Immunol 13:214–222, 2012.
6. Borghesi L, Milcarek C. From B cell to plasma cell: regulation of V(D)J recombination and antibody secretion. Immunol Res 36:27–32, 2006.
7. Zhang J, Quintal L, Atkinson A, Williams B, Grunebaum E, Roifman CM. Novel RAG1 mutation in a case of severe combined immunodeficiency. Pediatrics 116:445–449, 2005.
8. de Villartay JP, Lim A, Al-Mousa H, Dupont S, Déchanet-Merville J, Coumau-Gatbois E, et al. A novel immunodeficiency associated with hypomorphic RAG1 mutations and CMV infection. J Clin Invest 115:3291–3299, 2005.
9. Shiloh Y, Kastan MB. ATM: genome stability, neuronal development, and cancer cross paths. Adv Cancer Res 83:209–254, 2001.
10. Heyworth PG, Cross AR, Curnutte JT. Chronic granulomatous disease. Curr Opin Immunol 15:578–584, 2003.
11. Hart TC, Hart PS, Michalec MD, Zhang Y, Firatli E, Van Dyke TE, et al. Haim-Munk syndrome and Papillon-Lefevre syndrome are allelic mutations in cathepsin C. J Med Genet 37:88–94, 2000.
12. Lougaris V, Badolato R, Ferrari S, Plebani A. Hyper immunoglobulin M syndrome due to CD40 deficiency: clinical, molecular, and immunological features. Immunol Rev 203:48–66, 2005.
13. Freeman AF, Holland SM. The hyper-IgE syndromes. Immunol Allergy Clin North Am 28:277–291, 2008.
14. Ku CL, Picard C, Erdös M, Jeurissen A, Bustamante J, Puel A, et al. IRAK4 and NEMO mutations in otherwise healthy children with recurrent invasive pneumococcal disease. J Med Genet 44:16–23, 2007.
15. Khor CC, Chapman SJ, Vannberg FO, Dunne A, Murphy C, Ling EY, et al. A Mal functional variant is associated with protection against invasive pneumococcal disease, bacteremia, malaria and tuberculosis. Nat Genet 39:523–528, 2007.
16. Hernandez PA, Gorlin RJ, Lukens JN, Taniuchi S, Bohinjec J, Francois F, et al. Mutations in the chemokine receptor gene CXCR4 are associated with WHIM syndrome, a combined immunodeficiency disease. Nat Genet 34:70–74, 2003.
17. Bruton OC. Agammaglobulinemia. Pediatrics 9:722–728, 1952.
18. Park JH, Resnick ES, Cunningham-Rundles C. Perspectives on common variable immune deficiency. Ann NY Acad Sci 1246:41–49, 2011.
19. Hill AVS. Evolution, revolution and heresy in the genetics of infectious disease susceptibility. Philos Trans R Soc Lond B Biol Sci 367:840–849, 2012.
20. Khor CC, Vannberg FO, Chapman SJ, Guo H, Wong SH, Walley AJ, et al. CISH and susceptibility to infectious diseases. N Engl J Med 362:2092–2101, 2010.
21. Cheng MH, Anderson MS. Monogenic autoimmunity. Annu Rev Immunol 30:393–427, 2012.
22. Jacobson DL, Gange SJ, Rose NR, Graham NM. Epidemiology and estimated population burden of selected autoimmune diseases in the United States. Clin Immunol Immunopathol 84:223–243, 1997.
23. Vaidya B, Kendall-Taylor P, Pearce SHS. The genetics of autoimmune thyroid disease. J Clin Endocr Metab 87:5385–5397, 2002.
24. Pietropaolo M, Barinas-Mitchell E, Kuller LH. The heterogeneity of diabetes: unraveling a dispute: is systemic inflammation related to islet autoimmunity? Diabetes 56:1189–1197, 2007.

25. Tsuji T, Imajo Y, Sawabe M, Kuniyuki S, Ishii M, Hamada T, et al. Acquired cutis laxa concomitant with nephrotic syndrome. Arch Dermatol 123:1211–1216, 1987.
26. Kleopa KA. Autoimmune channelopathies of the nervous system. Curr Neuropharmacol 9:458–467, 2011.
27. Cotsapas C, Voight BF, Rossin E, Lage K, Neale BM, Wallace C, et al. Pervasive sharing of genetic effects in autoimmune disease. PLoS Genet 7:e1002254, 2011.
28. Miller DM, Rossini AA, Greiner DL. Role of innate immunity in transplantation tolerance. Crit Rev Immunol 28:403–439, 2008.
29. Anderson M, Su M. AIRE and T cell development. Curr Opin Immunol 23:198–206, 2011.
30. Oursler JR, Labib RS, Ariss-Abdo L, Burke T, O'Keefe EJ, Anhalt GJ. Human autoantibodies against desmoplakins in paraneoplastic pemphigus. J Clin Invest 89:1775–1782, 1992.

Chapter 8

Cancer

8.1 RARE CANCERS ARE FUNDAMENTALLY DIFFERENT FROM COMMON CANCERS

"The most savage controversies are those about matters as to which there is no good evidence either way."

—Bertrand Russell

It is worthwhile listing the differences between rare cancers and common cancers. In subsequent sections of this chapter, we will explore the biological basis of these differences, and why these differences have led to major advances in the prevention and treatment of all cancers.

1. Just a few types of common cancers account for the majority of occurrences of cancer.
2. Most of the different types of cancers are rare cancers.
3. Virtually every common cancer is composed of cells derived from the ectodermal or the endodermal layers of the embryo (see Glossary items, Ectoderm, Endoderm). Rare cancers derive from all three germ layers, but the majority of rare cancers derive from the mesoderm.
4. All of the childhood cancers are rare cancers.
5. All the advanced stage cancers that we can currently cure are rare cancers, and most of the curable rare cancers are cancers that occur in children.
6. Inherited syndromes that cause rare cancers are often associated with increased risk for developing common cancers; hence, the causes of rare cancers are related to the causes of common cancers.
7. Rare cancers are genetically simpler than common cancers (i.e., have fewer mutations). In many cases, we know the underlying mutation that leads to the development of rare cancers. We do not know the underlying mutation(s) that leads to common cancers.
8. Common cancers are genetically heterogeneous and may contain one or more rare types of cancer having the same clinical phenotype as the common cancer.
9. Most of what we know about the pathogenesis of cancer has come from observations on rare cancers.

Rare Diseases and Orphan Drugs. http://dx.doi.org/10.1016/B978-0-12-419988-0.00008-0

10. The rare cancers serve as sentinels for environmental agents that can cause various types of cancer; either rare or common. Common cancers cannot serve as sentinels.
11. Treatments developed for the rare cancers will almost certainly apply to the common cancers.

In Sections 8.2 through 8.4, we will try to explain why these 12 assertions are true, and what they teach us about the biology of cancer.

8.2 THE DICHOTOMOUS DEVELOPMENT OF RARE CANCERS AND COMMON CANCERS

"The digestive canal represents a tube passing through the entire organism and communicating with the external world, i.e. as it were the external surface of the body, but turned inwards and thus hidden in the organism."

—Ivan Pavlov

8.2.1 Just a few types of common cancers account for the majority of occurrences of cancer

Though there are thousands of types of human cancer, the bulk of cancer cases in humans are accounted for by just a few, under a dozen, types of cancer. The two most commonly occurring cancers of humans are basal cell carcinoma of skin and squamous cell carcinoma of skin. Together, these two tumors account for about 1.2 million new cancers each year in the U.S., nearly equal to the number of all the other types of cancers combined. These tumors are so common that, frequently, more than one basal cell carcinoma or squamous cell carcinoma will occur in the same individual. Fortunately for us, these two tumors seldom cause deaths; most cases are cured by simple excision. Cancer registries do not bother to collect records on these two cancers, and the published data on cancer incidence, compiled from registries and surveillance databases, typically ignores these two tumors. Nonetheless, we will see later in this chapter that basal cell carcinoma of skin and squamous cell carcinoma of skin tell us much about the biology of cancer in humans.

In Section 2.1, we discussed Pareto's principle, wherein a few common items account for the majority of instances of any collection. Cancer obeys Pareto's principle: a few cancers account for most cases of cancer occurring in humans. Collected U.S. data for the year 2008 indicate that, after excluding basal cell carcinomas and squamous cell carcinomas of skin, there were 1,437,180 new cancers. In the same year, there were 565,650 cancer deaths, of which 161,840 individuals died of lung cancer [1]. The percentage of U.S. cancer deaths from lung cancer was 28.6% (161,840/565,650). Also in 2008, there were 49,960 deaths from colorectal cancer, accounting for 8.8% of U.S. cancer deaths (49,960/565,650). Just two cancers (lung and colorectal) accounted for 37.4% of deaths from cancer in the U.S. When age-adjusted data are examined, the top five cancer killers (lung, colon, breast, pancreas, and prostate) account for 57% of all cancer deaths [1] (see Glossary item, Age-adjusted).

Observing that a few types of cancers account for the bulk of human cancer deaths, funding agencies have concentrated their efforts on finding cures for the most common cancers. Just seven types of common cancer, out of about 6000 known cancers, account for over 36% of cancer funding [2]. The justification for distributing cancer research funding toward research in the common cancers is simple. If cures can be found for the most common cancers, we could drastically reduce the number of cancer deaths in the U.S. and in the world. Curing a rare cancer that might affect a few hundred people worldwide would seem to be an ill-advised investment of our limited resources. Hence the rare cancers receive relatively little cancer funding compared with the common cancers.

The drawback to this straightforward approach is that it has failed. Despite decades of funding, we still do not know how to cure common cancers when they are diagnosed at an advanced disease stage. New discoveries in cancer genetics have highlighted the incredible complexity of the commonly occurring cancers. The complexity of the common cancers has been a seemingly insurmountable barrier blocking the development of simple and effective cures. Despite the long-term efforts of an army of cancer researchers, the age-adjusted death rate from cancers in the year 2000 was about the same as it was in 1975. A significant drop in the cancer death rate since the year 2000 is largely attributed to smoking cessation and other preventive measures; not due to effective new cures [3].

8.2.2 Most of the types of cancers are rare cancers

There are about 6000 types of cancer that have been assigned names by pathologists [4–6]. About a dozen of these cancers are common diseases. The remaining cancers (i.e., about 6000 entities) comfortably qualify as "rare" under U.S. Public Law 107-280, the Rare Diseases Act of 2002 [7]. Consequently, healthcare workers must somehow come to grips with 6000 types of rare cancers. Moreover, the variety of rare cancers is increasing rapidly. As we learn more and more about the genetics of cancers, we find that the common cancers can be subtyped into genetically distinct groups. Furthermore, we are finding an increasing number of alternate alleles and heterogeneous genes that account for rare diseases. Hence, the trend is leading us to divide the common cancers into genetically distinct subtypes that qualify as rare cancers, and to divide the known rare cancers into ultra-rare subtypes.

8.2.3 Virtually every common cancer is composed of cells derived from the ectodermal or the endodermal layers of the embryo. Rare cancers derive from all three germ layers, but the majority of rare cancers derive from the mesoderm

8.2.1 Rule—Most common cancers are caused by environmental agents.

Brief Rationale—The vast majority of cancers occur at body sites that are directly exposed to chemical, physical, or biological agents delivered by food, water, and air. The tissues that receive the highest levels of exposure are the same tissues that yield the highest number of tumors. Tissues of the body that are not directly exposed to outside agents (e.g., muscle, connective tissues) are not sites at which common cancers develop.

The human body is like a doughnut, with an outside, where the glaze is found, and an inside, the dough. Our skin would represent the outer-edge surface of the doughnut. Our intestines would represent the inner-edge surface of the doughnut, the glazed part that lines the doughnut hole. Our outer-edge surface is lined by epithelial squamous cells of skin epidermis. Our inner-edge surface is lined by epithelial enterocytes of the gastrointestinal mucosa.

Our connective tissues, muscle, adipose tissue, vessels, and bones would correspond to the doughy part of the doughnut. All the "inside" tissues are derived from the mesoderm, the embryonic layer that is sandwiched between the ectoderm, which gives rise to the epidermis and skin appendages (e.g., hair, sebaceous glands), and the endoderm, which gives rise to the gut and derivative organs (e.g., pancreas, liver, lungs).

Exposure to toxic and carcinogenic chemicals takes place on the doughnut surfaces (skin and gastrointestinal tract), and the epithelial organs that bud off these surfaces. Because ectoderm and endoderm, corresponding to the glazed part of the doughnut, are exposed to carcinogens delivered through the air, water, and food, all of the common cancers occurring in humans derive from ectodermal and endodermal tissues (e.g., skin cancers, lung cancer, colorectal cancer, prostate cancer, breast cancer). Because the cells deriving from the mesoderm (i.e., the connective tissues) are not exposed to high levels of environmental carcinogens, all of the mesodermal tumors (e.g., fibrosarcoma, rhabdomyosarcoma, liposarcoma, osteosarcoma, angiosarcoma) are rare cancers.

The skin, being on the outside surface of the body, is exposed daily to ultraviolet light, often for prolonged periods. Hence, it is no surprise that basal cell carcinoma and squamous cell carcinoma of skin, both of which preferentially occur on sun-exposed skin surfaces, are the two most common cancers of humans.

Additional factors contribute to the enormous disparity between cancers deriving from ectoderm/endoderm (i.e., the doughnut glaze) and cancers deriving from mesoderm (i.e., the dough). Cells lining the surface of the body are constantly dividing and sloughing into the atmosphere, in the case of epidermal squamous cells, or the gut lumen, in the case of enterocytes. With the exception of bone marrow cells, and the osteoblasts in growing bones, cells derived from mesoderm have a low level of cell division. Dividing cells are targets for the early mutational steps of carcinogenesis, because dividing cells can pass unrepaired mutations to their progeny (see Glossary item, Mutagen). This is one more reason why rapidly dividing ectodermal and endodermal cells of the human body are much more likely to give rise to cancers than are slowly dividing mesodermal cells.

8.2.2 Rule—In adults, diseases of cells derived from ectoderm or from endoderm typically have an environmental cause.

Brief Rationale—Tissues deriving from ectoderm and endoderm are exposed to toxins at higher levels than are the tissues that derive from mesoderm. When a disease targets ectodermal- or endodermal-derived cells, in adults, it is likely to have a toxic etiology. Cells of mesodermal origin (i.e., the inside cells) are typically spared, because they are less exposed to the environment.

When we examine the diseases known to be caused by environmental toxins, most target tissues of ectodermal or endodermal origin. We have previously discussed ectodermal (i.e., skin) tumors induced by ultraviolet light. In addition to basal cell carcinoma and squamous cell carcinoma of skin, melanoma is another cancer that arises almost exclusively from skin, particularly sun-exposed skin. Smoking causes lung cancer; the cancer most likely to kill humans. The lung is an ectodermally-derived organ. Examples of other toxin-caused diseases targeting ectodermal or endodermal tissues includes emphysema (cigarette toxins acting on lung), asthma (allergens acting on lung), type 1 diabetes (antibodies acting on endodermally-derived islet cells), and cirrhosis (alcohol acting on endodermally-derived hepatic cells).

8.2.3 Rule—Most of the metabolism of foreign compounds entering the human body is handled by cells derived from endoderm or ectoderm.

Brief Rationale—It stands to reason that the cells that receive the brunt of environmental toxins will be the cells that are adapted to detoxify exogenous chemicals.

The liver, of endodermal origin, is the principal metabolizing organ in the human body. The gut absorbs the complex chemicals in our food and sends the compounds via the portal vein to be processed in the liver. The hepatocyte is designed to receive, metabolize, and detoxify exogenous chemicals, using an enriched set of metabolic enzymes within an expanded smooth endoplasmic reticulum. Any molecule that can be metabolized by human cells can be metabolized by the liver.

8.2.4 Rule—Most chemical carcinogens need to be metabolized before they are converted to an active (i.e., mutagenic) molecular form.

Brief Rationale—Activated carcinogens are highly reactive molecules that can bind to just about any kind of molecule. Naturally active carcinogens would react with, and be neutralized by, non-genetic molecules before they could reach

DNA. Highly carcinogenic molecules exist as stable, inactive molecular species that are metabolized within cells to active molecules that react with DNA.

Many carcinogens are activated by enzymes (e.g., the cytochrome p450 pathway) within the smooth endoplasmic reticulum. Polymorphisms in the cytochrome P450 CYP2D6 gene influence the tissue targets of carcinogens [8].

8.3 THE GENETICS OF RARE CANCERS AND COMMON CANCERS

"The cell is basically an historical document, and gaining the capacity to read it (by the sequencing of genes) cannot but drastically alter the way we look at all of biology."

—Carl Woese [9]

8.3.1 All of the childhood cancers are rare cancers; hence, none of the common cancers are childhood cancers

8.3.1 Rule—Virtually all cancers of childhood have a germline genetic component to their pathogenesis.

Rationale—The common cancers have multi-step etiologies, requiring many years to develop, and occurring in adults. Children simply do not have the opportunity to express diseases that involve repeated exposures to commonly occurring environmental agents. Hence, cancers in children develop from inborn mutations. Cancer-causing germline mutations are rare; hence, childhood cancers are rare.

The earliest occurring childhood cancers are the congenital tumors (i.e., tumors present at birth). Obviously, congenital tumors are constrained to have a 9-month pathogenesis (i.e., the gestation period). Gestational choriocarcinoma is a cancer that arises from the so-called extra-embryonic cells of the developing organism, the cells that give rise to the fetal placenta. Many choriocarcinomas are known to arise from zygotes having a double set of paternal chromosomes. Studies of mouse embryogenesis have demonstrated that both maternal and paternal chromosomes are necessary for normal embryonic development. When there is a deficiency of maternal chromosomes and a redundancy of paternal chromosomes in a mouse conceptus, growth of extra-embryonic tissues (trophectoderm) is exaggerated. This is the presumed genetic mechanism accounting for a portion of the observed gestational choriocarcinomas [10,11].

The tumors that occur later in childhood may have acquired some, but not all, of the requisite genetic damage leading to the development of a cancer. We learn much by careful observations of inherited and non-inherited (i.e., sporadic) retinoblastoma, a rare tumor of primitive retinal cells (see Glossary items, Non-inherited genetic disease, Sporadic). Inherited retinoblastomas are often bilateral and occur in children. Non-familial retinoblastomas are unilateral and occur in an older age group. The reason for this difference is that children who develop retinoblastoma are born with a heterozygous mutation of the RB gene. To develop cancer, they need only acquire one additional mutation inactivating the alternate allele, and yielding a homozygous loss of gene function. Because only one additional mutational event is required in the inherited syndrome, the tumors occur often (two tumors in a single patient) and early. In adults who develop non-familial cases of retinoblastoma, two mutational events (one for each allele) must accumulate in cells of the retinal lineage, sometime during the life of the patient. This usually takes a long time to occur (accounting for the late onset) and occurs only rarely (accounting for only one occurrence of a tumor in an affected individual). The theoretical underpinnings of these observations were made by Knudsen and preceded the discovery of tumor suppressor genes and the RB gene mutation (see Glossary item, Tumor suppressor gene) [12,13].

Observations on retinoblastomas indicate that the same genetic lesion that occurs in the germline of rare childhood tumor syndromes may be acquired in the somatic cells of sporadically occurring retinoblastomas in adults. This observation foreshadows a theme that will be expanded in Section 12.1, "Shared Genes," that rare cancers account for subsets of common cancers.

8.3.2 All the advanced stage cancers that we can currently cure are rare cancers, and most of the curable rare cancers are cancers that occur in children

The cancer death rates shown here are stratified by age, and all races, males and females are included [14]. Rates are per 100,000 population and are age-adjusted to the 2000 U.S. population.

Age group	1950	1978	2005
0–4	11.1	4.6	2.2
5–14	6.7	4.1	2.5
15–24	8.6	6.1	4.1
25–34	20.4	14.2	9.1
35–44	63.6	50.7	32.8
45–54	174.2	179.6	118.3
55–64	391.3	428.9	329.7
65–74	710.0	803.4	748.8
75–84	1167.2	1204.1	1265.1
85 and over	1450.7	1535.3	1643.7
All ages	195.4	204.4	184.0

When you look at the different age groups, the biggest drops in cancer deaths occur in pediatric ages. For older individuals, aged 65 and up, the cancer death rate has been rising. Cancer deaths among adults are about 500 times more numerous than cancer deaths among children. We have made remarkable progress in treating childhood cancer, but dramatic advances in the treatment of childhood cancers do not significantly improve the aggregate death rate from cancer among all age groups. Because all childhood cancers are always rare, we can say that the greatest advances in cancer research have involved rare cancers.

8.3.3 Inherited syndromes that cause rare cancers are often associated with increased risk for developing common cancers; hence, the causes of rare cancers are related to the causes of common cancers

Many of the greatest advances in our understanding of common cancers have come through the study of rare familial cancer syndromes in which common types of cancer occur. Here are a few common cancers and the familial syndromes that account for a small percentage of cases.

- Colon tumors (benign and malignant)
  - Colorectal cancer hereditary non-polyposis
  - Polyposis syndrome, mixed hereditary
  - Turcot syndrome (central nervous system cancer and familial polyposis of the colon)
  - Mismatch repair gene pmsl1 colorectal cancer hereditary, non-polyposis type 3 included
  - Checkpoint kinase 2 *S. pombe* homologue of breast and colorectal cancer susceptibility
  - Colorectal adenomatous polyposis autosomal recessive
  - Oligodontia–colorectal cancer syndrome
  - Juvenile polyposis/hereditary hemorrhagic telangiectasia syndrome
  - Adenomatous polyposis of the colon (APC)
  - Peutz–Jeghers syndrome
  - Colorectal cancer hereditary non-polyposis type 2
  - Colorectal cancer susceptibility on chromosome 9

- Lung cancer
  - Lung cancer 1
  - Lung cancer, alveolar cell carcinoma included

- Breast cancer
  - Brca1 breast cancer type 1
  - Breast cancer 11–22 translocation associated
  - Brca2 breast cancer type 2
  - Brca3 breast cancer type 3

- Basal cell carcinoma of skin (see Glossary item, Basal cell carcinoma)
  - Basal cell carcinomas with milia and coarse sparse hair
  - Basal cell nevus syndrome
  - Basal cell carcinoma, multiple
  - Basaloid follicular hamartoma syndrome (see Glossary item, Hamartoma)
  - Basal cell carcinoma with follicular differentiation
  - Xeroderma pigmentosum complementation group b
  - Xeroderma pigmentosum 1
- Renal cell carcinoma
  - Renal carcinoma, familial associated 1 included
  - Renal cell carcinoma, papillary
  - Non-papillary renal carcinoma 1
  - Renal cell carcinoma, papillary 3
  - Leiomyomatosis and renal cell cancer hereditary
- Thyroid cancer
  - Thyroid carcinoma, familial medullary
  - Familial non-medullary thyroid cancer
  - Papillary thyroid microcarcinoma
  - Thyroid carcinoma, papillary with papillary renal neoplasia
  - Thyroid carcinoma, non-medullary 1
  - Thyroid carcinoma, Hürthle cell
  - Thyroid carcinoma, follicular
- Ovarian cancer
  - Epithelial ovarian cancer
  - Ovarian cancer, epithelial, susceptibility to
- Melanoma
  - Melanoma, cutaneous malignant 4
  - Melanoma, cutaneous malignant 3
  - Familial atypical multiple mole melanoma-pancreatic carcinoma syndrome
  - Dysplastic nevus syndrome, hereditary b-k mole syndrome
- Prostate cancer
  - Prostate cancer, hereditary x-linked
  - Prostate cancer, hereditary 1
  - Prostate cancer, hereditary 20
  - Prostate cancer, hereditary 7
  - Prostate cancer, hereditary 3
  - Prostate cancer/brain cancer, susceptibility

When we look at individual inherited cancer syndromes, we see that both rare and common cancers may result. Here is the list of different types of cancer associated with the Li–Fraumeni syndrome [15]. The syndrome-associated cancers are divided into common and rare cancers.

- Common tumors associated with Li–Fraumeni syndrome
  - Breast cancer
  - Lung adenocarcinoma
  - Colon cancer
  - Pancreatic cancer
  - Prostate cancer
- Rare tumors associated with Li–Fraumeni syndrome
  - Soft tissue sarcomas
  - Osteosarcomas
  - Brain tumors
  - Acute leukemias
  - Adrenocortical carcinomas
  - Wilms tumor
  - Phyllodes tumor of breast

It is worth noting that the common cancers associated with rare cancer syndromes have a similar morphologic appearance as their sporadic counterparts. This suggests that regardless of underlying genetic cause, the pathogenesis of each named common cancer tends to converge to its characteristic phenotype.

8.3.4 Rare cancers are genetically simpler than common cancers (i.e., have fewer mutations)

8.3.2 Rule—Rare tumors are much more likely to have a single cause, a single carcinogenic pathway, a single inherited gene, or a single acquired marker, than are any of the common tumors.

Brief Rationale—Many different factors can lead to a common cancer; this is why the cancer is common. Only very specific and highly unlikely factors (e.g., genetic mutation) lead to rare cancers; this is why they are rare.

Here are some simple gene alterations encountered in rare neoplasms (see Glossary item, Neoplasm).

- Cylindroma, CYLD1 tumor suppressor gene [16]
- Gastrointestinal stromal tumor, c-KIT mutation [17]
- Polycythemia vera, JAK2 mutation [18]
- Pilomatrixoma, beta-catenin [19]
- Pleomorphic adenoma, PLAG1 [20]
- Ovarian granulosa-stromal cell tumors, diploid or trisomy 12 [21]
- Clear cell hidradenoma of the skin, TORC1-MAML2 gene fusion [22]
- Lipoblastoma, rearrangements of chromosome bands 8q11-13 possibly involving PLAG1 [23]

- Lipomas, HMGIC
- Hemangioma, HMGIC
- Chondroid hamartoma, HMGIC

There are few causes for any particular rare tumor, and the gene that is known to cause a rare tumor will likely be the cause of most or all instances of the same rare tumor. The cause might be an inherited mutation, as is the case for inherited retinoblastoma. The cause might be a single exposure to an identified carcinogen at a documented moment in time, as in gestational exposure to diethylstilbestrol resulting in clear cell adenocarcinoma of the cervix in adolescent girls. **Because rare cancers often have a single cause, it seems reasonable to suppose that rare cancers can be easily treated or prevented by targeting one altered gene or one altered pathway, or by avoiding one specific carcinogenic event.**

8.3.5 Common cancers are heterogeneous and may contain biologically distinctive subsets of cancers that are rare

For example, microsatellite instability characterizes about 15% of colorectal cancers (see Glossary items, Microsatellite, Microsatellite instability). Lynch syndrome is an inherited cancer syndrome associated with a variety of cancers, including colorectal cancer. All Lynch syndrome cases of colorectal cancer have microsatellite instability. Lynch syndrome cases account for about 3% of the colorectal cancers that exhibit microsatellite instability. The remaining 12% of colorectal cancers that have microsatellite instability are sporadic. Hence, common colorectal cancers have a subset of cases that can be biologically distinguished by microsatellite instability, and a subset of this subset is familial and caused by a germline mutation that renders microsatellites unstable [24].

Additional examples abound. A rare subset of lung cancers is caused by a rearrangement in the NUT gene. Secretory breast carcinoma, formerly known as juvenile carcinoma of breast, is a rare type of invasive ductal breast cancer that is characterized by a specific fusion gene [25]. Myelodysplastic syndrome, a preleukemic condition for which the preponderance of cases occur in elderly individuals, is known to occur in children who inherit a predisposition to losing chromosome 7 in somatic blood forming cells [26,27].

8.3.3 Rule—In a tumor that can occur as a rare inherited form, or as a common sporadic form, we always learn the most by studying the rare inherited form and later extending our gained knowledge to the common sporadic form.

Brief Rationale—Only the subset of cases arising from an inherited germline mutation can be studied in affected and unaffected relatives.

Research depends on controlled experiments in which two groups are identical, with the exception of one perturbing factor. In the case of inherited diseases, molecular biologists can look for a gene that distinguishes two groups that are the same in nearly every way except that one group has a disease-causing gene in its germline, and the other does not. In the case of sporadically occurring tumors, there is no control group, and nothing to learn by comparing the genome of germlines. Much of what we know about tumors has come from studying familial cases, and then testing to see if the same gene that caused the familial cases is also present in the sporadic cases. Here are a few examples:

- Germline mutations of the p53 tumor suppressor gene are present in the rare Li–Fraumeni syndrome. A somatic p53 mutation is present in about half of all human cancers [28].
- Families with germline mutations of the KIT gene develop gastrointestinal stromal tumors (GISTs). Somatic mutations of KIT occur in the majority of sporadic GIST tumors.
- Germline RET gene mutations occur in familial medullary carcinoma of thyroid, and in most cases of sporadic medullary carcinoma of thyroid [29,30].
- Germline RB1 gene mutations occur in familial retinoblastoma syndrome and in sporadic cases of retinoblastoma [31].
- Germline patched (ptc) gene mutations occur in basal cell nevus syndrome and in sporadically occurring basal cell carcinomas [32].
- Germline PTEN mutations occur in Cowden syndrome and Bannayan–Riley–Ruvalcaba syndrome, two inherited disorders associated with a high rate of endometrial carcinomas. PTEN mutations are found in 93% of sporadically occurring endometrial carcinomas [15].

In tumor after tumor, the genetic lesion present in sporadically occurring cancers would not have been found without prior knowledge of the syndromic gene (the gene responsible for the rare inherited condition).

8.3.4 Rule—If you look hard enough, you can usually find examples of syndromic disorders accounting for what might otherwise be considered a sporadic or non-syndromic childhood cancer.

Brief Rationale—A germline mutation having the biological power to cause cancer might be expected to produce some additional phenotypic effects in the organism.

What must we assume if we believe that a germline mutation, found in every somatic cell in the body, will yield one or more types of cancer, and no other abnormality? We would need to assume that genes exist whose only function relates to the development of cancer. Actually, the known cancer genes (tumor suppressor genes and oncogenes) play fundamental roles in human development and in the activities of differentiated cells (see Glossary items, Oncogene, Tumor suppressor gene). The germline mutations that raise the incidence of cancer may be associated with various abnormalities; hence, they are syndrome-forming mutations. We find such associations among pediatric tumors. Examples include:

inguinal hernia and Ewing sarcoma; prune belly syndrome and congenital kidney cancer; Dubowitz syndrome (microcephaly, growth retardation, and a characteristic facial appearance) and rhabdomyosarcoma; Schinzel–Giedion syndrome (a multi-system developmental abnormality with dysmorphism of skull) and sacrococcygeal teratoma [33].

8.3.5 Rule—There is no such thing as a mutation that is necessary and sufficient, by itself, to cause cancer.

Brief Rationale—In the worst of the inherited cancer syndromes, tumors do not occur in every organ, or even in every individual who carries the cancer-causing mutation. The empiric absence of a 100% penetrant cancer mutation (i.e., one that always causes cancer) suggests that more than one event or condition must prevail during carcinogenesis.

With the exception of the extremely rare congenital tumors, cancers caused by germline mutations are not present at birth. They require events to play out over time. Furthermore, in those organs in which germline mutations produce a cancer in an at-risk organ, most of the cells of the organ (i.e., greater than 99.999% of the cells of the organ) do not become cancerous. Cancer is a rare cellular process, even in inherited syndromes that confer a high risk of developing cancer.

8.3.6 Rule—In contrast to rare cancers, common cancers are characterized by many different mutations in many different genes, and the affected genes will vary from patient to patient and from tumor sample to tumor sample within the same patient.

Brief Rationale—Common cancers are genetically unstable.

Genetic instability, resulting from any of many possible gene mutations, is found in virtually every common cancer, and the number of resulting gene mutations and splice variants carried by a common cancer continuously increases as the cancer grows (see Glossary item, Gene mutation rate, Genetic instability) [34–38]. Every cancer is composed of emerging subclones having new mutations, and some of these subclones will inevitably demonstrate growth advantages over other subclones, thus producing increasingly aggressive subclonal outgrowths in a heterogeneous tumor (see Glossary items, Tumor heterogeneity, Cancer progression) [39].

8.4 USING RARE DISEASES TO UNDERSTAND CARCINOGENESIS

"It is once again the vexing problem of identity within variety; without a solution to this disturbing problem there can be no system, no classification."

—Roman Jakobson

8.4.1 Most of what we know about the pathogenesis of cancer has come from observations on rare cancers

Like most other biological processes (e.g., blood coagulation, inflammation, protein synthesis), cancer proceeds in steps. The purpose of this section is to describe what we think we know about the biological steps leading to cancer, and how we might understand these steps much better, if we apply ourselves to studying rare cancers and rare cancer syndromes.

8.4.1 Rule—Carcinogenesis, the pathogenesis of tumors, is a multi-step process.

Brief Rationale—Interventions can stop the process of carcinogenesis at various points in tumor development (e.g., the precancer stage), indicating the presence of multiple biological steps, each with characteristic properties and vulnerabilities.

Until the middle of the twentieth century, little was known about what caused cancer and how cancers developed in the body. As we found an increasing number of chemical agents that caused cancers, and after DNA's role in the genetic code was discovered, a rather simple scenario emerged that seemed to explain the way that cancers developed. First, a carcinogen mutated a normal cell, rendering it cancerous. Over time, the cell multiplied at a rate faster than its normal counterparts. Eventually, the population produced from the original cancer cell became large enough to visualize grossly. The visible tumor could be excised, examined microscopically, and assigned a name, based on the normal tissue it resembled morphologically. This is a one-step model. A one-time interaction between a cell and a carcinogen produces a mutant cancer cell, and this cancer cell divides to produce a mass of cells that we call a cancer. By the early 1960s, this simplistic hypothesis was fully discredited. Leslie Foulds' groundbreaking book *Neoplastic Development* firmly established that cancer develops through multiple biological steps [40].

What evidence proves that cancer is a multi-step process? First, we know that very long latencies follow the exposure of a person to a carcinogen and the emergence of a cancer. In the case of asbestos-induced mesotheliomas, there are many reported cases wherein navy shipyard workers, and the members of their families, were exposed to asbestos for a very brief time during World War II, only to develop cancers 20–40 years later. If a cancer cell were to begin growing in 1944 at a rate similar to the growth rate we observe in clinically encountered mesotheliomas, it would have killed the patient within a few months or years (i.e., in about 1945 or 1946). Assuming that a carcinogen can create a cancer cell via a mutational event, an assumption that is totally incorrect, then the tumor cell created back in 1945 must have either lain dormant or must have grown as a clonal community at a rate much slower than that observed in mesotheliomas. In either case, the original tumor cell must have been biologically different from the tumor cells observed in mesotheliomas. Therefore, the original tumor cell must have undergone a biological step that changed its phenotype. Hence, carcinogenesis of asbestos-induced mesotheliomas must involve multiple steps.

Precancers serve as the morphologic proof of the multi-step process of pathogenesis. Precancers are non-invasive lesions that precede the emergence of invasive cancers (see Glossary item, Invasion). After a time, invasive cancers develop from the precancer. If we knew nothing about the biological events that precede the development of precancers, we could nonetheless infer that carcinogenesis is a multi-step process because it passes through a precancerous stage.

8.4.2 Rule—Each step in carcinogenesis is a potential target of cancer prevention.

Brief Rationale—The key thing to know about carcinogenesis is that it occurs in steps. Because there are multiple steps in carcinogenesis, there are multiple opportunities for blocking the progression of cancer [41,3].

In 2006, cancer epidemiologists noticed that the incidence of new breast cancer cases in the U.S. dropped 15% from August 2002 to December 2003. In July 2002, a report was published indicating that hormone treatments for menopause raised the risk of developing breast cancer [42]. As a result, beginning in 2002, millions of women whose ages correlated with the peak years of developing breast cancer abandoned hormone treatment. The subsequent unexpected drop in breast cancer incidence was attributed to the cessation of hormonal supplementation in menopausal women [42].

There is a serious problem with the kind of reasoning that inferred that hormone treatment cessation, beginning in July 2002, can produce a sustained drop in the incidence of breast cancer, beginning several months later. We know that carcinogenesis is a process that occurs over several decades. A tumor that occurs in a woman in the year 2003 would have been initiated way back in 1990 or earlier. If we wanted to know the intervention that dropped the 2003 cancer rate, should we not look at events occurring in 1990 and thereabouts? Likewise, can we not infer that withdrawal of a cancer-causing agent in 2002 would produce a drop of breast cancer occurrences in 2015 not 2003? What is wrong with our thinking?

It is a mistake to assume that interventions that reduce cancer incidence must operate at the first step of carcinogenesis, when carcinogens interact with normal cells. **The rapid drop in breast cancer incidence following withdrawal of hormone replacement implies that carcinogenesis is a multi-step process, and hormone withdrawal blocked a late step in carcinogenesis**. Epidemiologic observations confirm that carcinogenesis is a multi-step process. Using similar epidemiologic reasoning, British cancer epidemiologist Richard Peto proved that smoking-induced lung cancer is a multi-step process [43].

What do we currently know, or think that we know, about the mysterious process known as carcinogenesis? A commonly held view is that carcinogenesis is a long process that involves the accumulation of genetic and epigenetic alterations that confer the malignant phenotype onto a clone of cells. The envisioned sequence of events that comprise carcinogenesis begins with initiation, wherein a carcinogen damages the DNA of a cell, producing a mutant founder cell that produces a clone of cells that have one or more subtle (i.e., morphologically invisible) differences from the surrounding cells (see Glossary item, Initiation). We can only speculate on the alterations found in initiated cells, but these might include mutations that cause cells to be less likely to senesce and die, or be more likely divide, or be less genetically stable, or better able to survive in a hypoxic environment. In the cases of the inherited cancers occurring in children or in young adults, an inborn mutation begins the process early in life. After a time, which could easily extend into years, subclones of the original clone emerge that have additional properties that are conducive to the emergence of the malignant phenotype (e.g., new mutations that confer additional growth or survival advantages, or greater ability to grow under the available conditions). The process of continual subclonal selection continues, usually for a period of years, until a morphologically distinguishable group of cells appears; the precancer. Subclonal cells from the precancer eventually emerge with the full malignant phenotype (i.e., the ability to invade surrounding tissues and metastasize to distant sites; see Glossary item, Invasion).

The process of carcinogenesis proceeds stepwise, over time, from initiation (i.e., the first cellular event in the pathogenesis of cancer, which is presumed to involve the production of a heritable abnormality in a somatic cell), to latency (i.e., the long period of carcinogenesis during which cellular changes do not yield morphologically visible abnormalities), to the precancer stage, characterized by nuclear atypia (see Glossary item, Nuclear atypia), to invasive cancer, to progression, wherein the tumor accumulates additional genotypic and phenotypic abnormalities. Left untreated, malignant cells eventually metastasize and grow at distant sites (see Glossary item, Dormancy). Most deaths from cancer arise from complications caused by metastatic disease.

The steps of carcinogenesis unfold over decades. Hence, invasive cancers are a disease of middle-aged and elderly individuals, but carcinogenesis is a disease process occurring in a relatively young population.

8.4.3 Rule—Rare cancers and rare cancer syndromes have helped us to dissect the various steps of carcinogenesis.

Brief Rationale—We see rare cancers and rare cancer syndromes having inherited defects involving a specific step in carcinogenesis. These would include polymorphisms in genes that metabolize carcinogens at the time of initiation, that repair DNA (e.g., xeroderma pigmentosum), that preserve the integrity of DNA replication, that control microsatellite stability (e.g., hereditary nonpolyposis colon cancer syndrome), that control apoptosis, that activate tumor suppressor genes (e.g., Li–Fraumeni syndrome) and tumor oncogenes (BCR/ABL fusion gene in chronic myelogenous leukemia), that drive hyperplasia of particular cell types (e.g., c-KIT gastrointestinal stromal tumors), and so on.

As in all multi-step processes that occur in discrete biological steps, the rare cancers give us an opportunity to dissect the process, step by step, by examining the inherited mutations that operate specific steps in the process. The general approach to dissecting cellular pathways by studying the rare diseases will be described in Section 10.4.

8.4.2 The rare cancers serve as sentinels for environmental agents that can cause various types of cancer, rare and common. Common cancers cannot serve as sentinels

When we look at past triumphs in cancer epidemiology, most are examples in which a small population was exposed to a specific carcinogen that produced a rare tumor.

Epidemiologic triumphs in cancer research:

- Hepatic angiosarcoma (thorotrast exposure in the 1940s, polyvinyl chloride exposure in the 1970s)
- Mesothelioma (World War II asbestos exposure by U.S. Navy shipbuilders, with mesotheliomas developing 20+ years after exposure)
- Scrotal cancer (British chimney sweeps in the eighteenth century)
- Oral cancer in teenagers (tobacco chewers, late twentieth century)
- AIDS-related cancers (HIV infections, beginning about 1980)
- Lung cancer (a rarity in the nineteenth century, lung cancer became the number one cancer killer of humans, with the popularization of cigarette smoking)
- Leukemia (increased in workers exposed to benzene)
- Thyroid cancer in Chernobyl (pulsed exposure to radiation)

8.4.3 Treatments developed for the rare cancers will apply to common cancers

8.4.4 Rule—Rare cancers are easier to cure than common cancers.

Brief Rationale—The malignant phenotypes of rare cancers are often driven by a single genetic alteration or a single cellular pathway. It is feasible to target and inhibit a single pathway with a single drug. Common cancers are driven by hundreds or thousands of aberrant pathways. We currently have no way of inhibiting all of the possible pathways that drive the malignant phenotype in common cancers.

The list of cancers that can be cured in an advanced stage of growth consists exclusively of rare cancers [44]:

- Choriocarcinoma
- Acute lymphocytic leukemia of childhood
- Burkitt lymphoma
- Hodgkin lymphoma
- Acute promyelocytic leukemia
- Large follicular center cell (diffuse histiocytic) lymphoma

- Embryonal carcinoma of testis
- Hairy cell leukemia
- Seminoma

When we learn how to cure the rare tumors, we can extend our knowledge to treat common cancers whose key pathways are also found in rare, curable cancers. The topic of sharing cures among rare cancers and common cancers will be revisited in Section 13.3.

REFERENCES

1. Seer. Estimated New Cancer Cases and Deaths for 2008. Available from: http://seer.cancer.gov/csr/1975_2005/results_single/sect_01_table.01.pdf, viewed December 2, 2013.
2. Funding for Various Research Areas. Available from: http://obf.cancer.gov/financial/historical.htm.
3. Berman JJ. Precancer: The Beginning and the End of Cancer. Jones & Bartlett, Sudbury, 2010.
4. Berman JJ. Developmental Lineage Classification and Taxonomy of Neoplasms. Available from: http://www.julesberman.info/neoclxml.gz.
5. Berman JJ. Neoplasms: Principles of Development and Diversity. Jones & Bartlett, Sudbury, 2009.
6. Berman JJ. Tumor classification: molecular analysis meets Aristotle. BMC Cancer 4:10, 2004. Available from: http://www.biomedcentral.com/1471-2407/4/10.
7. Rare Diseases Act of 2002, Public Law 107–280, 107th U.S. Congress, November 6, 2002.
8. Wolf CR, Smith CA, Gough AC, Moss JE, Vallis KA, Howard G, et al. Relationship between the debrisoquine hydroxylase polymorphism and cancer susceptibility. Carcinogenesis 13:1035–1038, 1992.
9. Woese CR. Bacterial evolution. Microbiol Rev 51:221–271, 1987.
10. McGrath J, Solter D. Completion of mouse embryogenesis requires both the maternal and paternal genomes. Cell 37:179–183, 1984.
11. Barton SC, Surani MAH, Norris ML. Role of paternal and maternal genomes in mouse development. Nature 311:374–376, 1984.
12. Knudson AG Jr, Hethcote HW, Brown BW. Mutation and childhood cancer: a probabilistic model for the incidence of retinoblastoma. Proc Natl Acad Sci USA 72:5116–5120, 1975.
13. Knudson AG. Mutation and cancer: statistical study of retinoblastoma. Proc Natl Acad Sci USA 68:820–823, 1971.
14. Seer Cancer Statistics Review 1975–2005. Table I-2 56-year trends in U.S. cancer death rates. Available from: http://seer.cancer.gov/csr/1975_2005/results_merged/topic_historical_mort_trends.pdf.
15. Omim. Online Mendelian Inheritance in Man. Available from: http://omim.org/downloads, viewed June 20, 2013.
16. Biggs PJ, Chapman P, Lakhani SR, Burn J, Stratton MR. The cylindromatosis gene (cyld1) on chromosome 16q may be the only tumour suppressor gene involved in the development of cylindromas. Oncogene 12:1375–1377, 1996.
17. Berman J, O'Leary TJ. Gastrointestinal stromal tumor workshop. Hum Pathol 32:578–582, 2001.
18. Zhang L, Lin X. Some considerations of classification for high dimension low-sample size data. Stat Methods Med Res 2011. Nov 23. Available from: http://smm.sagepub.com/content/early/2011/11/22/0962280211428387.long, viewed January 26, 2013.

19. Chan EF, Gat U, McNiff JM, Fuchs E. A common human skin tumour is caused by activating mutations in beta-catenin. Nature Genet 21:410–413, 1999.
20. Kandasamy J, Smith A, Diaz S, Rose B, O'Brien C. Heterogeneity of PLAG1 gene rearrangements in pleomorphic adenoma. Cancer Genet Cytogenet 177:1–5, 2007.
21. Fletcher JA, Gibas Z, Donovan K, Perez-Atayde A, Genest D, Morton CC, et al. Ovarian granulosa-stromal cell tumors are characterized by trisomy 12. Am J Pathol 138:515–520, 1991.
22. Behboudi A, Winnes M, Gorunova L, van den Oord JJ, Mertens F, Enlund F, et al. Clear cell hidradenoma of the skin—a third tumor type with a t(11;19)-associated TORC1-MAML2 gene fusion. Genes Chromosomes Cancer 43:202–205, 2005.
23. Sciot R, Akerman M, Dal Cin P, De Wever I, Fletcher CD, Mandahl N, et al. Cytogenetic analysis of subcutaneous angiolipoma: further evidence supporting its difference from ordinary pure lipomas: a report of the CHAMP Study Group. Am J Surg Pathol 21:441–444, 1997.
24. Boland CR. Clinical uses of microsatellite instability testing in colorectal cancer: an ongoing challenge. J Clin Oncol 25:754–756, 2007.
25. Tognon C, Knezevich SR, Huntsman D, Roskelley CD, Melnyk N, Mathers JA, et al. Expression of the ETV6-NTRK3 gene fusion as a primary event in human secretory breast carcinoma. Cancer Cell 2:367–376, 2002.
26. Lizcova L, Zemanova Z, Malinova E, Jarosova M, Mejstrikova E, Smisek P, et al. A novel recurrent chromosomal aberration involving chromosome 7 in childhood myelodysplastic syndrome. Cancer Genet Cytogenet 201:52–56, 2010.
27. Shannon KM, Turhan AG, Chang SS, Bowcock AM, Rogers PC, Carroll WL, et al. Familial bone marrow monosomy 7. Evidence that the predisposing locus is not on the long arm of chromosome 7. J Clin Invest 84:984–989, 1989.
28. Royds JA, Iacopetta B. p53 and disease: when the guardian angel fails. Cell Death Differ 13:1017–1026, 2006.
29. Marshall E. Genetic testing. Families sue hospital, scientist for control of Canavan gene. Science 290:1062, 2000.
30. Vezzosi D, Bennet A, Caron P. Recent advances in treatment of medullary thyroid carcinoma. Ann Endocrinol (Paris) 68:147–153, 2007.
31. Blanquet V, Turleau C, Gross-Morand MS, Senamaud-Beaufort C, Doz F, et al. Spectrum of germline mutations in the RB1 gene: a study of 232 patients with hereditary and nonhereditary retinoblastoma. Hum Mol Genet 4:383–388, 1995.
32. Johnson RL, Rothman AL, Xie J, Goodrich LV, Bare JW, Bonifas JM, et al. Human homolog of patched, a candidate gene for the basal cell nevus syndrome. Science 272:1668–1671, 1996.
33. Mehes K, Kosztolanyi G. Clinical manifestations of genetic instability overlap one another. Pathol Oncol Res 10:12–16, 2004.
34. Greenman C, Stephens P, Smith R, Dalgliesh GL, Hunter C, Bignell G, et al. Patterns of somatic mutation in human cancer genomes. Nature 446:153–158, 2007.
35. O'Driscoll L, McMorrow J, Doolan P, McKiernan E, Mehta JP, Ryan E, et al. Investigation of the molecular profile of basal cell carcinoma using whole genome microarrays. Mol Cancer 5:74, 2006.
36. Gorringe KL, Chin S, Pharoah P, Staines JM, Oliveira C, Edwards PAW, et al. Evidence that both genetic instability and selection contribute to the accumulation of chromosome alterations in cancer. Carcinogenesis 26:923–930, 2005.
37. Srebrow A, Kornblihtt AR. The connection between splicing and cancer. J Cell Sci 119:2635–2641, 2006.

38. Wang Z, Cummins JM, Shen D, Cahill DP, Jallepalli PV, Wang TL, et al. Three classes of genes mutated in colorectal cancers with chromosomal instability. Cancer Res 64:2998–3001, 2004.
39. Swanton C. Intratumor heterogeneity: evolution through space and time. Cancer Res 72:4875–4882, 2012.
40. Foulds L. Neoplastic Development. Academic Press, New York, 1969.
41. Alberts DS. Reducing the risk of colorectal cancer by intervening in the process of carcinogenesis: a status report. Cancer J 8:208–221, 2002.
42. Kolata G. Reversing trend, big drop is seen in breast cancer. New York Times, December 15, 2006.
43. Peto R. Epidemiology, multistage models and short-term mutagenicity tests. In Origins of Human Cancer. Hiatt HH, Watson JD, Winsten JA, eds. Cold Spring Harbor Publications, New York, pp. 1403–1428, 1977.
44. Holland Frei Cancer Medicine. Kufe D, Pollock R, Weichselbaum R, Bast R, Gansler T, Holland J, Frei E, eds. BC Decker, Ontario, Canada, 2003.

Part III

Fundamental Relationships between Rare and Common Diseases

Chapter 9

Causation and the Limits of Modern Genetics

9.1 THE INADEQUATE MEANING OF BIOLOGICAL CAUSATION

"In most cases, the molecular consequences of disease- or trait-associated variants for human physiology are not understood."

—Teri A. Manolio and co-authors [1]

Let's start our discussion of biological causation with a simple disease whose cause is understood by laypersons and clinicians alike. The common gastric ulcer is an erosion of the stomach wall caused by excess acid eating through the stomach lining (see Figure 9.1). This definition, or something much like it, appears throughout the medical literature [2]. The definition takes a plausible cause and juxtaposes it against an immediate result; acid eats through stomach lining causing ulcer. The problem here is that this explanation ignores the biological steps that lead from acid production to ulcer formation, thereby eliminating much chance of rationally preventing, treating, or even understanding the pathogenesis of gastric ulcers.

Here is how ulcers of the stomach, or ulcers of just about any mucosal lining, are created. The cells that line the uppermost layer of a mucosal lining normally die and slough off the mucosa. In the case of the gut, the superficial cells slough into the gut lumen and are excreted in the feces. The sloughed cells are replaced by cells regenerated from dividing cells located at the bottom layer of the mucosa. When the mucosal surface is exposed to a toxic chemical agent (such as stomach acid), cells on the surface die and slough at a higher rate than normal. When this happens, the regenerative cells divide more frequently, achieving a new steady state in which the acid-exposed superficial cells are replaced at a higher rate than normal. This condition can continue for a very long time; damaged superficial cells are replaced by overworked regenerative cells, with no net erosion of the mucosal lining. If the toxic exposure increases to a level beyond regenerative capacity, then the lining of the mucosa thins. At some point, the lining erodes completely, exposing the submucosa to the direct

Rare Diseases and Orphan Drugs. http://dx.doi.org/10.1016/B978-0-12-419988-0.00009-2

FIGURE 9.1 Gross specimen of opened stomach, with a large, round, ulcer (upper left). The margins of the ulcer are well demarcated, as if pressed by a cookie cutter. The ulcer is a focus wherein the surface mucosa is eroded, and the underlying tissue is inflamed. (*Source: MacCallum WG. A textbook of pathology* [3].)

effects of acid. This usually results in inflammation. Eventually, an ulcer may erode into a large vessel or through a tissue wall, creating a surgical emergency.

When cause is separated from effect by an intervening series of pathogenetic steps, we begin to understand the different factors that might influence development of a gastric ulcer. Basically, a gastric ulcer is the final event in a typically long process in which bottom layer regenerative cells balance top layer cell losses, until such time as they yield to exhaustion. If an individual were nutritionally deficient, his ability to regenerate sloughed mucosal cells might be diminished. Likewise, if the individual were taking a drug that inhibited cell division, then the ability to compensate for cell losses would be diminished. Either condition might hasten the development of an ulcer. If an individual had an atrophic (i.e., thinned) lining of the mucosal surface, there might be diminished ability to compensate for any cell loss. In the case of gastric ulcers, this might occur in a low-acid environment. Chronic conditions that exacerbate the turnover of lining cells, such as chronic bacterial infection, might also lead to ulcer production.

Understanding the many biological steps that lead to ulcer formation will help us find alternative approaches to preventing and treating ulcers. In the case of gastric ulcers, knowing the underlying cause is not as helpful as understanding the pathogenesis.

In the field of medicine, we often cannot assign a specific cause to a particular disease without seriously misleading ourselves. For example, what is the cause of rheumatic fever? Rheumatic fever is an autoimmune process that targets the heart. Rheumatic fever occurs in people who have been infected with a Group A strain of *Streptococcus pyogenes*. The infection, which usually presents as a pharyngitis, elicits an immune response against a bacterial antigen. The antibody species that target the bacterial antigen happen to cross-react with proteins in normal heart and vessels. These cross-reacting antibodies damage the heart and vessels to produce rheumatic fever. Rheumatic fever is one of the most thoroughly studied and best understood diseases known to man. Knowing all that we know about the pathogenesis, pathology, and clinical features of rheumatic fever, it should be easy to specify the cause of the disease. Alas, this is not the case. For example, we cannot assert that rheumatic fever is caused by *Streptococcus pyogenes* because not all cases of infection lead to rheumatic fever, and because the clinical features of the disease are not actually caused by the infection. Likewise, we cannot assert that rheumatic fever is an autoimmune disease because it does not result from a defect in the autoimmune response. Basically, rheumatic fever involves a normal immune response to a foreign antigen (i.e., a protein of *Streptococcus pyogenes* bacteria) that happens to cross-react with the heart proteins. Furthermore, we cannot claim that rheumatic fever is caused by a heart defect; the heart is an innocent bystander in a process that evolved over time in tissues other than the heart (i.e., the pharynx and other tissues in which immunocytes reside). Despite everything we know about rheumatic fever, it is difficult to specify its underlying cause.

Consider what may happen when the pathogenesis of a disease is unknown, but an army of geneticists is eager to find the cause. The following discussion is loosely adapted from an essay written by Geoffrey Rose in 1985 [4], and described in one of my previous works [5]. Smokovia is an imaginary land where the most important crop is tobacco, and everyone in the country smokes cigarettes. Mothers pass out cigarettes to the whole family after lunch and dinner. Students are encouraged to smoke in the classroom. Cigarettes are smoked in every hospital corridor.

In Smokovia, there is a very high rate of lung cancer. One out of five Smokovians dies of the disease. Nobody has a clue why this might be. Because there is no subpopulation of Smokovians who do not smoke, there is no indication that the risk of lung cancer might be reduced in non-smokers.

The Smokovian scientists decide that there must be some defect in the gene pool that predisposes some Smokovians to cancer. They have compared the genomes of Smokovians who develop lung cancer from those Smokovians who have lived a full life without developing lung cancer. Every patient with lung cancer had a characteristic gene variation that was lacking in the cancer-free population. The scientists developed a method for detecting the cancer-prone gene variant from a simple blood test. They suggested that, as a stop-gap solution, Smokovians carrying the lung cancer gene might choose not to procreate. In the meantime, Smokovian scientists would work on a technique to replace the cancer-prone gene with the cancer-resistant gene in early embryos.

While this work was proceeding, Smokovia was hit by a terrible tobacco blight. In a matter of weeks, the entire crop of tobacco was destroyed. It would take scientists years and years to develop a blight-resistant tobacco. Because tobacco was the most important crop in Smokovia, all scientists were inducted into the service of the tobacco industry.

Decades passed. As a nation, Smokovians endured the agony of cigarette withdrawal. Economic times were bad, but the farmers eventually switched to alternate crops. Smokovia was slowly recovering, and people noticed that the rate of lung cancer was dropping and dropping. After about 50 years, lung cancer had become a rare disease. At first, people thought that their genetic selection project had paid off. But a genetic census of Smokovians showed that lung cancer had vanished even in the small subpopulation of Smokovians who had the cancer-prone gene. Nobody could understand what had happened.

Fifty years after the blight, Smokovian scientists found a blight-resistant tobacco plant. Once more, tobacco was planted, cigarettes were produced, and Smokovians resumed their national past-time. High lung cancer rates returned. Again, tumors occurred in the subpopulation of Smokovians with the cancer-prone gene.

The Smokovian scientists were preoccupied by "cause" when they should have been thinking in terms of the processes of pathogenesis. When everyone smoked, there was little hope pinning the blame on tobacco. Their search for a causal gene was scientifically valid, but futile. Their plan to breed out cancer-prone individuals was neither helpful nor fair. Had they studied the temporal changes in bronchi that led to the development of cancer, they may have had a better chance to infer that toxic and mutagenic alterations in bronchial epithelial cells were caused by an environmental agent. Their biggest mistake was simply a failure to study the pathogenesis of the disease.

Smokovia is an imaginary realm. Surely here in the real world, we have practical ways of assigning causation. Or perhaps not. Consider the process whereby physicians determine causes of death. An individual has smoked three packs of cigarettes every day of his life, beginning at the age of 15 years. At age 50, he develops chronic obstructive pulmonary disease, and at age 60, he develops lung cancer. At age 63, with carcinoma extending to both lobes of the lung, the patient develops pneumonia. Soon thereafter, while he gasps for breath, his heart stops. The patient's personal physician pronounces that the patient is dead, and now she must complete the patient's death certificate. She writes as the cause of death "cardiopulmonary arrest," a diagnosis that carries no clinical value, as it merely conveys that at the time of death, the patient's lungs ceased breathing, and his heart stopped beating. The doctor confused mode of death with cause of death, a common error. The underlying process that led to the patient's demise was cigarette abuse or perhaps cigarette addiction, but knowing that the patient was hooked on cigarettes does not tell us that the patient developed lung cancer and chronic obstructive pulmonary disease. Knowing that the patient had lung cancer and chronic obstructive pulmonary disease does not convey that the patient developed pneumonia (see Figure 9.2).

You would think that doctors would know how to correctly determine the cause of death, but numerous studies conducted in many different nations all show that cause of death records have many deficiencies [6–12]. The sticking point seems to be the word "cause." A death certificate should contain a list that

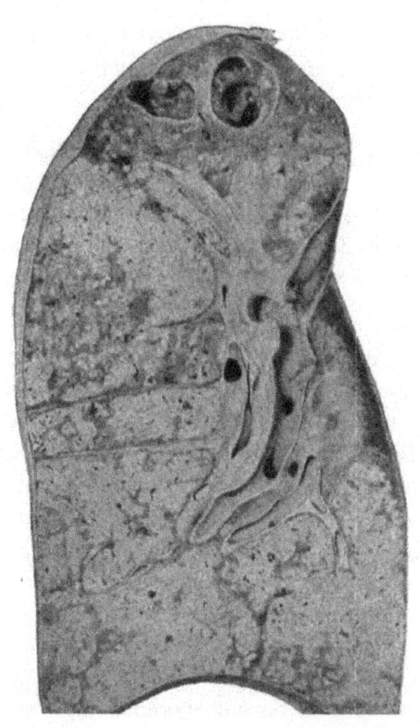

FIGURE 9.2 A cross-section of a gross specimen of lung, cut apex to base. The lung is involved entirely by a consolidating tuberculous pneumonia. Notice a few round blebs in the upper lobe. See color plate at the back of the book. (*Source: MacCallum WG. A textbook of pathology* [3].)

includes the underlying cause of death, followed by the conditions that occurred as a result of the underlying cause of death, followed by the proximate cause of death (i.e., the condition that led directly to the death of the patient and immediately preceded the death). A separate listing may include clinically significant conditions that did not directly lead to the death of the patient. If the patient had several underlying conditions that would be expected to lead to death, then the physician should apply the "but-for" test to choose among the underlying causes. From the field of law, the "but-for" test attempts to determine whether a sequence of actions leading to an event could have happened without the occurrence of a particular underlying action or condition. In terms of the death certificate, "but-for" the chosen underlying cause of death, the sequence of events leading to the individual's death would not have occurred (see Glossary items, But-for, Causes of death, Cause of death error, Proximate cause, Underlying cause of death).

Just as it is difficult to provide a scientific definition for the term "cause," it is likewise difficult to provide a scientific definition for "sporadic," a term that conveys an opposite process. The term "sporadic" describes a case occurrence of a disease in an individual with no special risk factors, or without any discernible cause, as though by random chance. The term "sporadic," like the term "cause," seldom means what we intend it to mean.

9.1.1 Rule—Sporadic diseases are non-sporadic diseases that we do not understand.

Brief Rationale—A sporadic disease, by definition, occurs randomly, with no known cause. Diseases do not occur at random and without cause. Once the cause is understood, the sporadic disease becomes non-sporadic.

Schizophrenia is a common disease with a prevalence of about 1.1%. This translates to about 51 million individuals worldwide who suffer from this mental disorder. Many cases of schizophrenia occur in families and such cases are considered to be inherited and non-sporadic. Other cases seem to have no familial association and are considered sporadic. Are these sporadic cases caused by environmental factors, or are they caused by *de novo* mutations that arose in the affected individuals (see Glossary item, *De novo* germline mutation)? Recent evidence would suggest that many of the so-called sporadic cases arise from new mutations in affected individuals [13].

It can be difficult to make the distinction between sporadic and non-sporadic disease. In the case of cancer, the occurrence of a tumor that is thought to be sporadic may turn out to be non-sporadic, when a specific cause is found. Likewise, a tumor thought to be non-sporadic might revert back to being sporadic when its presumed cause is shown not to apply.

For example, an individual who develops breast cancer may have a strong family history of breast cancer, and affected family members may all carry a predisposing BRCA gene mutation. In this case, the patient's tumor is considered to be non-sporadic. If the patient is tested and found to lack a BRCA mutation, then the tumor must be considered to be sporadic in this case, despite any family predisposition. If additional studies show that the patient has an inherited cancer-causing gene different from the BRCA gene, then the tumor would revert back to being non-sporadic.

As a biological concept, the term "cause" has limited scientific value. It is far better to think in terms of "pathogenesis," the sequence of cellular processes that begins with some event and ends with a disease. Until this chapter, our discussion of rare diseases has focused on causes, usually monogenic, that are wholly responsible for the expression of a disease phenotype. Thinking in terms of causation can be helpful, because it simplifies our conception of biological reality, and focuses our attention on a "but-for" event that is responsible for the eventual expression of disease. **When we go beyond causation and begin to study pathogenesis, we find a sequence of processes that apply to the general development of diseases. Understanding these processes, and controlling their expression, can lead to effective treatments of classes of diseases that share one or more pathogenetic steps.**

9.2 THE COMPLEXITY OF THE SO-CALLED MONOGENIC RARE DISEASES

"How is it that you keep mutating and can still be the same virus?"

—Chuck Palahniuk, in his novel *Invisible Monsters*

Monogenic diseases, caused by an aberration in a single gene, account for the majority of rare diseases. Monogenic diseases are the simplest diseases occurring in any organism, and one would think that once we identify the protein coded by the mutant gene, we would fully understand the disease. Nothing could be further from the truth. Knowing the gene, and its protein, may bring us to the root of the disease, but it does not explain how the disease develops; it does not explain the pathogenesis of the disease.

In point of fact, we know very little about the pathogenesis of most of the rare, monogenic diseases. When we try to study the pathogenesis of rare diseases, we quickly learn that they are much more complex than we had imagined. The complexity derives from the general biological properties of genes and of the cellular processes that influence the expression of genes. Let us examine a partial list of factors that add to the complexity of monogenic diseases.

1. A single gene may produce a protein product whose function varies depending on the specific site and type of mutation in the gene. Hence, variations in a gene can produce different diseases.

For example, different mutations of the same gene, desmoplakin, cause the following diseases:

- Arrhythmogenic right ventricular dysplasia 8
- Dilated cardiomyopathy with woolly hair and keratoderma
- Lethal acantholytic epidermolysis bullosa
- Keratosis palmoplantaris striata II
- Skin fragility-woolly hair syndrome

There are hundreds of examples of single genes that can produce more than one disease (see Glossary item, 1-gene-to-many-diseases). **Appendix I contains a list of approximately 170 genes, with each gene known to be the underlying cause of more than one listed genetic disease.**

The disease caused by a gene may change depending on whether the gene is expressed as a germline mutation or a somatic mutation. In the case of the MYCN gene, a germline mutation resulting in MYCN gene haploinsufficiency (i.e., for which one gene is non-functional while the gene on its matching chromosome expresses a normal gene product) may produce Feingold syndrome, a developmental disorder characterized by microcephaly, limb malformations, esophageal and duodenal atresias, and other developmental alterations (see Glossary items, Malformation, Haploinsufficiency). The same gene, occurring in somatic cells (i.e., as new mutations in tissue cells of adult organisms) as an amplified gene, is associated with neuroblastoma formation.

In some cases, the diseases produced by a specific genetic mutation will change depending on the mutation's parental lineage. Prader–Willi syndrome is a genetic disease characterized by growth disorders (e.g., low muscle tone, short stature, extreme obesity, and cognitive disabilities). Angelman syndrome is a genetic disease characterized by neurologic disturbances (e.g., seizures, sleep disturbances, hand-flapping), and a typifying happy demeanor. Both diseases can occur in either gender and both diseases are caused by the same microdeletion at 15q11-13 (see Glossary items, Microdeletion, Contiguous gene deletion syndrome, Contig disease, Interstitial deletion, Imprinting). When the microdeletion occurs on the paternally derived chromosome, the disease that results is Prader–Willi syndrome. When the microdeletion occurs on the maternally derived chromosome, the disease that results is Angelman syndrome.

In some cases, variation in the sites of mutations in a gene does not produce different diseases, but may account for one disease with different levels of severity. For example, in the case of Wiskott–Aldrich syndrome, discussed in Section 7.1, mutations that truncate the protein product of the WAS gene will produce severe disease, while mutations that produce changes in single amino acids, without changing the length of the protein, will tend to produce mild disease [14].

In some cases, the gain or loss of methylation at a gene site may produce disorders of nearly opposite clinical features. For example, the H19 differentially methylated region is a site on chromosome 11p15.5 in which microdeletions occur in some cases of Beckwith–Wiedemann syndrome and Russell–Silver syndrome. Opposite methylation patterns in the H19 differentially methylated region will cause Beckwith–Wiedemann syndrome when there is gain-of-methylation and Russell–Silver syndrome when there is loss-of-methylation (see Glossary item, Gain-of-function) [15]. Beckwith–Wiedemann syndrome is characterized by tissue overgrowth and tumor formation [16]. Russell–Silver syndrome is characterized by dwarfism. The role of methylation in epigenetic regulation will be described in further detail in Section 10.2.

2. A single gene may encode a regulatory protein that affects many other proteins to produce a disease that affects many different tissues through unrelated mechanisms. It may be difficult or impossible to determine all the different proteins and pathways that are altered by a defective regulatory gene.

In general, diseases due to genes encoding transcription factors are characterized by multiple anomalies of development and growth. Transcription factors are proteins that bind to specific DNA sequences to control the transcription of DNA to RNA. A mutation in a single transcription factor can produce a phenotypically complex syndrome. For example, a mutation in the gene encoding transcription factor TBX5 causes Holt–Oram syndrome, consisting of hand malformations, heart defects, and other malformations (see Glossary item, Homeobox).

3. A protein with a single function may exert a pleiotypic response in different types of cells and tissues, causing many different phenotypic changes

in tissues to produce a seemingly complex disease phenotype. Consider the example of the rare disease ligneous conjunctivitis. Ligneous conjunctivitis is caused by a deficiency of a single protein, plasminogen. Plasmin, the activated form of plasminogen, breaks down fibrin, a protein produced during coagulation and clot formation. In the absence of plasminogen, fibrin accumulates in various sites, and the accumulating fibrin dries out as a hard material. On the surface of the eyes, dried fibrin elicits inflammation, leading to a thick, hardened focus of conjunctivitis (i.e., ligneous conjunctivitis). Accumulating fibrin in the middle ear and the tracheo-bronchial mucosa (of the lungs) leads to inflammation at these sites. In the brain, an occlusive hydrocephalus may occur, due to fibrin deposits blocking the normal flow and clearance of cerebrospinal fluid in the brain ventricles. In retrospect, the pathogenesis of ligneous conjunctivitis is simple to understand. All of the pleiotropic effects are the result of a deficiency of a single protein, with a single function, that happens to be expressed in several different organs to produce a variety of clinical conditions that are closely related to one another; but not obviously so. Ligneous conjunctivitis is an example of the simplest form of pleiotropism, wherein seemingly unrelated phenotypes result from an alteration in a single expressed protein and a single functional pathway.

Another example of pleiotypia resulting from a gene with a single function is found in the WHIM syndrome. WHIM is an acronym for warts, hypogammaglobulinemia, infections and myelokathexis (congenital leukopenia and neutropenia). We now know that WHIM is a combined immunodeficiency disease caused by an alteration in the chemokine receptor gene CXCR4 [17]. Warts result from a lowered immune repression of papillomaviruses. Likewise, the other phenotypic components of the disease arise from the aberrant chemokine. Though the altered CXCR4 gene produces a syndrome with a complex phenotype, it does so through the action of one protein with one function.

4. A protein with a single function may exert a single type of response, but that response may depend on the genetic and epigenetic conditions under which the protein is expressed. Hence, different individuals, each with their own unique genome and epigenome, will respond differently to the same genetic aberration.

If a disease were truly caused by an aberration of a single gene, then all of the consequences of the genetic aberration would be identical in every person with the gene. In fact, some monogenic diseases have remarkably uniform clinical phenotypes in affected populations (e.g., sickle cell disease). What would happen if the same genetic aberration were recapitulated in a mouse? If the mouse homologue served the same purpose as the human gene, and if the gene were the sole cause of the disease, then you might expect the disease to be the same in man and mouse.

Lesch–Nyhan disease is a rare syndrome caused by a deficiency of HGPRT (hypoxanthine-guanine phosphoribosyl transferase), an enzyme involved in purine metabolism. In humans, HGPRT deficiency results in high levels of uric acid, with resultant renal disease and gout. A vast array of neurologic and psychologic signs accompanies the syndrome, including self-mutilation. Neurologic features tend to increase as the affected child ages. The same HGPRT deficiency of humans can be produced in mice. Mice with HGPRT deficiency do not have disease. As far as anyone can tell, mice with HGPRT deficiency are totally normal [18]. How can this be? **A single gene cannot cause a disease all by itself. Every monogenic disease is expressed in a complex system wherein the defective gene is a participant in various pathways that eventually lead to a disease**. The mouse, evidently, has a set of pathways that compensates for the deficiency in HGPRT.

Diabetes is usually a common polygenic disease. There are rare subtypes of type 2 diabetes that have a monogenic origin. As you would expect, these rare subtypes arise in children, and have a Mendelian pattern of inheritance. One such monogenic form of diabetes is MODY-8 (maturity-onset diabetes of the young), caused by a mutation in the carboxyl-ester lipase gene. This same mutation was delivered to a transgenic mouse, intended as an animal model for MODY (see Glossary item, Transgenic). Mice carrying the same altered gene as the human failed to develop any signs of diabetes, or pancreatic damage, or any dysfunction caused by the mutated gene [19].

Though there is often striking phenotypic homogeneity among humans with the same genetic defect, there are many exceptions. Modifier genes can influence the time of onset of disease, the severity of disease, and the clinical phenotype of genetic diseases [20].

5. A single protein encoded by a single gene may have many different biological effects and functions, and these functions may differ based on the cell type in which the protein is expressed, the stage of development in which the protein is expressed, and the cellular milieu (e.g., concentrations of substrate or protein inhibitors) for a given cell type at a particular moment in time. Hence, a specific aberration in a single gene may produce different diseases, depending on factors that are difficult to anticipate or analyze.

Sometimes, one gene may code for a protein that has multiple different roles, thus producing diseases of widely disparate clinical phenotypes. For example, nuclear lamina (lamin a/c) has several biological roles: controlling nuclear shape; influencing transcription; and organizing heterochromatin. Mutations in the LMNA gene cause more than 10 different clinical syndromes, including neuromuscular and cardiac disorders, premature aging disorders, and lipodystrophy. Likewise, the polyfunctional TP53 gene has been linked to 11 clinically distinguishable cancer-related disorders [21].

9.2.1 Rule—A single pleiotropic gene is likely to be associated with several phenotypically unrelated diseases.

Brief Rationale—Genes with pleiotropic pathological effects, and genes that alter a pathway that operates in many different types of cells, are likely to play a role in the pathogenesis of more than one disease, simply because they perturb many different cellular processes.

A bull in a china shop will do more damage than a mouse in a china shop. For example, the APOE gene encodes apolipoprotein E, which is involved in the synthesis of lipoproteins. One common allele of the APOE locus, e4, increases the risk of two common diseases with no obvious biological relationship: Alzheimer disease and heart disease [22,23]. A rare locus of APOE is associated with longevity [24].

6. The pathogenesis of a monogenic disease may be complex, requiring many events to occur in a particular sequence over a period of time, culminating in a disease phenotype. Deviations from the usual steps in pathogenesis may delay or eliminate the occurrence of disease.

Many of the rare monogenic diseases express a characteristic clinical phenotype at birth (e.g., birth defects), or in early childhood. A minority of rare, monogenic diseases are not expressed until adulthood. What can we infer from this observation?

9.2.2 Rule—Monogenic rare diseases that express in late adolescence, or in adulthood, are likely to require additional events (i.e., somatic genetic mutations, toxic exposures, or the accumulation of molecular species or cellular alterations caused by the original genetic defect) that occur over time.

Brief Rationale—If this were not the case, every inherited genetic defect would be expected to express itself clinically at birth or in early childhood.

The many inherited cancer syndromes produce tumors in a younger age group than the same tumors that occur sporadically. Still, these inherited tumors tend to occur in early or mid-adulthood, not at or near birth (see Chapter 8). Cancer is a multi-step process. An inherited mutation that accounts for one step in the process may shorten the time for development of the cancer, but it cannot eliminate the remaining steps.

Huntington disease is a rare monogenic inherited disease that usually begins in adults between 35 and 45 years of age. It is caused by a CAG triplet repeat inside the Huntington gene (see Glossary item, Trinucleotide repeat syndrome). The mutant gene slowly poisons brain cells, particularly neurons in the caudate nucleus, putamen, and substantia nigra. The toxic effects of the mutant protein are slow to cause injury, hence the late onset of disease.

Cardiofaciocutaneous syndrome is a rare monogenic inherited disorder characterized by a set of distinctive congenital abnormalities involving the face, heart, and other organs. It is caused by mutations in any of several different genes, including BRAF. In a zebrafish model of cardiofaciocutaneous syndrome, fish embryos express the BRAF disease allele. Treatment of the affected embryos with inhibitors of the pathway affected by the BRAF mutation will restore normal development in these fish [25]. The inhibitors needed to be administered in a window of time when the BRAF mutation exerted its teratogenic effect. In this case, the pathogenesis of disease could be interrupted by an additional event occurring at a crucial moment in time.

9.3 ONE MONOGENIC DISORDER, MANY GENES

"Make everything as simple as possible, but not simpler."

—Albert Einstein

In the previous section of this chapter, we saw how one gene can cause several phenotypically distinctive diseases. In this section, we shall review the mechanisms whereby one disease can be caused by any one of several different genes. When one clinical phenotype is caused by any one of several different genes, the phenomenon is referred to as locus heterogeneity.

Here are a few examples wherein rare, monogenic diseases can be caused by errors in any one of several different genes:

- **Tuberous sclerosis** is an inherited monogenic rare syndrome that produces multiple benign hamartomas, as well as certain types of cancers. The genetic basis of tuberous sclerosis involves bi-allelic inactivation of either of two unlinked genes that seem to have equivalent pathogenic roles. The genes are TSC1 (encoding hamartin) and TSC2 (encoding tuberin). In this disease, the hamartin and tuberin genes lock together in a protein complex. A defect in either gene disrupts the same pathway [26].
- **Bardet–Biedl syndrome** is characterized by rod-cone dystrophy, obesity, polydactyly, and a variety of organ abnormalities. The various forms of Bardet–Biedl syndrome are accounted for by mutations in one of at least 14 different genes. Although the underlying pathogenesis of Bardet–Biedl syndrome is yet to be clarified, there is evidence to suggest that each of the gene mutations known to cause Bardet–Biedl produce a defect in the basal body of ciliated cells [27]. Such defects produce the pleiotropic phenotype that characterizes Bardet–Biedl syndrome.
- **Li–Fraumeni syndrome** is an inherited cancer syndrome characterized by an increased risk of developing such common cancers as breast cancer, lung cancer, colon cancer, pancreatic cancer, and prostate cancer. Various types of rare cancers associated with the Li–Fraumeni syndrome include soft tissue sarcomas, osteosarcomas, brain tumors, acute leukemias, adrenocortical carcinomas, Wilms tumor, and phyllodes tumor of breast. The observation that common cancers and rare cancers having a common underlying genetic cause would seem to indicate that a rare genetic cause of a common disease can sometimes occur within a gene that is known to cause a rare disease. This theme will be further developed in Section 12.1.

  Li–Fraumeni syndrome was originally believed to be caused exclusively by mutations in the TP53 gene encoding protein p53. TP53 is an example of a tumor suppressor gene. The absence of a tumor suppressor reduces the cell's normal ability to suppress cellular events that increase the susceptibility of cells to cancer. In the case of the p53 gene, loss of activity reduces the ability of cells to undergo apoptosis, a process by which cells commit suicide following DNA damage. By continuing to survive and divide, damaged cells contribute to a subpopulation of cells at risk for progressing through the stages of carcinogenesis. As it turns out, mutations in genes other than TP53 can produce a syndrome similar to, if not indistinguishable from, Li–Fraumeni syndrome. In addition to TP53, the genes that produce forms of Li–Fraumeni syndrome include CHEK2 and BRCA1 [28]. In all three cases, the resulting syndrome results in a very high risk for breast cancer [29]. All three genes have similar functions: controlling whether cells live or die following DNA damage.

- **Retinitis pigmentosa** is a group of inherited conditions characterized by the progressive loss of photoreceptor cells in the retina. Rhodopsin consists of the protein moiety opsin and a reversibly covalently bound cofactor, retinal [30]. More than 100 mutations in the rhodopsin gene account for about 25% of cases. About 150 mutations have been reported in the opsin gene. Other mutated genes causing variants of retinitis pigmentosa involve pre-mRNA splicing factors, as well as post-translational errors in protein folding and other errors of chaperone proteins (see Glossary item, Alternative RNA splicing, Post-translational protein modification). Mutations in any one of more than 35 different genes can cause variant forms of retinitis pigmentosa. Retinitis pigmentosa is unusual for being a disease that can be inherited as an autosomal dominant, autosomal recessive, or X-linked disorder. Digenic and mitochondrial forms of retinitis pigmentosa have been described, and the disease can appear as a solitary disorder or as part of a multi-organ syndrome (e.g., NARP syndrome of neuropathy, ataxia, and retinitis pigmentosa caused by a mutation in the mitochondrial DNA gene MT-ATP6).

  Why there are so many forms of retinitis, with such a large repertoire of disease-causing genes, is somewhat of a mystery. Most of the genes causing various forms of retinitis pigmentosa express constituents of specialized photoreceptors found exclusively in retinal photoreceptor cells (e.g., rhodopsin). Other genes that cause retinitis pigmentosa are active in many different cells (e.g., splicing factors). The outer segment of rod photoreceptors are continuously shed from the tips of cells and replaced by new segments. Rods are extraordinarily dependent on maintaining a high rate of self-renewal, and small deficiencies in cell synthesis may precipitate the loss of these cells [31–33].

- **Epidermolysis bullosa** is an inherited disease characterized by blistering of the skin and mucosal membranes (e.g., mouth). It is always caused by a defect causing the epidermis to be poorly anchored to the underlying dermis. Over 300 gene defects can result in epidermolysis bullosa. Depending on the variant form of the disease, any of several different genes may serve as the underlying cause (e.g., COL, PLEC, Desmoplakin genes). There is also an autoimmune form of epidermolysis bullosa acquisita, wherein antibodies target Type VII collagen, a component of the basement membrane glue that helps bind epidermis with dermis. Epidermolysis bullosa will be discussed again in Section 10.1 as an example of disease convergence (see Glossary item, Convergence).

There are also instances in which a rare phenotypic condition occurs as a component of multiple syndromes, each caused by a different genetic mutation. For example, inherited hemophagocytic lymphohistiocytosis is a component of Chediak–Higashi syndrome and of Griscelli syndrome. Hemophagocytosis is the pathological phagocytosis (i.e., engulfment) of red blood cells by macrophages. Acquired hemophagocytic lymphohistocytosis can occur in Letterer–Siwe disease [34]. In all cases, the final pathogenetic steps of these phenotypically related diseases involves the hypersecretion of cytokines by lymphocytes and macrophages, precipitating a severe, and life-threatening, inflammatory response that includes hemophagocytosis.

In instances where a combined gene deficiency is found, the root cause may be a microdeletion that deletes multiple genes at once. Alternately, a combined deficiency may be caused by a pleiotropic gene that controls the synthesis of several different proteins. In combined factor V and factor VIII clotting factor deficiency, a defect in either the LMAN1 or MCFD2 genes results in diminished transport of factor V and factor VIII from the endoplasmic reticulum to the Golgi apparatus. Hence, the post-translational processing of both these factors is incomplete, and a combined deficiency results. The gene products of MCFD2 and LMAN1 form a cargo receptor complex that acts on a similar set of proteins. Hence, mutations in either gene can produce the same combined deficiency of factor V and factor VIII [35].

The number of rare genetic syndromes that can be caused by any one of several different genes is quite long. A few additional examples are listed here:

- Autosomal dominant cutis laxa can be caused by a mutation of the elastin gene or the fibulin-5 gene.
- Hypotrichosis simplex of the scalp can be caused by mutation in the CDSN gene or the KRT74 gene.
- Oguchi disease can be caused by a mutation in the arrestin gene or the rhodopsin kinase gene.
- Autosomal dominant form of throbocytopenia can be caused by a mutation in the ANKRD26 gene or the cytochrome c gene.

The diseases discussed in this section are examples of disease convergence, in which different underlying processes eventually converge to a common phenotype. The concept of phenotype convergence in the presence of genotypic diversity will be explored in Section 10.2.

9.4 GENE VARIATION AND THE LIMITS OF PHARMACOGENETICS

"That all science is description and not explanation, that the mystery of change in the inorganic world is just as great and just as omnipresent as in the organic world, are statements which will appear platitudes to the next generation."

—Karl Pearson in 1899 [36]

The good news is that we're learning a great deal about the genetics and biology of genetic diseases, and our knowledge is increasing at an enormous rate. The bad news is that the genome is more complex than anyone had ever imagined. The upshot is that the future of pharmacogenetics and personalized medicine will be slow in coming.

Pharmacogenetics applies advances in genetics to drug discovery, development, and individualized response prediction (i.e., who will benefit from a drug, based on genetic testing or profiling of the individual). The term "pharmacogenomics" has no standard definition, but it usually applies to drug-related biological assessments based on whole genome or multi-gene tests (see Glossary item, Pharmacogenomic). To simplify the issue, the word "pharmacogenetics" will be used consistently herein to cover all possible uses of either term.

In the early days of molecular genetics, it was hoped that there would be a one mutation/one disease correlation for every disease that was caused, in whole or in part, by an altered gene. If there were diseases caused by more than one altered gene working in concert, it was assumed that the number of genes would be rather small, maybe two to five. This was a reasonable assumption at the time, because if there were many genes involved in a disease, then that would imply that there were many differences in sequence and structure of disease-causing genes from one person to the next. Very few scientists were expecting to find the richly polymorphic nature of the human genome.

As it turns out, there is a great deal of polymorphism in the human population and in the genes that cause human diseases. It is estimated that there are about 10 million single nucleotide polymorphisms (SNPs) in the human population, and a SNP occurs about once in every 300 nucleotides [37]. This may be an underestimate. It seems that the more individuals you sample for an SNP database, the more variations you can find. Nelson and colleagues, in a 2012 paper, examined gene variations among 14,002 people in 202 genes selected as potential drug targets [38]. They found variants in one of every 17 nucleotides. The authors concluded that, with such high frequencies of genetic variations, we will never be able to catalogue all the sequenced variants in a single database.

Knowing that there is a great deal of variation in genomes of populations of humans, can we safely infer that there is also great variation in the genes that cause human diseases? In 1987, the first mutation in the PAH gene, responsible for causing phenylketonuria, was reported. A few months later, a second mutation was found. The mutations kept coming. There are now more than 500 different mutations of the PAH gene that are known to cause phenylketonuria [20]. When different mutations of a gene can cause the same disease, then the disease is said to have allelic heterogeneity. Many other monogenic diseases have abundant allelic heterogeneity. The CFTR (transmembrane conductance regulator) gene that causes cystic fibrosis has more than 1000 gene variants [20]. Likewise, there are over 1500 mutations found in the von Hippel–Lindau tumor suppressor gene [39].

In addition to allelic heterogeneity (i.e., variations on a single gene causing the same disease), we must somehow deal with locus heterogeneity (i.e., mutations in different genes that can lead to the same disease. The retinal dystrophies, including the various forms of retinitis pigmentosa, are caused by various mutations in any one of at least 100 different genes [40].

Because the human genome has millions of polymorphisms, and because a single disease gene may have hundreds of different pathologic variants, the simplest pharmacogenetics test is fraught with difficulty. Consider screening individuals for BRCA gene mutations that would indicate a high risk for breast cancer. On one study, 300 women from high risk breast cancer families were negative for BRCA1 and BRCA2 mutation on the first screening. On rescreening these women, an additional 12% were found to be positive when tested against a larger library of known BRCA1 and BRCA2 gene mutations [29]. Likewise, a laboratory at the University of Alabama that has tested over 5000 individuals for mutations causing neurofibroamatosis type 1 reports that it is continually encountering previously unknown mutations [41]. Nobody knows if or when we will come to a time when we can say that all of the genetic causes of neurofibromatosis type 1 are known.

It is easy to recite monogenic diseases that can be caused by any one of hundreds of gene variants (e.g., epidermolysis bullosa, retinitis pigmentosum, and the retinal dystrophies, collagenopathies, inherited arrhythmias). It would seem impossible to screen for every possible gene variation that can cause any disease or that can increase an individual's likelihood of developing any disease.

Aside from the aforementioned false-negative issue (i.e., missing disease-causing variants that have not been previously encountered), the field of pharmacogenetics labors under a serious false-positive issue. **Not every individual who carries a disease-causing gene will develop the predicted disease**. Apparently, a disease gene may be a necessary and underlying cause of a disease, but it may not be sufficient to produce the disease absent other conditions in the organism. In these cases, the disease gene is said to have incomplete penetrance. The quantitative measure of penetrance is the percentage of individuals who carry the gene and who develop the disease divided by the total number of people who carry the gene. Epistasis may modify penetrance; one gene may be influenced by a particular allele of another gene (see Glossary item, Epistasis). Environmental and epigenetic factors can also influence gene function. Such factors may influence whether a disease-causing mutation is expressed as disease (i.e., penetrance); and may also influence the age at which the disease emerges, or the severity of the disease, or phenotype of the disease (i.e., which clinical problems will develop). Because the factors that influence penetrance are many, and because we do not fully understand how these factors operate, and because we cannot know all the factors that may influence penetrance, we can expect to be misled, at times, when pharmacogenetic testing tells us that we harbor a disease-causing mutation.

If we cannot find all the mutations that cause simple, monogenic diseases, is there any reason to expect that we will find every mutation in every gene that contributes to common, polygenic diseases?

9.4.1 Rule—At least for the near future, personalized medicine will be impossible to attain, except for a small and select set of diseases and genotypes.

Brief Rationale—The genotypes of the simplest diseases are highly complex. Factors that modify genotype and phenotype (e.g., genome, epigenome, cellular physiology, and environmental agents) are even more complex. Because we cannot cope with these complexities, we cannot reliably diagnose every individual with a genetic disease, nor can we rationally predict how therapeutic responses will differ from individual to individual.

9.5 ENVIRONMENTAL PHENOCOPIES OF RARE DISEASES

"[C]ommon disease phenotypes may comprise, to some currently undefined extent, a heterogeneous collection of individually rare, genetically distinct disorders [42], the common phenotype being explained by involvement of different disease-associated genes in related physiologic pathways."

—J. Crow [43]

Phenocopy diseases are medical conditions that closely mimic a genetic disease, but are caused or triggered by an environmental factor. In many cases, phenocopy diseases are non-hereditary and acute. In some cases, the phenocopy disease is reversible when the environmental trigger is removed or when an appropriate treatment is applied.

Here are just a few well-known examples of phenocopy diseases, listing the phenocopy disease, followed by its rare disease equivalent:

- **Acquired conduction defect—inherited conduction defect**

Disorders of the electrical systems in humans that defects of ion flux across membranes are known as channelopathies. The inherited cardiac conduction channelopathies were discussed in Section 5.3.

Because the anti-arrhythmogenic and anti-epileptic drugs typically target ion channels, they are the drugs most likely to produce, as an adverse side effect, disorders of cardiac conduction. For example, rufinamide, an oral antiepileptic drug, has been reported to cause QT-interval shortening [44]. Quinidine, disopyramide, and procainamide have been reported to produce QT prolongation [45].

Several channelopathies can be acquired as autoimmune diseases, in which antibodies react with ion channels, or related cellular components upon which the ion channels depend (e.g., myasthenia gravis, Lambert–Eaton myasthenic syndrome, cerebellar ataxia associated with VGCC antibodies, acquired neuromyotonia, Morvan fibrillary chorea, limbic encephalitis) [46].

Progressive familial heart block type IA is a genetic disorder of the cardiac conduction system. Clinically similar conditions can be acquired when the tissues of the conduction systems are damaged, as in: myocardial infarct, conduction system ischemia (i.e., lack of blood flow to components of the conduction system, particularly the His–Purkinje conduction tissue), age-related degeneration of conduction system, and complications of procedures (i.e., insertion of wires or lines into the heart chambers) [47].

- **Acquired porphyria cutanea tarda—inherited porphyria cutanea tarda**

The porphyrias are a group of disorders caused by deficiencies of enzymes involved in the synthesis of heme, a constituent of hemoglobin, p450 liver cytochromes, catalase, peroxidase, and myoglobin. Two tissues account for the bulk of heme synthesis in the body: erythropoietic cells and liver cells. All of the porphyrias are characterized by excessive production of porphyrin molecules.

Porphyria cutanea tarda is caused by a deficiency of uroporphyrinogen decarboxylase, an enzyme involved in heme synthesis within liver cells. The disease is manifested as blistering and discoloration in sun-exposed areas of the skin. It is surmised that sunlight reacts with porphyrins to produce toxic oxygen radicals.

About 20% of the cases of porphyria cutanea tarda are inherited as a uroporphyrinogen decarboxylase deficiency. Many individuals with inherited deficiencies will never experience any of the skin manifestations of the disease (see Glossary item, Penetrance). Most of the remainder of cases of porphyria cutanea tarda are due to acquired liver damage, such as that produced by long-term alcohol production or hepatitis C infection. The damaged liver cells have a defective heme pathway leading to an excess of porphyrin.

- **Acquired von Willebrand disease—inherited von Willebrand disease**

Von Willebrand factor is a complex protein, the largest protein found in plasma, and is required for platelet adhesion. Reduction in von Willebrand factor results in a clotting disorder. Von Willebrand disease can result from inherited deficiency or it can be acquired through several mechanisms. In an autoimmune variant of the disease, antibodies reacting with the factor produce a protein complex that is rapidly cleared, effectively producing a deficiency. As a large, complex molecule, von Willebrand factor is particularly vulnerable to mechanical disruption. Artificial heart valves have been observed to produce von Willebrand disease. In cases of thrombocythemia (i.e., increased numbers of platelets in blood), excess platelets can absorb the von Willebrand factor to produce a functional deficiency.

- **Aminoglycoside-induced hearing loss—inherited mitochondriopathic deafness**

For reasons that are not fully understood, cochlear cells of the middle ear tend to die off when their mitochondria are damaged; more so than most other cells of the body. It has been hypothesized that defects in mitochondria trigger apoptosis of cochlear cells, and this may lead to permanent hearing impairment [48,49], a topic that will be revisited in Section 10.1. Subtle mitochondriopathic events accumulating over time may contribute to age-related hearing loss (i.e., presbyacusis). Not surprisingly, inherited mutations of mitochondria account for syndromic (i.e., multisystem) and non-syndromic hearing loss [50,51].

Non-syndromic hearing loss may occur as an acquired condition following treatment with ototoxic antibiotics that target mitochondria. The ability of bacterial antibiotics to cause mitochondrial toxicity is linked to an evolutionary event that occurred 2 or 3 billion years ago. Evolutionary biologists believe that the very first eukaryote came fully equipped with one or more mitochondria and that these mitochondria were trapped bacteria, one of only two types of organism available to the primordial eukaryote. The Alpha Proteobacteria are characterized by their small size, their Gram negativity, and their intimate associations with eukaryotic cells. The Alpha Proteobacteria live as symbionts, endosymbionts, or as intracellular parasites in eukaryotic cells. This close relationship between Alpha Proteobacteria and Class Eukaryota may extend back to the very first eukaryotic cell. Based on sequence similarities between the Alpha Proteobacteria and eukaryotic mitochondria, it has been proposed that eukaryotic mitochondria evolved from an endosymbiotic member of Class Alpha Proteobacteria. All existing eukaryotic organisms (including humans) descended from ancestors that contained mitochondria. Furthermore, all existing eukaryotic organisms, even the so-called amitochondriate classes (i.e., organisms without mitochondria), contain vestigial forms of mitochondria (i.e., hydrogenosomes and mitosomes) [52–55].

Because mitochondria are a type of bacteria permanently trapped within our cells, it is no wonder that antibacterial agents are capable of inducing mitochondrial toxicity that mimics rare, inherited causes of deafness. The aminoglycosides are particularly prone to producing ototoxicity [56].

- **Antabuse (disulfiram) treatment—inherited alcohol intolerance**

Individuals with alcohol dehydrogenase deficiency experience unpleasant physical reactions after ingesting even small amounts of alcohol. Most prominent are flushing and systemic vasodilation. This inherited syndrome is very common among individuals of Asian descent. It is caused by mutations in the ADH1C gene, encoding a subunit of alcohol dehydrogenase.

Disulfiram (trade name Antabuse) inhibits aldehyde dehydrogenase, another enzyme involved in alcohol metabolism. Disulfiram toxicity mimics inherited alcohol intolerance, causing flushing and systemic vasodilation. Other symptoms may include nausea, hypotension, and difficulty breathing. Disulfiram is administered as a discouragement to individuals suffering from alcohol abuse.

- **Drug-induced methemoglobinemia—inherited methemoglobinemia**

Methemoglobin is hemoglobin containing iron in its ferric [$Fe3^{+}$] ion form, rather than its normal ferrous [$Fe2^{+}$] ion form. The ferric form is less able to release oxygen from red cells than the ferrous form, producing generalized hypoxia, sometimes turning affected individuals blue. Methemoglobin is a normal byproduct of hemoglobin synthesis. An enzyme system in red blood cells keeps the normal methemoglobin concentration low. An inherited deficiency of methemoglobin reductase leads to methemoglobinemia.

A variety of drugs and chemicals mimic inherited methemoglobinemia by oxidizing blood, thereby accelerating the rate of formation of methemoglobin, and overwhelming the enzyme system that would normally maintain safe methemoglobin levels. Such oxidizing agents include trimethoprim, sulfonamides, dapsone, local anesthetics such as articaine and prilocaine, aniline dyes, and nitrates such as bismuth nitrates.

- **Fetal exposure to methotrexate—Miller syndrome**

Fetal exposure to methotrexate produces a similar set of malformation to that found in Miller syndrome, also known as postaxial acrofacial dystosis. Anatomic abnormalities in Miller syndrome include micrognathia (i.e., small jaw), cleft lip, cleft palate, and hypoplasia or aplasia of the postaxial elements of the limbs (i.e., the medial side of arms and forearms, and the lateral side of the legs and thighs). Miller syndrome is caused by an alteration in the gene encoding dihydroorotate dehydrogenase, an enzyme involved in pyrimidine synthesis. Methotrexate interferes with purine synthesis and, to a lesser extent, pyrimidine synthesis [57].

- **Methylmalonic acidemia caused by severe deficiency of vitamin B12—inherited methylmalonic acidemia**

Methylmalonic acidemia is caused by any one of about a half dozen genes encoding proteins that contribute to the pathway whereby precursor molecules are converted into succinyl coA (e.g., MMAA, MMAB, MMACHC, MMADHC, LMBRD1, or MUT genes). The genetic disorder is diagnosed in neonates who typically present with hyperammonemia and encephalopathy. Judging by the number of different pathway enzymes whose deficiency results in the same or similar clinical phenotype, it would seem that methylmalonic acidemia is a good example of a convergent pathway disease; any deficiency in the pathway leads to the same clinical phenotype. This generalization extends to at least one non-genetic cause: vitamin B12 deficiency. Vitamin B12 is required for the last step of the pathway leading to succinyl coA formation. A deficiency of vitamin B12 in neonates will produce a phenocopy of methylmalonic acidemia [58].

- **MPTP toxicity and drug-induced Parkinsonism—Parkinson disease**

Parkinson disease is a neurodegenerative disease that targets the dopamine-generating cells in the substantia nigra. Most cases of this disease have no known genetic cause, but a small subset of Parkinson cases are caused by monogenic mutations (e.g., SNCA, PRKN, LRRK2, PINK1).

Various drugs can preferentially kill neurons in the substantia nigra and about 7% of individuals with Parkinson syndrome have drug-acquired disease. Antipsychotic drugs are particularly prone to producing Parkinson-like side effects.

One agent, MPTP, has gained notoriety for its ability to produce rapid, irreversible Parkinsonism, with a single injection [59]. The lipophilic drug crosses the blood–brain barrier and is activated in the brain to toxic metabolites that have high specificity for dopamine-producing neurons in the substantia nigra [59].

- **Osteolathyrism and scurvy—inherited collagenopathies**

Collagen production is a relatively complex process, and you might expect that external agents can interfere with its synthesis. The toxin beta-aminopropionitrile is synthesized in certain plants of Genus *Lathyrus*, which includes sprouts of pea and lentils. Ingestion of beta-aminopropionitrile inhibits lysyl oxidase, required for proper cross-linking of procollagen. The resulting condition may mimic inherited collagenopathies, sometimes producing skeletal deformities, aortic dissection, and a host of abnormalities resulting from weakened connective tissues (see Glossary item, Collagenopathy).

Scurvy is caused by insufficient vitamin C in the diet. Vitamin C is a required co-factor in the synthesis of procollagen, specifically in the hydroxylation of proline. When vitamin C is not present in the diet, a collagenopathy ensues.

Scurvy was virtually unknown prior to the era of sea exploration. In 1497, Vasco da Gama sailed from Lisbon to Calcutta. The then-traditional diet of sailors excluded fruits and green vegetables. His adventure was the first recorded voyage sufficiently long enough to induce scurvy, which killed three-fifths of his men. In 1593, Sir Richard Hawkins cured scurvy with oranges and lemons. Over the next several centuries, the cure for scurvy was repeatedly lost and rediscovered by later generations of explorers (e.g., captains of the Dutch East India Company in the 1600s, James Lind in 1747), but scurvy persisted as a common scourge of long-distance sailors well into the nineteenth century. In 1848, the exploring ships *Erebus* and *Terror*, on their way to find the Northwest Passage to the Pacific Ocean, were trapped in ice. One hundred and five men reached shore, but they neglected to salvage their lemon juice. The men died of scurvy. Today, synthesized vitamin C is cheap and plentiful; dietary insufficiency is uncommon.

Scurvy is an example of a rare disease that suddenly became a somewhat common disease, staying common for over 300 years before retreating into rarity.

The list of acquired conditions that mimic rare, genetic diseases is too numerous to discuss each one in detail. Listed here are a few additional phenocopy diseases and readers are encouraged to pursue these topics in external resources:

- Alcohol-induced sideroblastic anemia—inherited sideroblastic anemia
- B12 deficiency—inherited pernicious anemia
- Cardiomyopathy due to alcohol abuse—inherited dilated cardiomyopathy

- Lead-induced encephalopathy—inherited tau encephalopathy [60]
- Myopathy produced by nucleoside analogue reverse transcriptase inhibitors (i.e., HIV drugs)—inherited mitochondrial myopathy [61]
- Pseudo-Pelger–Huet anomaly—inherited Pelger–Huet anomaly [62]
- Thalidomide-induced phocomelia—Roberts syndrome and SC pseudothalidomide syndrome
- Warfarin embryopathy—brachytelephalangic chondrodysplasia punctata [63]

9.5.1 Rule—Phenocopy diseases are typically mimics of rare diseases, not common diseases.

Brief Rationale—The prototypical phenocopy disease involves a single agent having a specific effect on a single pathway in a limited number of cell types. In theory, any pathway can be altered by a drug to produce a phenotype that mimics a monogenic disease. A simple interruption of normal cellular function of a gene or a pathway is consistent with what we see in rare diseases and in phenocopy diseases, and lacks the cumulative acquisition of multiple genetic or cellular aberrations that typically characterize the common diseases.

Phenocopy diseases provide important clues to the pathogenesis of rare and common diseases for the following reasons:

1. There is usually one pathway involved, often found in a limited number of cell types, and the phenocopy disease teaches us how this pathway operates and how it can be disrupted.
2. The pathway disrupted in the phenocopy disease is almost always the same pathway that is disrupted in the rare genetic disease. Hence, the phenocopy tells us how the rare disease expresses itself, and this is something that we can seldom infer from our knowledge of the gene mutation associated with the rare disease.
3. When the genetic cause of the rare disease is unknown, the careful study of its phenocopy will always yield a set of candidate genes that may operate in the rare disease.
4. Pharmacologic treatments for the phenocopy disease may apply to pathways operative in the genetic form of the disease or in the common diseases.
5. The pathway involved in a phenocopy disease can contribute to the pathogenesis of a common disease. Hence, understanding the phenocopy diseases brings us a little closer to understanding common diseases [60]. This topic will be discussed further in Chapter 10.
6. Recognizing the cause of a phenocopy disease may curtail potential environmental catastrophes.

The phenocopy diseases help us to focus on the cellular pathways leading to disease. If you exclusively study the genetics of disease, you will likely miss the cellular pathways that link rare diseases with common diseases. **The phenocopy diseases remind us that you can have a disease without a gene, but you cannot have a disease without a pathway**.

REFERENCES

1. Manolio TA, Collins FS, Cox NJ, Goldstein DB, Hindorff LA, Hunter DJ, et al. Finding the missing heritability of complex diseases. Nature 461:747–753, 2009.
2. Ajeigbe KO, Olaleye SB, Oladejo EO, Olayanju AO. Effect of folic acid supplementation on oxidative gastric mucosa damage and acid secretory response in the rat. Indian J Pharmacol 43:578–581, 2011.
3. MacCallum WG. A Textbook of Pathology, 2nd edition. WB Saunders Company, Philadelphia and London, 1921.
4. Rose G. Sick individuals and sick populations. Int J Epidemiol 14:32–38, 1985.
5. Berman JJ. Neoplasms: Principles of Development and Diversity. Jones & Bartlett, Sudbury, 2009.
6. Documentation for the Mortality Public Use Data Set, 1999. Mortality Statistics Branch, Division of Vital Statistics, National Center for Health Statistics, 1999.
7. Frey CM, McMillen MM, Cowan CD, Horm JW, Kessler LG. Representativeness of the surveillance, epidemiology, and end results program data: recent trends in cancer mortality rate. JNCI 84:872, 1992.
8. Ashworth TG. Inadequacy of death certification: proposal for change. J Clin Pathol 44:265, 1991.
9. Kircher T, Anderson RE. Cause of death: proper completion of the death certificate. JAMA 258:349–352, 1987.
10. Walter SD, Birnie SE. Mapping mortality and morbidity patterns: an international comparison. Intl J Epidemiol 20:678–689, 1991.
11. Berman JJ. Methods in Medical Informatics: Fundamentals of Healthcare Programming in Perl, Python, and Ruby. Chapman and Hall, Boca Raton, 2010.
12. Berman JJ. Biomedical Informatics. Jones & Bartlett, Sudbury, MA, 2007.
13. Xu B, Roos JL, Dexheimer P, Boone B, Plummer B, Levy S, et al. Exome sequencing supports a de novo mutational paradigm for schizophrenia. Nat Genet 43:864–868, 2011.
14. Jin Y, Mazza C, Christie JR, Giliani S, Fiorini M, Mella P, et al. Mutations of the Wiskott-Aldrich Syndrome Protein (WASP): hotspots, effect on transcription, and translation and phenotype/genotype correlation. Blood 104:4010–4019, 2004.
15. Soejima H, Higashimoto K. Epigenetic and genetic alterations of the imprinting disorder Beckwith-Wiedemann syndrome and related disorders. J Hum Genet 58:402–409, 2013.
16. Weksberg R, Shuman C, Beckwith JB. Beckwith-Wiedemann syndrome. Eur J Hum Genet 18:8–14, 2010.
17. Hernandez PA, Gorlin RJ, Lukens JN, Taniuchi S, Bohinjec J, Francois F, et al. Mutations in the chemokine receptor gene CXCR4 are associated with WHIM syndrome, a combined immunodeficiency disease. Nat Genet 34:70–74, 2003.
18. Engle SJ, Womer DE, Davies PM, Boivin G, Sahota A, Simmonds HA, et al. HPRT-APRT-deficient mice are not a model for Lesch-Nyhan syndrome. Hum Mol Genet 5:1607–1610, 1996.
19. Raeder H, Vesterhus M, El Ouaamari A, Paulo JA, McAllister FE, Liew CW, et al. Absence of diabetes and pancreatic exocrine dysfunction in a transgenic model of carboxyl-ester lipase-MODY (maturity-onset diabetes of the young). PLoS One 8:e60229, 2013.
20. Nebert DW, Zhang G, Vesell ES. From human genetics and genomics to pharmacogenetics and pharmacogenomics: past lessons, future directions. Drug Metab Rev 40:187–224, 2008.
21. Vogelstein B, Lane D, Levine AJ. Surfing the p53 network. Nature 408:307–310, 2000.
22. Pritchard JK, Cox NJ. The allelic architecture of human disease genes: common disease-common variant…or not? Human Mol Gen 11:2417–2423, 2002.

23. Corder EH, Saunders AM, Strittmatter WJ, Schmechel DE, Gaskell PC, Small GW, et al. Gene dose of apolipoprotein E type 4 allele and the risk of Alzheimer's disease in late onset families. Science 261:921–923, 1993.
24. Beekman M, Blanch H, Perola M, Hervonen A, Bezrukov V, Sikora E, et al. Genome-wide linkage analysis for human longevity: genetics of healthy aging study. Aging Cell 12:184–193, 2013.
25. Anastasaki C, Estep AL, Marais R, Rauen KA, Patton EE. Kinase-activating and kinase-impaired cardio-facio-cutaneous syndrome alleles have activity during zebrafish development and are sensitive to small molecule inhibitors. Hum Mol Genet 18:2543–2554, 2009.
26. van Slegtenhorst M, Nellist M, Nagelkerken B, Cheadle J, Snell R, van den Ouweland A, et al. Interaction between hamartin and tuberin, the TSC1 and TSC2 gene products. Hum Mol Genet 7:1053–1057, 1998.
27. Ansley SJ, Badano JL, Blacque OE, Hill J, Hoskins BE, Leitch CC, et al. Basal body dysfunction is a likely cause of pleiotropic Bardet-Biedl syndrome. Nature 425:628–633, 2003.
28. Silva AG, Ewald IP, Sapienza M, Pinheiro M, Peixoto A, de Nobrega AF, et al. Li-Fraumeni-like syndrome associated with a large BRCA1 intragenic deletion. BMC Cancer 12:237, 2012.
29. Walsh T, Casadei S, Coats KH, Swisher E, Stray SM, Higgins J, et al. Spectrum of mutations in BRCA1, BRCA2, CHEK2, and TP53 in families at high risk of breast cancer. JAMA 295:1379–1388, 2006.
30. Hubbard R, Wald G. The mechanism of rhodopsin synthesis. Proc Natl Acad Sci USA 37:69–79, 1951.
31. Faustino NA, Cooper TA. Pre-mRNA splicing and human disease. Genes Dev 17:419–437, 2003.
32. Korenbrot JI, Fernald RD. Circadian rhythm and light regulate opsin mRNA in rod photoreceptors. Nature 337:454–457, 1989.
33. Tanackovic G, Ransijn A, Thibault P, Abou Elela S, Klinck R, Berson EL, et al. PRPF mutations are associated with generalized defects in spliceosome formation and pre-mRNA splicing in patients with retinitis pigmentosa. Hum Mol Genet 20:2116–2130, 2011.
34. Dufourcq-Lagelouse R, Pastural E, Barrat FJ, Feldmann J, Le Deist F, Fischer A, et al. Genetic basis of hemophagocytic lymphohistiocytosis syndrome (review). Int J Mol Med 4:127–133, 1999.
35. Zhang B, McGee B, Yamaoka JS, Guglielmone H, Downes KA, Minoldo S, et al. Combined deficiency of factor V and factor VIII is due to mutations in either LMAN1 or MCFD2. Blood 107:1903–1907, 2006.
36. Pearson K. The Grammar of Science. Adam and Black, London, 1900.
37. Genetics Home Reference. National Library of Medicine. July 1, 2013. Available from: http://ghr.nlm.nih.gov/handbook/genomicresearch/snp, viewed July 6, 2013.
38. Nelson MR, Wegmann D, Ehm MG, Kessner D, St Jean P, Verzilli C, et al. An abundance of rare functional variants in 202 drug target genes sequenced in 14,002 people. Science 337:100–104, 2012.
39. Nordstrom-O'Brien M, van der Luijt RB, van Rooijen E, van den Ouweland AM, Majoor-Krakauer DF, Lolkema MP, et al. Genetic analysis of von Hippel-Lindau disease. Hum Mutat 31:521–537, 2010.
40. Glockle N, Kohl S, Mohr J, Scheurenbrand T, Sprecher A, Weisschuh N, et al. Panel-based next generation sequencing as a reliable and efficient technique to detect mutations in unselected patients with retinal dystrophies. Eur J Hum Genet 2:99–104, 2013.

41. Olson S, Beachy SH, Giammaria CF, Berger AC. Integrating Large-Scale Genomic Information into Clinical Practice: Workshop Summary. The National Academies Press, Washington, DC, 2012.
42. McClellan J, King M. Genomic analysis of mental illness: a changing landscape. JAMA 303:2523–2524, 2010.
43. Crow YJ. Lupus: how much "complexity" is really (just) genetic heterogeneity? Arthritis Rheum 63:3661–3664, 2011.
44. Schimpf R, Veltmann C, Papavassiliu T, Rudic B, Göksu T, Kuschyk J, et al. Drug-induced QT-interval shortening following antiepileptic treatment with oral rufinamide. Heart Rhythm 9:776–781, 2012.
45. Yap YG, Camm AJ. Drug induced QT prolongation and torsades de pointes. Heart 89:1363–1372, 2003.
46. Kleopa KA. Autoimmune channelopathies of the nervous system. Curr Neuropharmacol 9:458–467, 2011.
47. Park DS, Fishman GI. The cardiac conduction system. Circulation 123:904–915, 2011.
48. Chen H, Tang J. The role of mitochondria in age-related hearing loss. Biogerontology November 8, 2013.
49. Raimundo N, Song L, Shutt TE, McKay SE, Cotney J, Guan MX, et al. Mitochondrial stress engages E2F1 apoptotic signaling to cause deafness. Cell 148:716–726, 2012.
50. Kokotas H, Petersen MB, Willems PJ. Mitochondrial deafness. Clin Genet 71:379–391, 2007.
51. Yamasoba T, Tsukuda K, Suzuki M. Isolated hearing loss associated with T7511C mutation in mitochondrial DNA. Acta Otolaryngol Suppl:13–18, 2007.
52. Stechmann A, Hamblin K, Perez-Brocal V, Gaston D, Richmond GS, van der Giezen M, et al. Organelles in Blastocystis that blur the distinction between mitochondria and hydrogenosomes. Curr Biol 18:580–585, 2008.
53. Tovar J, Leon-Avila G, Sanchez LB, Sutak R, Tachezy J, van der Giezen M, et al. Mitochondrial remnant organelles of Giardia function in iron-sulphur protein maturation. Nature 426:172–176, 2003.
54. Tovar J, Fischer A, Clark CG. The mitosome, a novel organelle related to mitochondria in the amitochondrial parasite Entamoeba histolytica. Mol Microbiol 32:1013–1021, 1999.
55. Burri L, Williams B, Bursac D, Lithgow T, Keeling P. Microsporidian mitosomes retain elements of the general mitochondrial targeting system. PNAS 103:15916–15920, 2006.
56. Selimoglu E. Aminoglycoside-induced ototoxicity. Curr Pharm Des 13:119–126, 2007.
57. Ng SB, Buckingham KJ, Lee C, Bigham AW, Tabor HK, Dent KM, et al. Exome sequencing identifies the cause of a Mendelian disorder. Nat Genet 42:30–35, 2010.
58. Higginbottom MC, Sweetman L, Nyhan WL. A syndrome of methylmalonic aciduria, homocystinuria, megaloblastic anemia and neurologic abnormalities in a vitamin B12-deficient breast-fed infant of a strict vegetarian. N Engl J Med 299:317–323, 1978.
59. Watanabe Y, Himeda T, Araki T. Mechanisms of MPTP toxicity and their implications for therapy of Parkinson's disease. Med Sci Monit 11:17–23, 2005.
60. Zhu H-L, Meng S-R, Fan J-B, Chen J, Liang Y. Fibrillization of human tau is accelerated by exposure to lead via interaction with His-330 and His-362. PLoS ONE 6:e25020, 2011.
61. Chapplain JM, Beillot J, Begue JM, Souala F, Bouvier C, Arvicux C, et al. Mitochondrial abnormalities in HIV-infected lipoatrophic patients treated with antiretroviral agents. J Acquir Immune Defic Syndr 37:1477–1488, 2004.

62. Wang E, Boswell E, Siddiqi I, Lu CM, Sebastian S, Rehder C, et al. Pseudo-Pelger-Huet anomaly induced by medications: a clinicopathologic study in comparison with myelodysplastic syndrome-related pseudo-Pelger-Huet anomaly. Am J Clin Pathol 135:291–303, 2011.
63. Franco B, Meroni G, Parenti G, Levilliers J, Bernard L, Gebbia M, et al. A cluster of sulfatase genes on Xp22.3: mutations in chondrodysplasia punctata (CDPX) and implications for warfarin embryopathy. Cell 81:1–20, 1995.

Chapter 10

Pathogenesis: Causation's Shadow

10.1 THE MYSTERY OF TISSUE SPECIFICITY

"In mammals the genome is shaped by epigenetic regulation to manifest numerous cellular identities."

—David A. Khavari, George L. Sen, and John L. Rinn [1]

The genetic diseases are characterized by germline gene alterations that are present in every nucleated cell of the body. Nonetheless, the clinical phenotypes that are produced by these omnipresent genes are restricted to one or a few organs or cell types. What is the mechanism whereby a defect that is present in every cell of the body is manifested in only a few organs?

It should come as no surprise that many different mechanisms account for this fascinating phenomenon. **In almost every case, our understanding of tissue specificity has come from astute observations on rare diseases.** Here is a list of the currently studied biological processes that restrain the phenotypic expression of genetic diseases.

- **Cell-specific gene expression**

The most obvious reason why certain tissues and not others are affected by genetic diseases relates to cell-type specific gene expression. For example, congenital hypothyroidism is a primary disease of the thyroid. It cannot be a primary disease of any other organ because the thyroid gland is the only organ that produces thyroid hormones. The indirect effects of hypothyroidism are found in those organs that respond to thyroid hormones. The primary site of disease and the secondarily affected organs are predetermined by normal endocrine physiology.

Sometimes cell-specific gene expression plays a transient role during development. In blepharophimosis, ptosis, and epicanthus inversus syndrome type I (BPES1), a FOXL2 mutation causes eyelid abnormalities and reduces the number of ovarian follicles in the fetal organism. Expression of the FOXL2 gene during development is found primarily in the ovaries and the eyelids, thus explaining the cell-type-specific disease phenotype [2,3].

Rare Diseases and Orphan Drugs. http://dx.doi.org/10.1016/B978-0-12-419988-0.00010-9

Of course, if we fully understood the mechanisms that control cell-type-specific gene expression, we might be able to recruit various types of cells to produce proteins that are normally restricted to another cell type. For example, if insulin secretion were deficient due to developmental absence of pancreatic islets, then it might be advantageous to activate insulin synthesis in hepatocytes or gut lining cells.

The cell-type-specific control of gene expression is highly complex (see Glossary items, Enhancer, Trans-acting, Cis-acting, Promoter). For every complex biological mystery, the clues are found within a rare disease. Cornelia de Lange syndrome is an inherited disease associated with many different developmental defects, including intellectual disability, skeletal abnormalities, small stature, and gastrointestinal dysfunction. Cornelia de Lange syndrome is caused by a mutation in the Nipbl gene, which loads cohesin at promoter sites. Cohesin occupies different promoters on different types of cells, thus mediating cell-type-specific gene expression [4]. As we expected, a defect in the process through which cell-type gene expression is controlled adversely influenced the development of many different tissues.

- **Cell-type specificity of pathways**

Processes that occur exclusively in one cell type will obviously produce a primary disease restricted to one tissue or organ. For example, every genetic disorder of hemoglobin synthesis produces a primary disease of red blood cells, the only cells in the body that synthesize hemoglobin. Because hemoglobin is an essential protein involved in oxygen exchange, deficiencies in hemoglobin will produce widespread secondary changes in many different organs.

10.1.1 Rule—Mutations of pathways in unspecialized cells tend to produce disorders that are found in multiple organs.

Brief Rationale—If a cellular function is general (i.e., occurring in many different types of cells), a defect in the cellular function is likely to cause dysfunction in many different organs.

Cystic fibrosis is caused by an inherited deficiency of the cystic fibrosis transmembrane conductance regulator (CFTR). CFTR regulates the movement of chloride and sodium ions across epithelial membranes. Ion exchange across epithelial tissues is deployed by many different kinds of cells, including virtually every type of duct-lining cell, and every mucus-producing epithelial cell, such as the mucus-producing cells of the lung and gastrointestinal tract. In cases of cystic fibrosis, mucus-producing cells produce a thick, viscous product that cannot be easily cleared from ducts (e.g., pancreas) and organ conduits (e.g., bronchi, gastrointestinal tract, seminal vesicles). A defect in a generally necessary cellular function will produce primary disease in many different organs.

Mutations of pathways in highly specialized cells sometimes produce isolated (i.e., non-syndromic) disease. For example, a variant of non-syndromic deafness is maternally inherited and is associated with mutation in mitochondrial DNA. Mitochondrial DNA mutations are expressed in every cell of the body, with the exception of red blood cells. One might assume that a mitochondrial defect would produce pathological changes in many different types of cells. Not always. In the case of non-syndromic hearing loss, mitochondrial mutations that result in deafness produce a highly specific loss of cells of the inner ear, without producing known deficits in other organs. Raimundo and coworkers have shown that there is a stress pathway leading to cell death that is preferentially expressed in inner ear neurons [5]. A mitochondrial mutation associated with non-syndromic hearing loss, the A1555G mtDNA mutation, activates this pathway, causing cells to die in the inner ear, and producing progressive deafness [5].

- **Cell-type disease specificity determined by the weakest link**

There are many instances wherein a mutation affects pathways in many different cells, but it only produces a clinical phenotype in a small subset of affected tissues. In many of these instances, the clinical phenotype arises in cells that are least able to cope with the mutational effects.

Here are a few examples of weakest link tissue specificity:

1. **Vitamin B12 deficiency.** Vitamin B12 is required for efficient DNA synthesis and cell division. Cells that have the highest cell division rate are the same cells that are most affected by B12 deficiency. Bone marrow has a very high rate of cell division. As expected, B12 deficiency results in anemia (i.e., a reduction in circulating red blood cells) due to a defect in maturation of the red cell lineage.

   In addition to its weakest link toxicity for bone marrow cells, vitamin B12 deficiency exerts a specific toxic effect on the nervous system via a cell-type-specific mechanism. As described in Section 9.5, vitamin B12 is required for the last step of the pathway leading to succinyl coA formation. A deficiency of vitamin B12 produces elevated levels of methylmalonic acid. Over time, methylmalonic acid incorporates into myelin and destabilizes the myelin sheath, required for the normal fast conduction of impulses along neuronal axons. The resultant myelinopathy produces sensory and motor neuron deficiencies and subacute combined (i.e., posterior and lateral column) degeneration of the spinal cord. Alterations in mentation have also been observed.

2. **Retinitis pigmentosa.** As described in Section 9.3, retinitis pigmentosa is one of the most genetically heterogeneous of the inherited diseases. It can occur as a solitary disease, or it can occur in syndromic conditions (e.g., Usher syndrome, which is characterized by losses of hearing and vision).

   A non-syndromic form of retinitis pigmentosa is caused by a mutation in any one of four genes coding for splicing factors (splicing factor mutations) (i.e., PRPF31, PRPF3, PAP1, and PRPF8). These splicing factors are found in virtually every type of cell in the body. Why would deficits in any of these ubiquitous factors lead to one specific defect? Tanackovic and coworkers have suggested that the protein processing demands in retinal cells are extremely high, due to the high turnover of rhodopsin molecule [6]. The same splicing defects that are tolerated in other types of cells are destructive to retinal photoreceptor cells. This particular form of retinitis pigmentosa is an example of weakest link cell specificity.

3. **Leukoencephalopathy with vanishing white matter.** Leukoencephalopathy with vanishing white matter is a leukodystrophy (i.e., a degenerative disease of the white matter of the brain), caused by loss of function mutation in any of five genes encoding subunits of the translation initiation factor EIF-2B. The gene product initiates protein synthesis in cells throughout the body, but its clinical expression is often isolated to two cell types in the brain: oligodendrocytes and astrocytes (see Glossary items, Initiation factor, Translation factor). These cells have a particularly high rate of protein synthesis, and it is this heightened requirement for the gene product that seems to render these cells sensitive to the mutation. The disease is triggered or exacerbated by certain types of stress to the central nervous system. Susceptible individuals who carry the mutation may develop acute symptoms of leukodystrophy (cerebellar ataxia, spasticity, optic atrophy) after head trauma.

It should be noted that a weakest link mechanism may apply in instances for which no differences in pathway requirements can be measured in the affected tissues. The reason that a weakest link cause for tissue specificity cannot always be determined by any measurable test is subtle, and best explained with an economic analogy. Suppose everyone in a population receives a weekly salary of $200. One person has expenses of $199 per week, and manages to save a $1 each week. Another person has expenses of $201 per week, and is continually in debt. Over time, he loses his house and cannot provide for his own welfare. The difference in requirements between the two individuals is just $2, an insignificant quantity that might evade an accountant's inspection. **The lesson here is that a cell that is particularly susceptible to disease may have a pathway requirement not measurably different from the pathway requirement of cells that do not express disease susceptibility**.

- **Convergence**

As applied to diseases, convergence occurs when different genes, cellular events, exposures, and pathogenetic mechanisms all lead to a similar clinical phenotype. Convergence is found in common diseases and in rare diseases. In the case of systemic responses to injury, convergence may have an evolutionary origin. For example, humans have evolved to respond in an orchestrated way to a variety of pathologic stimuli. Various antigens can stimulate an orchestrated acute allergic response that may be identical for a wide variety of antigens (hives, bronchial constriction, puffy eyes). Likewise, humans have evolved to a systemic response to local infection that is specific for our species [7].

Convergence is observed in all the rare diseases that have genetic heterogeneity, either allelic heterogeneity or locus heterogeneity (see Section 9.3). In these cases, many underlying genetic causes yield the same clinical phenotype.

10.1.2 Rule—Regardless of the path taken, many pathologic processes will converge to the same pathologic condition.

Brief Rationale—There are a limited number of ways that the body can respond to malfunctions.

Think about all the things that can go wrong with your car. The engine can stop, the fuel system can be interrupted, the battery may die, the brakes may fail, any of the four tires can flatten, the headlights may not work, the electrical system may suffer a circuit shortage, and so on. It seems like a long list, but it is not. Maybe a dozen common problems account for the vast majority of car problems. Add these to a few dozen less likely problems, and you have a listing that would cover 99% of automobile repair issues. Every auto repairman knows that there are a limited number of systems in the car that can go bad. Repairs are relatively easy if the repairman can determine the system or part that is at fault.

Whereas the number of different auto problems is limited, the number of events that can lead to these problems is virtually infinite. An auto repairman knows that for every engine breakdown, there might be thousands of possible causes. A non-functioning engine can be corrected by taking out the bad engine and putting in a new engine. If he is a very good repairman, he will determine whether a problem in a different system (e.g., the fuel injector) was indirectly responsible for the engine failure. Diagnostic tools should determine when a defect in one system is responsible for a defect in another system.

Humans, like automobiles, are highly complex. Nonetheless, there are a limited number of problems that can occur in a complex organism. Heart attacks exemplify pathological convergence. Many different pathological processes can lead to the blockage of a coronary artery, such as: atherosclerotic plaque, hypertrophy of the arterial wall, spasms of the artery, acute infection of the artery, thrombus formation within the artery, arterial tear or dissection, developmental defects resulting in narrowing. Genes and environment contribute to these mechanisms. In the end, they can all produce one clinical phenotype; the all-too-common heart attack.

Hypertension is another excellent example of convergence toward a common phenotype. As discussed in Section 5.4, there are numerous genetic and environmental causes of hypertension. The causes of hypertension may include overactivity of the renin–angiotensin system, or channel defects at various sites of the renal tubule, or arterial wall pathology, or increased salt consumption. Regardless of the underlying cause of hypertension, all inherited and acquired forms of the disease converge onto one physiologic pathway: increased net salt balance leading to increased intravascular volume, leading to augmented cardiac output, leading to elevated blood pressure [8]. Regardless of the underlying mechanism leading to an individual's hypertension, diuretics such as hydrochlorothiazide or furosemide, which reduce the reabsorption of sodium in the kidneys, will almost always lower blood pressure. We see a similar phenomenon with rare and common causes of diabetes. Extremely rare single gene diabetes, including HNF1A MODY and permanent neonatal diabetes associated with the KCNJ11 and ABCC8 genes, is controlled with sulfonylurea, the same drug used to treat common type 2 diabetes. The cause of monogenic diabetes is quite different from the cause of common type 2 diabetes, but their pathways converge; and all these diseases respond to the same treatment [9].

Some of the rare diseases exhibit convergence with one another. For example, epidermolysis bullosa is an inherited disease characterized by blistering of the skin and mucosal membranes (e.g., mouth). It is always caused by a defect in the mechanism whereby the epidermis is anchored onto the underlying dermis. Blisters are formed in locations where the epidermis lifts off the dermis, usually at sites of friction. Over 300 gene defects can result in epidermolysis bullosa. Depending on the variant form of the disease, any of several different genes may serve as the underlying cause (e.g., COL, PLEC, Desmoplakin genes). There is also an autoimmune form of epidermolysis bullosa acquisita, wherein antibodies target Type VII collagen, a component of the basement membrane glue that lies between the epidermis and the dermis. Regardless of the underlying cause, all variants of epidermolysis bullosa converge to a blistering phenotype.

10.1.3 Rule—A large set of cellular defects accounts for a relatively small number of possible pathologic conditions.

Brief Rationale—In any complex system, there are a limited number of functional parts, but each functional part can break down due to a vast number of possible defects.

Convergence can be thought of as the opposite of pleiotropism. In the case of pleiotropism, one mutation produces a variety of physiological effects. In the case of convergence, a variety of mutations or pathologic processes lead to the same phenotype. Is there some physical law that explains why convergence is such a universal phenomenon in so many complex systems? Maybe it is all due to entropy. We can see that nothing in the universe is ever as chaotic as we might expect from the complexity of the individual elements of the system. Despite the enormous number of atoms in the universe, there seem to be just a few dozen types of cosmological bodies (e.g., stars, planets, black holes). These bodies assemble into galaxies that seem to have a relatively narrow array of shapes and sizes. In the case of biological systems, complex processes settle for a limited number of outcome categories.

10.1.4 Rule—Regardless of the complexity of a system, the outcomes are typically repeatable and stable.

Brief Rationale—All existing biological systems, despite their complexity, converge toward stability. If a biological system were unstable, it would cease to exist.

If a germline defect caused chaotic reactions in many different types of cells, the organism would never survive the developmental process. Death would ensue early *in utero*, and the disease would not manifest in a living individual. As it happens, *in utero* death is a frequent occurrence; nearly half of all embryos die (i.e., spontaneously abort). It is reasonable to assume that some of these deaths are caused by mutations that destabilize normal development.

The phenomenon of convergence may explain some of the genetic complexity that seems to characterize many, if not all, of the common diseases. When there are hundreds or thousands of gene variations that are associated with one disease, it is likely that all these different genes contribute to a limited range of available disease pathways. **In diseases that have a complex genetic etiology, it makes sense to examine the pathways that converge to a final clinical phenotype, rather than to try to understand the individual contribution from each variant gene.**

- **Vulnerability of post-mitotic cells**

The post-mitotic cell is fully differentiated and has lost the ability to divide (see Glossary item, Post-mitotic). As discussed in Section 4.4, much of the aging process consists of degenerative changes that occur in post-mitotic cells such as neurons, cartilage cells, and muscle cells (see Glossary item, Aging). When an injurious process affects many different cell types, it is likely to cause functional deficiencies in cell populations that cannot replace the injured cells with new, healthy cells.

For example, the tauopathies, discussed in Section 4.3, are neurodegenerative disorders wherein tau protein accumulates within the central nervous system. Tau proteins are involved in the stabilization of microtubules in many different types of cells, but the accumulated tau protein is specifically toxic to irreplaceable, post-mitotic neurons.

- **Co-conditional factors**

When you impose an increasing number of restrictions on a cellular process, you narrow the range of participating cell types. For example, in xeroderma pigmentosum, there is an inherited DNA repair deficiency in every cell of the body. Xeroderma pigmentosum cells are particularly sensitive to the toxic effects of DNA damage induced by ultraviolet light. Ultraviolet light cannot penetrate deeper than the skin. Hence, the clinical phenotype of xeroderma pigmentosum is restricted to the skin and the cornea.

- **Unsolved mysteries**

In animals, DNA sequences are not transcribed directly into full-length RNA molecules ready for translation into a final protein. There is a pre-translational process wherein transcribed sections of DNA, so-called introns, are spliced together, and a single gene can be assembled into alternative spliced products. Alternative splicing is one method whereby more than one protein form can be produced by a single gene [10,11]. Cellular proteins that coordinate the splicing process are referred to, in aggregate, as the spliceosome (see Glossary item, Spliceosome). **Errors in normal splicing can produce inherited disease, and it is estimated that 15% of disease-causing mutations involve splicing [12,13].**

One might expect that mutations in spliceosomes would cause deficiencies in diverse cell types with multi-organ and multi-system disease (e.g., syndromic disease). This is not the case. For example, spliceosome mutations account for a form of retinitis pigmentosa and of spinal muscular atrophy [10]. In both diseases, pathology is limited to a specific type of cell; retinal cells and their pigment layer in retinitis pigmentosa, and motor neuron cells in the spinal muscular atrophy.

Today, nobody can adequately explain the cell-type specificity of diseases that arise from a constitutive loss of function of essential splicing factors.

10.2 CELL REGULATION AND EPIGENOMICS

"I would be quite proud to have served on the committee that designed the E. coli genome. There is, however, no way that I would admit to serving on a committee that designed the human genome."

—David Penny [14]

In earlier chapters, we introduced the concept of the epigenome, without elaborating on its most general role in the pathogenesis of disease. In this section, we will step back and describe the role of the epigenome in the pathogenesis of rare diseases, and we shall see how this knowledge may help us to understand and treat common diseases, such as aging and cancer.

10.2.1 Rule—The genome establishes the identity of an organism; the epigenome establishes the behavior of the individual cells within the organism.

Brief Rationale—Each terrestrial organism can be distinguished from every other terrestrial organism by its unique sequence of DNA. Furthermore, each organism can be identified as a member of a particular species based on DNA sequences inherited from a common ancestor. Within the organism, there are many different cell types, each with a specific behavior (e.g., hepatocyte, neuron, keratinocyte). Because each cell of an organism contains the same DNA, cell-type behavior cannot be determined by the genome. Because cell type is inherited (i.e., a hepatocyte produces another hepatocyte, never a keratinocyte),

there must be a heritable component of somatic cells that acts on the genome to direct its activities. We call this the epigenome.

At a minimum, the epigenome consists of the non-sequence modifications to DNA that control the expression of genes. These modifications include DNA methylations, as well as histone, and non-histone chromatin complexes and modifications thereof (see Glossary item, DNA methylation). Beyond this minimalist definition, there are expanded versions of the definition that would include non-coding RNAs, RNA splicing factors, and non-sequence modifications of RNA that might influence gene expression [15]. As used in this book, the terms "epigenome" and "epigenetics" apply exclusively to non-sequence alterations in chromosomes that are heritable among somatic cell lineages. In particular, the epigenome controls differentiation and the biological behavior of the different somatic cell types of the body.

The most common form of methylation in DNA occurs on cytosine nucleotides, most often at locations wherein cytosine is followed by guanine. These methylations are called CpG sites. CpG islands are concentrations of CpG dinucleotides that have a GC content over 50% and that range from 200 base pairs (bp) to several thousand bp in length. There are about 29,000 to 50,000 CpG islands [16]. The patterns of methylation are inherited from somatic cell to somatic cell and alterations in methylation patterns are epimutations that persist through a cell's lineage [17].

Various proteins bind specifically to CpG sites. For example, MECP2 is a chromatin-associated protein that modulates transcription. MECP2 binds to CpGs; hence, alterations in CpG methylation patterns can alter the functionality of MECP2. Mutations in MECP2 cause RETT syndrome, a progressive neurologic developmental disorder and a common cause of mental retardation in females. It has been suggested that the MECP2 mutation disables normal protein–epigenome interactions [18].

10.2.2 Rule—The epigenome influences which tissues will be affected by a disease-causing gene mutation.

Brief Rationale—In Section 10.1, we described how the expression of genes in different types of cells accounts for much of the tissue specificity of genetic diseases. We can now recognize that the difference of gene expression in different types of cells is determined by the epigenome.

In addition to influencing tissue specificity of disease, the epigenome influences the phenotype of a disease.

10.2.3 Rule—Conditions of the epigenome can determine which disease, among several, may result from a gene mutation.

Brief Rationale—The epigenome determines gene expression; an unexpressed protein-coding gene, regardless of the mutations it may contain, cannot cause

disease. The epigenome determines which genes are expressed, and in which tissues.

In Section 9.2, we discussed the epigenetic dichotomy between Prader–Willi syndrome and Angelman syndrome, both caused by a microdeletion at 15q11-13. When the microdeletion occurs on the paternally derived chromosome, the disease that results is Prader–Willi syndrome. When the microdeletion occurs on the maternally derived chromosome, the disease that results is Angelman syndrome. The explanation for this peculiar dichotomy lies in the parental pattern of epigenetic modification. As background for this topic, there are a few concepts that must be introduced. Soon after fertilization, embryonic cells erase their inherited epigenetic modifications. The process must occur so that primitive embryonic cells start with a clean slate, adding epigenetic modifications to cells, as appropriate, to guide normal development. In every human genome, there is a small set of genes that avoids erasure. Such genes maintain the epigenetic profile of the parental genes, and we say that such genes are "imprinted" (see Glossary items, Imprinting, Loss of imprinting, Erasure).

The microdeletion that causes Prader–Willi syndrome and Angelman syndrome at 15q11-13 contains about 4 million bp covering many different genes on one chromosome. Let us imagine that the microdeletion is maternally inherited (i.e., lies on the chromosome contributed by the mother). The paired chromosome, contributed by the father, lacks the 15q11-13 microdeletion. In theory, the paternal chromosome could compensate for the lost genes in the microdeletion area. However, the paternal chromosome has a gender-typical pattern of gene silencing, and certain genes on the undeleted paternal chromosome will be unexpressed, resulting in the absence of expression for such genes in the haploid set of chromosomes. Had the 15q11-13 microdeletion occurred on the paternal chromosome instead of the maternal chromosome, the mother's gender-typical pattern of gene silencing would result in the lack of expression of a different set of genes. You can see that the gender of the parent who passes the microdeletion to the offspring will determine which genes are unexpressed in the offspring; hence, which disease occurs. Diseases whose expression is caused by imprinting do not obey a Mendelian pattern of inheritance. If the 15q11-13 had obeyed Mendelian inheritance, offspring with the microdeletion would all have the same disease traits, regardless of the gender of parent who passed the microdeletion.

10.2.4 Rule—In general, epigenetic hypermethylation leads to gene suppression. Hypomethylation promotes gene expression.

Brief Rationale—As an empirical observation, hypermethylated chromosomal regions, such as the X-chromosome Barr body, are genetically hypoactive.

Females are born with two X-chromosomes, but nature seems to set the maximum at one. In every somatic cell, one of the two X-chromosomes is inactive, and the inactive chromosome has a different morphology than the active chromosome. The inactive chromosome is shrunken, compact, hypermethylated, and usually sticks to the edge of the nucleus (i.e., adjacent to the nuclear envelope), where there is minimal transcriptional activity. This inactive X-chromosome is called the Barr body. The choice of which X-chromosome is inactivated (i.e., paternal X or maternal X) occurs seemingly randomly in different cells' early development, but after the germline is established. Hence, every genetically normal female is a somatic mosaic, composed of patches of cells clonally descended from an embryonic cell that had an active paternally inherited X-chromosome or an active maternally inherited X-chromosome (see Glossary item, Somatic mosaicism).

Aside from the observed hypermethylation in inactive X-chromosomes, the presence or absence of methylation can account for clinically opposite disease phenotypes. As described in Section 9.2, opposite patterns of methylation in a microdeletion region in 11p15.5 account for some cases of two clinically near-opposite diseases. Hypermethylation causes Beckwith–Wiedemann syndrome, a syndrome that produces tissue overgrowth. Hypomethylation produces Russell–Silver syndrome, a cause of dwarfism (see Glossary item, Russell–Silver syndrome).

10.2.5 Rule—The disease phenotype (i.e., the clinical presentation) of rare diseases caused by epigenetic mutations is impossible to predict at present.

Brief Rationale—Cell regulation via the epigenome is extremely complex. A simple epigenome modification will have pleiotypic effects that are beyond our ability to measure. Hence, the best we can do at present is to observe the changes in trait that occur when the epigenome is altered, and try to make some sense of our observations.

Though we cannot predict the changes that result from an epigenetic alteration, we can do a bit of experimentation to see the range of effects that might occur when we alter the methylation patterns in cells. Such patterns can be altered with methylating agents or with hypomethylating agents. One of the most impressive demonstrations of induced heritable epigenetic change occurs when pregnant mice are fed a diet supplemented with a methylating agent. The offspring inherit the viable yellow phenotype, associated with large size, obesity, hyperinsulinism, heightened susceptibility to cancer, and a shorter average lifespan [19]. This observation applies to a particular strain of mice that are known to show imprinted heritability of the viable yellow trait. Nonetheless, the lesson holds in this special case: methylation via an environmental additive can produce a heritable epimutation that permanently affects multiple tissues in the offspring.

Cancer is a disease characterized by changes in just about every organelle, pathway, and molecule in cancer cells. The epigenome is changed drastically in cancer, and epigenetic instability is as much a feature of cancer progression as is genetic instability (see Glossary items, Epigenetic instability, Mutator phenotype). A range of genes have been shown to be hypermethylated in cancer; hypermethylation has also been observed the CpG islands in cancer cells (see Glossary item, CpG island). It is tempting to ask whether hypomethylation can reverse some of the malignant phenotype attributed to hypermethylation. This is the case at least for myelodysplastic syndrome, a preleukemic condition [20]. Two hypomethylating agents, 5-azacytidine and 5-aza-2-deoxycytidine, have already won U.S. FDA approval for the treatment of myelodysplastic syndrome [20,21]. In addition to hypomethylation agents, the histone deactylase inhibitors are being examined as modulators of the cancer epigenome, and as candidate chemotherapeutic agents.

Of course, there is an interplay between epigenome and genome in cancers. In acute promyelocytic leukemia, a gene translocation produces the PML/RAR(alpha) fusion protein [22]. Normally, promyelocytes differentiate to become non-dividing myelocytes (neutrophils). Neutrophils are the major circulating nucleated cell and play a crucial role in inflammation and the body's defenses against infections. The PML/RAR(alpha) fusion protein causes the promyelocyte to divide, producing more promyelocytes and fewer neutrophils. Eventually, the population of clonal promyelocytes arising from the neoplastic progenitor cell will attain a sufficiently large number to be recognized clinically as a promyelocytic leukemia.

Acute promyelocytic leukemia is one of the few cancers that can achieve clinical remission without treatment with cytotoxic agents. Remission is achieved with all-trans retinoic acid. Treated promyelocytic cancer cells differentiate and become non-dividing mature myelocytes [23]. The mechanism by which the neoplastic fusion protein, PML/RAR(alpha), induces a neoplastic phenotype, and the mechanism whereby all-trans retinoic acid reverses the neoplastic phenotype seems to be mediated through the epigenome. It is hypothesized that PML/RAR(alpha) modifies histone deacetylase complexes resulting in the inappropriate transcriptional repression of genes that would normally inhibit promyelocyte proliferation. All-trans retinoic acid is thought to reverse this effect [22]. If this turns out to be the case, promyelocytic leukemia would serve as an example of a gene mutation that employs epigenetic alterations to sustain a neoplastic phenotype.

At present, rare hematologic malignancies seem to be most responsive to targeted epigenetic therapy. Time will tell whether the same approach will be effective against common cancers.

10.2.6 Rule—The epigenome, unlike the genome, is constantly changing in every type of cell throughout development and throughout adult life.

Brief Rationale—Cells have elaborate DNA repair systems that do a fairly good job at maintaining an unchanging sequence of nucleotides throughout life.

The epigenome can be easily altered. We see the effects of changes in the epigenome when we look at the process of cellular differentiation, which is based on epigenomic modification.

There is every indication to believe that the longer we live, the greater is the deviation from our neonatal epigenome. Much of our knowledge comes from observations on monozygotic (i.e., identical) twins.

It is a common observation that monozygotic twins look alike at birth, often growing into early adulthood as a pair of strikingly similar individuals. But as the decades go by, identical twins begin to diverge in clinical appearance. They seem to age differently, and they develop different diseases, with the greatest discordance in the common, acquired diseases of adulthood.

Fraga and coworkers found that monozygotic twins are born with nearly identical epigenetic patterns of DNA methylation and histone acetylation. This near-identity persisted in the early years, but in later years, monozygotic twins had widely divergent patterns of DNA methylation and histone acetylation. These divergent patterns were accompanied by discordances in gene expression. The discordance in diseases occurring in monozygotic twins is, an observation that you would not expect to see if disease susceptibility was determined by genes [24,25].

10.2.7 Rule—Children are afflicted by diseases with a strong genetic influence, while the elderly are afflicted by diseases with a strong epigenetic influence.

Brief Rationale—Children have not lived long enough to accumulate epigenetic changes. The elderly have lived so long that any genetic disease would have manifested decades earlier.

Logically, the following rule should also apply.

10.2.8 Rule—Acquired epigenetic alterations play a much greater role in the common diseases than in the rare diseases.

Brief Rationale—The rare diseases are typically diseases of children, which are driven by genetics. The common diseases happen to be diseases of adults and the elderly, who have lived long enough to accumulate epigenetic alterations.

In general, children are born into their diseases, while the elderly "earn" their diseases through a lifetime of toxic exposures and accumulated cellular damage.

10.3 DISEASE PHENOTYPE

"Individuals do not belong in the same taxon because they are similar, but they are similar because they belong to the same taxon."

—George Gaylord Simpson [26]

10.3.1 Rule—Diseases with the same clinical phenotype tend to exhibit similar aberrant metabolic pathways regardless of any differences in the underlying defects that caused the diseases.

Brief Rationale—Earlier in this chapter, we discussed the following rule: "Regardless of the path taken, many pathologic processes will converge to the same pathologic condition." Here, we approach the phenomenon of disease convergence from the opposite direction. If we have two conditions that have converged to the same clinical phenotype, can we assume that both conditions employ the same cellular pathways? Probably so, because diseases are manifestations of pathologic conditions in specific types of cells. Cells of a specific type are highly restrained to express a limited set of cell-type-specific pathways.

Assuming the rule is true, we would expect any disease that causes premature aging, regardless of its underlying cause (e.g., chromatin instability in Hutchinson–Gilford progeria, DNA instability in Werner syndrome, mitochondrial degeneration in Wolfram syndrome, telomere shortening in dyskeratosis congenita), to produce cellular lesions wherein the dominant cellular pathways are similar. Why would we expect this to be true? Basically, cells of any given cell type (e.g., neurons, gut lining cells, white blood cells, etc.) are programmed to express a certain limited number of physiologic pathways (i.e., there is a limit to each cell's physiologic options). When cells of a given type behave in a similar pathologic fashion (i.e., increased fragility, reduced respiratory capacity, increased apoptosis, uncontrolled cellular division, etc.), there are a restricted number of pathways that can produce these specific disease phenotypes. Therefore, you might expect that common pathology phenotypes share common aberrant metabolic pathways.

In the case of aging, there is some experimental evidence to suggest that this rule is true. Lafferty-Whyte and coworkers have shown that the gene expression patterns and pathways that produce the senescent phenotype are similar among various aging diseases regardless of the underlying causes [27].

10.3.2 Rule—When a disease phenotype is fully accounted for by the function of one pathway, then we can expect every disease having the same phenotype to have alterations in the same pathway.

Brief Rationale—If alternate single pathways or alternate sets of pathways could account for the equivalent phenotype produced by a single pathway, then we would expect to have encountered some instances of this phenomenon. No such instances are known. In the absence of such encounters, a single pathway for all instances of the clinical phenotype seems likely.

If we are lucky enough to know that a phenotype arises from a perturbation of a known pathway, then it does not make much difference which pathway enzymes are targeted. The resulting phenotype will be determined by the overall activity of the pathway. There are many examples wherein this rule holds true. The biology of GIST (gastrointestinal stromal tumor) is one such example.

GIST arises from the interstitial cells of Cajal, located in the gut wall. An activated tyrosine kinase molecule drives the growth of the interstitial cells of Cajal and leads to tumor formation. Early studies indicated that most GISTs had one of several different mutations of c-KIT, an activator of tyrosine kinase that was overexpressed in the tumor. Moreover, a new drug, imatinib mesylate, was found to inhibit tyrosine kinase. By inhibiting tyrosine kinase, imatinib mesylate arrested the activation of the tyrosine kinase growth factor and produced a dramatic reduction in the size of GISTs. The successful treatment of GISTs by imatinib mesylate was the first example of a minimally toxic drug targeted against a specific molecule, resulting in the rapid reduction in size of a solid tumor [28].

A lingering mystery at the time related to c-kit negative tumors; tumors that seemed to harbor no mutations in the c-KIT gene. Some researchers believed that every GIST harbored a c-kit mutation, but that there were technical limits to identifying all the mutations that existed. Other researchers believed that there might be alternate pathways to the development of GIST, and these alternate pathways awaited discovery [29]. As it turns out, there was at least one more mutation in another gene that led to a minority of cases of GIST, but this gene product participated in the same pathway as the c-kit protein. Mutations in either protein led to the activation of the same tyrosine kinase pathway that was blocked by imatinib mesylate.

This second protein is the PDGFR-alpha (platelet-derived growth factor receptor alpha). Mutations in the genes coding for PDGFR-alpha or c-kit produced identical types of tumors responsive to the same targeted therapy [30]. **Because different genes coding for different constituents of a pathway will tend to produce the same disease, it makes little sense to develop drugs that target each gene product. A drug that is active against the altered protein product of one mutated gene may not be active against all the variant mutant forms of the gene or against alternate proteins of the same pathway. It makes much more sense to develop drugs that target the key enzymes that drive the pathway under normal conditions**. This concept of pathway-targeted drug development is discussed more fully in Section 13.2.

10.3.3 Rule—An agent that causes a common disease will also cause multiple rare diseases.

Brief Rationale—Common diseases are complex, and an agent that produces common disease must exert biological effects on many different pathways. Some of these pathways are likely to produce rare clinical phenotypes.

For example, there are several common diseases caused by smoking. These include bronchogenic lung cancer, emphysema, and chronic obstructive pulmonary disease (see Glossary item, Bronchogenic carcinoma). Likewise, there are rare diseases that are caused by smoking, including: Buerger disease, also known as thromboangiitis obliterans, a progressive vasculopathy involving small and medium arteries and veins; and Warthin tumor, a benign salivary gland tumor [31].

Alcohol is known to produce liver cirrhosis and pancreatitis, both common diseases (see Figures 10.1 and 10.2). Alcohol also produces Marchiafava–Bignami disease, characterized by demyelination of the corpus callosum. Another neurologic sequelae of alcoholism is Wernicke–Korsakoff syndrome, characterized by vision changes, ataxia, and impaired memory resulting from brain atrophy involving the mammillary bodies, thalamus, periaqueductal gray, 3rd and 4th ventricles, and the cerebellum. Alcoholism also triggers or aggravates various metabolic diseases (e.g., gout, gall stones) and is thought to increase the incidence of several types of cancers (e.g., esophageal cancer, liver cancer, oral cancer).

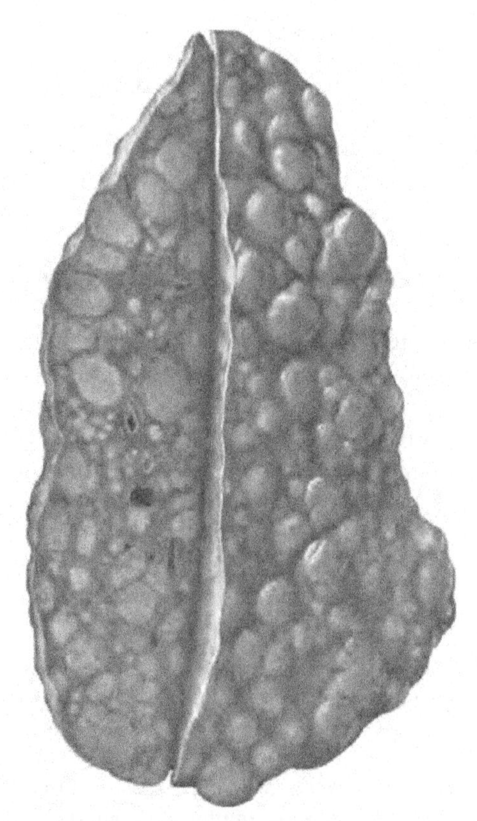

FIGURE 10.1 Gross specimen of liver involved by cirrhosis. Round nodules have replaced normal smooth parenchyma, and the tissues between the nodules are invested with bands of fibrous tissue. (*Source: MacCallum WG. A Textbook of Pathology* [32].)

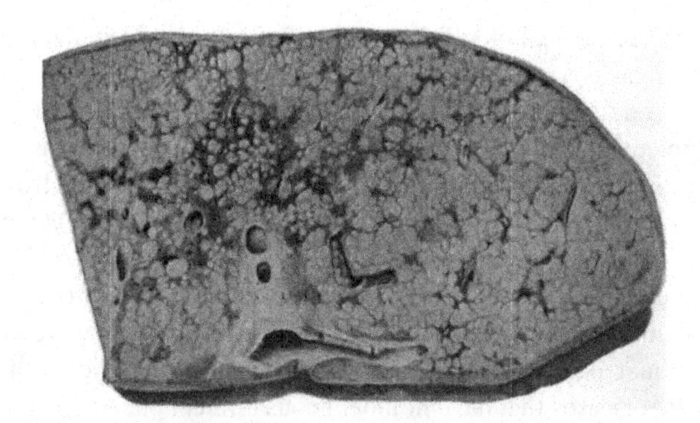

FIGURE 10.2 **Cross-section of gross specimen of liver involved by cirrhosis.** In this specimen, the nodules are all small, but of varying size (i.e., each nodule has a slightly different size from its neighbor nodules). Fibrous tissue fills the spaces between nodules, and the capsule covering the liver is thickened. See color plate at the back of the book. (*Source: MacCallum WG. A Textbook of Pathology* [32].)

10.4 DISSECTING PATHWAYS USING RARE DISEASES

"Intellectual history is a relay race, not a 100-yard dash."

—Susan Jacoby

When we learn that a particular inherited disease is caused by a specific gene defect, we like to believe that the disease phenotype results directly from the gene defect. This is almost never the case, even for the simplest monogenic diseases wherein a gene codes for a functional protein. In general, functional proteins are mere cogs in a complex system that eventually leads to a cellular outcome, such as blood coagulation, respiration, collagen synthesis, faithful DNA replication, and so on. Because many different proteins participate in virtually every cellular function, a disease of similar clinical phenotype can be produced by defects of any of the participating factors in the pathway.

10.4.1 Rule—Complex physiological pathways that are unique to humans (e.g., coagulation pathways, immune pathways, neural pathways, metabolism pathways) can only be understood by studying rare diseases affecting steps in the pathway.

Brief Rationale—Researchers have had great success collecting cases of rare deficiencies, and then piecing together a plausible and testable complex pathway.

A key property of many inherited pathway disorders lies in the ability of mixing studies to complement individual gene deficiencies. For example, suppose you know that there are multiple steps in blood coagulation, and you have access to two volunteers, each with a coagulation deficiency. If you mix their blood together and add a coagulant, you might find that their mixed blood sample clots, like normal blood. Each blood sample provided a clotting protein that the other lacks, and the end result was blood that clots. The two blood samples are said to be complementary. More technical experiments may reveal which complementary factors precede others in the clotting cascade. As an experimentalist, you would want to have access to blood samples from all of the many different inherited clotting disorders, so that you could eventually piece together the complete clotting pathway in proper order.

This aforementioned scenario sketches a time-honored approach to unraveling the functional pathways of genetic diseases. Most of what we know about complex cellular pathways has come from careful observations of heterogeneous genetic diseases that have homogeneous clinical phenotypes.

Here is a sampling of classes of genetic diseases and the normal cellular pathways they helped elucidate:

- Immune deficiencies → normal immune mechanisms
- Collagenopathies → collagen synthesis (see Glossary item, Collagenopathy)
- Sarcoglycanopathies → Sarcoglycan complex synthesis [33]
- Hemophilias → blood clotting cascade
- Inherited conduction disorders → cardiac conduction system (see Section 6.2)
- Glucose-6-phosphate dehydrogenase deficiency → hexose monophosphate shunt (see Section 11.4)
- Congenital disorders of glycosylation → post-translational modifications of proteins (see Glossary item, Congenital disorders of glycosylation)
- Leukocyte adhesion deficiencies → hematopoietic integrin pathway. Integrin activation enables circulating blood cells to modulate their affinity for endothelial ligands, thus promoting the extravasation of immune-response cells into sites of infection [34]
- Mitochondriopathies → cellular respiration (see Glossary item, Mitochondriopathy)
- Inborn errors of metabolism → pathways of cellular metabolism
- Dyslipidemias → lipid metabolism
- Peroxisome biogenesis disorders → assembly and function of peroxisomes (see Glossary item, Peroxisome biogenesis disorder) [35]
- Inherited disorders of DNA repair deficiencies → DNA repair mechanisms
- Inherited conditions that have phenocopy diseases → toxicologic mechanisms of disease (see Section 9.5)
- Vesicular transport disorders → mechanisms for transporting proteins to their functional location

After a protein molecule is translated from mRNA, it must undergo post-translational modifications and brought to its proper cellular location. Such post-translational steps are often divided into disorders of post-translational modification (e.g., congenital disorders of glycosylation) and protein transport disorders. For work on protein transport mechanisms, the 2013 Nobel Prize in Physiology or Medicine was awarded to James E. Rothman, Randy W. Schekman, and Thomas C. Sudhof. Much of the progress in this field was based on studies of inherited transport disorders in humans (see Glossary item, Vesicular trafficking disorder) [36].

As we develop new, powerful sequencing techniques, we shall learn more and more about the genetic alterations that influence the rare diseases and the common diseases. If we can link gene defects to pathway disturbances, we will be in a good position to treat groups of diseases with a single compound that targets their shared pathway. This topic will be discussed further in Section 13.3.

10.5 PRECURSOR LESIONS AND DISEASE PROGRESSION

"And what physicians say about disease is applicable here: that at the beginning a disease is easy to cure but difficult to diagnose; but as time passes, not having been treated or recognized at the outset, it becomes easy to diagnose but difficult to cure."

—Niccolo Machiavelli

Precancers were discussed in Section 8.4. Just as cancers are preceded by precancers, every disease with a complex pathogenesis, and this would include every common disease and most rare diseases, will have a precursor lesion that can be detected by morphologic, biochemical, or functional tests. Even in those diseases for which no precursor lesion has been defined or studied, we can infer the precursor's existence based on our conceptual understanding of pathogenesis. Pathogenesis is the series of cellular events that precedes the emergence of a disease; hence, there must be some set of events leading to some condition that precedes the clinical phenotype of the fully developed disease. Furthermore, we can infer a set of rules that must apply to precursors of common diseases.

10.5.1 Rule—There are monogenic precursors of common diseases.

Brief Rationale—Every disease begins with a single error.

We can presume that the source of these monogenic precursors will be gene variants that do not, by themselves, cause disease. These gene variants will be among the many gene polymorphisms that are found in any population of humans.

10.5.2 Rule—Precursor lesions are more common than the common diseases that they produce.

Brief Rationale—Precursor lesions are dependent upon additional events and processes that push pathogenesis forward. If these additional events and processes fail to occur, then the precursor cannot progress. Because every fully developed disease has a precursor, while not every precursor develops into a disease, we can infer that there must be more precursor lesions than developed diseases.

Much of the currently known genetic burden associated with common diseases derives from susceptibility genes; genes that do not cause disease, but that make the individual more sensitive to some disease-causing agent in the environment. Obviously, if the exposure to disease-causing agents does not occur, the disease will not develop.

10.5.3 Rule—Monogenic precursors are easier to treat than fully expressed common diseases.

Brief Rationale—A monogenic precursor lesion has one gene that is the underlying cause of its expressed phenotype, and the one gene may influence as few as one pathway. If the causal pathway can be restored to a normal level of activity, the precursor lesion may regress or otherwise fail to develop into a fully expressed disease phenotype.

10.5.4 Rule—More than one type of disease may have the same precursor lesion.

Brief Rationale—There are a limited number of pathologic pathways in cells, and these pathways tend to converge to a smaller number of pathways during pathogenesis. Hence, it seems likely that there will be some precursor lesions that are common to more than one disease.

Here, we can think about the phenocopy diseases. In almost every case, an agent alters a metabolic pathway that converges with pathways used in a genetic disease, eventually producing an equivalent clinical phenotype. Each environmental phenotype must have at least one precursor in common with its genetic counterpart. The same line of reasoning would apply to genetic diseases that have locus heterogeneity, and to common diseases with several distinct etiologic causes (e.g., cirrhosis caused by alcohol and cirrhosis caused by hemochromatosis).

10.5.5 Rule—As we learn more and more about the pathogenesis of diseases, new therapies will be targeted against the most sensitive precursor lesions, not against the fully developed disease.

Brief Rationale—When we successfully eliminate a precursor lesion, we eliminate all the diseases that may develop from the common precursor.

REFERENCES

1. Khavari DA, Sen GL, Rinn JL. DNA methylation and epigenetic control of cellular differentiation. Cell Cycle 9:3880–3883, 2010.
2. Fogli A, Rodriguez D, Eymard-Pierre E, Bouhour F, Labauge P, Meaney BF, et al. Ovarian failure related to eukaryotic initiation factor 2B mutations. Am J Hum Genet 72:1544–1550, 2003.
3. Crisponi L, Deiana M, Loi A, Chiappe F, Uda M, Amati P, et al. The putative forkhead transcription factor FOXL2 is mutated in blepharophimosis/ptosis/epicanthus inversus syndrome. Nat Genet 27:159–166, 2001.
4. Kagey MH, Newman JJ, Bilodeau S, Zhan Y, Orlando DA, van Berkum NL, et al. Mediator and cohesin connect gene expression and chromatin architecture. Nature 467:430–435, 2010.

5. Raimundo N, Song L, Shutt TE, McKay SE, Cotney J, Guan MX, et al. Mitochondrial stress engages E2F1 apoptotic signaling to cause deafness. Cell 148:716–726, 2012.
6. Tanackovic G, Ransijn A, Thibault P, Abou Elela S, Klinck R, Berson EL, et al. PRPF mutations are associated with generalized defects in spliceosome formation and pre-mRNA splicing in patients with retinitis pigmentosa. Hum Mol Genet 20:2116–2130, 2011.
7. Seok J, Warren HS, Cuenca AG, Mindrinos MN, Baker HV, Xu W, et al. Genomic responses in mouse models poorly mimic human inflammatory diseases. Proc Natl Acad Sci USA 110:3507–3512, 2013.
8. Lifton RP, Gharavi AG, Geller DS. Molecular mechanisms of human hypertension. Cell 104:545–556, 2001.
9. Klupa T, Skupien J, Malecki MT. Monogenic models: what have the single gene disorders taught us? Curr Diab Rep 12:659–666, 2012.
10. Faustino NA, Cooper TA. Pre-mRNA splicing and human disease. Genes Dev 17:419–437, 2003.
11. Sorek R, Dror G, Shamir R. Assessing the number of ancestral alternatively spliced exons in the human genome. BMC Genomics 7:273, 2006.
12. Pagani F, Baralle FE. Genomic variants in exons and introns: identifying the splicing spoilers. Nat Rev Genet 5:389–396, 2004.
13. Fraser HB, Xie X. Common polymorphic transcript variation in human disease. Genome Res 19:567–575, 2009.
14. Graur D, Zheng Y, Price N, Azevedo RB, Zufall RA, Elhaik E. On the immortality of television sets: "function" in the human genome according to the evolution-free gospel of ENCODE. Genome Biol Evol 5:578–590, 2013.
15. Jia G, Fu Y, Zhao X, Dai Q, Zheng G, Yang Y, et al. N6-methyladenosine in nuclear RNA is a major substrate of the obesity-associated FTO. Nat Chem Biol 7:885–887, 2011.
16. Bogler O, Cavenee WK. Methylation and genomic damage in gliomas. In Genomic and Molecular Neuro-Oncology. Zhang W, Fuller GN, eds. Jones & Bartlett, Sudbury, MA, pp. 3–16, 2004.
17. Lancaster AK, Masel J. The evolution of reversible switches in the presence of irreversible mimics. Evolution 63:2350–2362, 2009.
18. Amir RE, van den Veyver IB, Wan M, Tran CQ, Francke U, Zoghbi HY. Rett syndrome is caused by mutations in X-linked MECP2, encoding methyl-CpG-binding protein 2. Nat Genet 23:185–188, 1999.
19. Wolff GL, Kodell RL, Moore SR, Cooney CA. Maternal epigenetics and methyl supplements affect agouti gene expression in Avy/a mice. FASEB J 12:949–957, 1998.
20. Garcia-Manero G. Modifying the epigenome as a therapeutic strategy in myelodysplasia. Hematol Am Soc Hematol Educ Program 2007:405–411, 2007.
21. Raza A, Cruz R, Latif T, Mukherjee S, Galili N. The biology of myelodysplastic syndromes: unity despite heterogeneity. Hematol Rev 2:e4, 2010.
22. Nouzova M, Holtan N, Oshiro MM, Isett RB, Munoz-Rodriguez JL, List AF, et al. Epigenomic changes during leukemia cell differentiation: analysis of histone acetylation and cytosine methylation using CpG island microarrays. J Pharmacol Exp Ther 311:968–981, 2004.
23. Flynn PJ, Miller WJ, Weisdorf DJ, Arthur DC, Brunning R, Branda RF. Retinoic acid treatment of acute promyelocytic leukemia: in vitro and in vivo observations. Blood 62:1211–1217, 1983.
24. Chatterjee A, Morison IM. Monozygotic twins: genes are not the destiny? Bioinformation 7:369–370, 2011.

25. Wong AHC, Gottesman II, Petronis A. Phenotypic differences in genetically identical organisms: the epigenetic perspective. Human Mol Gen 14:R11–R18, 2005.
26. Simpson GG. Principles of Animal Taxonomy. Columbia University Press, New York, 1961.
27. Lafferty-Whyte K, Cairney CJ, Jamieson NB, Oien KA, Keith WN. Pathway analysis of senescence-associated miRNA targets reveals common processes to different senescence induction mechanisms. Biochim Biophys Acta 1792:341–352, 2009.
28. Fletcher CD, Berman JJ, Corless C, Gorstein F, Lasota J, Longley BJ, et al. Diagnosis of gastrointestinal stromal tumors: a consensus approach. Int J Surg Pathol 10:81–89, 2002.
29. Berman J, O'Leary TJ. Gastrointestinal stromal tumor workshop. Hum Pathol 32:578–582, 2001.
30. Burger H, den Bakker MA, Kros JM, van Tol H, de Bruin AM, Oosterhuis W, et al. Activating mutations in c-KIT and PDGFRalpha are exclusively found in gastrointestinal stromal tumors and not in other tumors overexpressing these imatinib mesylate target genes. Cancer Biol Ther 4:1270–1274, 2005.
31. Pinkston JA, Cole P. Cigarette smoking and Warthin's tumor. Am J Epidemiol 144:183–187, 1996.
32. MacCallum WG. A Textbook of Pathology, 2nd edition. WB Saunders Company, Philadelphia and London, 1921.
33. Duggan DJ, Hoffman EP. Autosomal recessive muscular dystrophy and mutations of the sarcoglycan complex. Neuromuscul Disord 6:475–482, 1996.
34. Bunting M, Harris ES, McIntyre TM, Prescott SM, Zimmerman GA. Leukocyte adhesion deficiency syndromes: adhesion and tethering defects involving beta 2 integrins and selectin ligands. Curr Opin Hematol 9:30–35, 2002.
35. Weller S, Cajigas I, Morrell J, Obie C, Steel G, Gould SJ, et al. Alternative splicing suggests extended function of PEX26 in peroxisome biogenesis. Am J Hum Genet 76:987–1007, 2005.
36. Gissen P, Maher ER. Cargos and genes: insights into vesicular transport from inherited human disease. J Med Genet 44:545–555, 2007.

Chapter 11

Rare Diseases and Common Diseases: Understanding their Fundamental Differences

11.1 REVIEW OF THE FUNDAMENTALS IN LIGHT OF THE INCIDENTALS

"In fact, diseases that exhibit simple Mendelian patterns of inheritance tend to be rare. Rather, complex diseases arise from numerous genetic and environmental factors working together."

—Johanna Craig [1]

Way back in Chapter 3, we discussed the "Six observations that distinguish common diseases from rare diseases." With the benefit of everything we have learned in subsequent chapters, we can revisit these same observations and provide a few additional insights.

11.1.1 Common diseases typically occur in adults; rare diseases are often diseases of childhood

The rare diseases are often inherited monogenic diseases. Consequently, every cell in the body contains the causal gene, and the gene has the opportunity to exert its effect throughout the period of embryonic development and into childhood. Hence, the rare diseases tend to occur in childhood. The common diseases are caused by multiple factors that accumulate throughout life. Hence, the common diseases tend to occur in adults. In general, the incidence of common diseases steadily increases with age (see Figure 11.1).

11.1.1 Rule—When you graph the frequency of occurrence of a rare disease against the age of the individuals that develop the disease, there is usually one clear peak.

Brief Rationale—Rare diseases often result from a single mutation that enters the germline at the time of conception. The process by which the gene mutation leads to a clinical disease will require roughly the same length of time in most affected individuals, producing a smooth, single peak when disease occurrences are graphed against age of occurrence.

Rare Diseases and Orphan Drugs. http://dx.doi.org/10.1016/B978-0-12-419988-0.00011-0

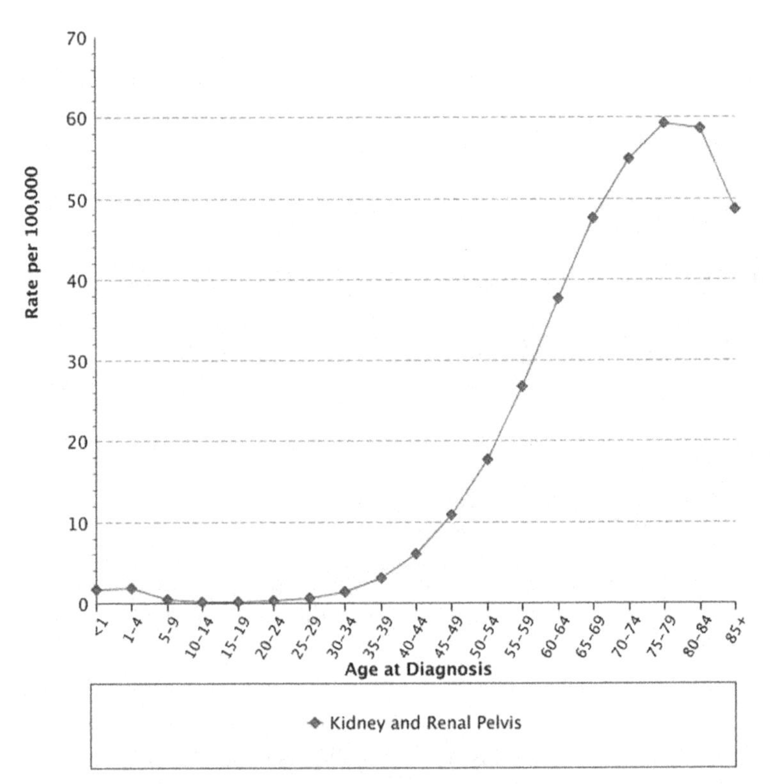

FIGURE 11.1 Graph showing incidence of kidney and renal pelvis cancer by age of occurrence of disease. The incidence of cancer of the kidney and renal pelvis rises steadily, as age increases, toward a single peak. The graph was generated at the National Institute's Surveillance, Epidemiology and End Results "Fast Stats" query site. The query input settings were: SEER incidence; Age-specific rates, 1992–2010, All races, Both sexes, All ages, Kidney and renal pelvis. See color plate at the back of the book. (*Available from: http://seer.cancer.gov/faststats/selections.php?series=cancer, viewed on November 29, 2013.*)

There are exceptions to the "one peak" rule. Some diseases have a bimodal distribution (i.e., two peaks). Distributions with more than two peaks are likely to occur, but the peaks in polymodal graphs run into one another and cannot, in general, be distinguished with certainty. Our ability to tease out polymodal data peaks may be improved somewhat as we become more adept at collecting information on large numbers of individuals, with verified, detailed quantitative feature data (i.e., age of occurrence of disease, gene mutations present in lesions, gene expression profiles).

11.1.2 Rule—Bimodality, when it occurs, is more often observed in the rare diseases than in the common diseases.

Brief Rationale—Because there are many occurrences of a common disease, second peaks (i.e., subpopulations with separate peak occurrence with age)

are likely to be masked by the large number of occurrences of the larger peak. Because the total number of individuals with a rare disease is small, a relatively small subpopulation with its own specific age of disease occurrence is likely to produce a visible second peak when the data are graphed.

For example, Hodgkin lymphoma, a rare tumor, has two peaks of disease occurrence (see Figure 11.2).

What does it mean when a rare disease breaks the "one peak" rule and demonstrates a bimodal age distribution? Here are a few possibilities:

1. Two different diseases, presumably with overlapping phenotypes, occur in two peak age groups, and are mistakenly assigned the same name.
2. A population is exposed to two environmental disease-causing agents, one working slower than the other.

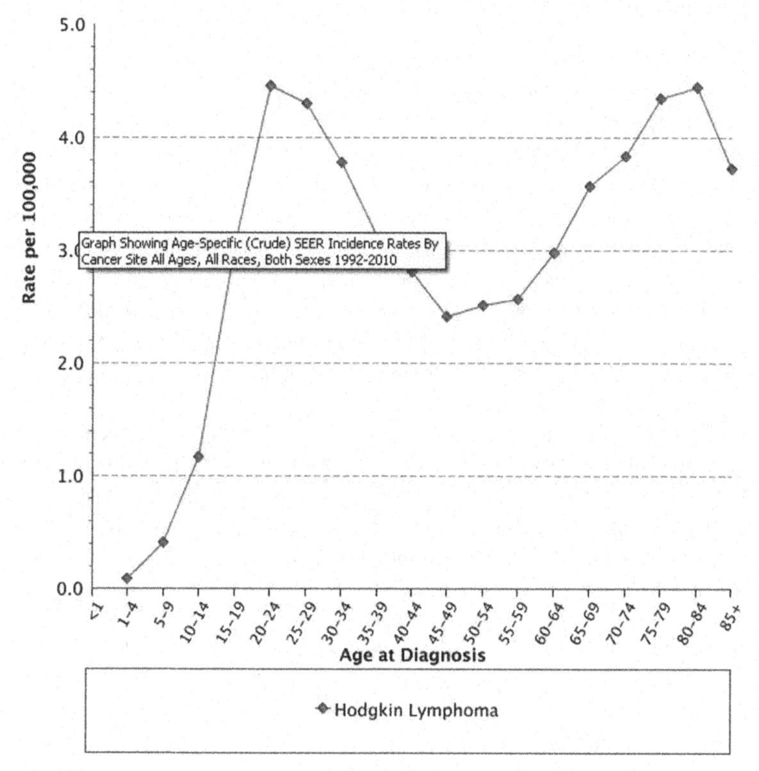

FIGURE 11.2 Graph showing incidence of Hodgkin lymphoma by age of occurrence of disease. There are two peaks in the graph. The first peak occurs in the early 20s. After the first peak, there is a trough, in the mid-40s, after which incidence increases steadily with age, toward a second peak. The graph was generated at the National Cancer Institute's Surveillance, Epidemiology and End Results "Fast Stats" query site. The query input settings were: SEER incidence; Age-specific rates, 1992–2010, All races, Both sexes, All ages, Hodgkin lymphoma. See color plate at the back of the book. (*Available from: http://seer.cancer.gov/faststats/selections.php?series=cancer, viewed on November 29, 2013.*)

3. A subpopulation is exposed to a different concentration of disease-causing agent, or at a different age, either resulting in disease occurring at a different average age for the subpopulation.
4. Two genetic causes for the same disease have different latencies (i.e., lengths of time for the disease to develop).
5. Two subpopulations have different disease modifiers (i.e., sets of genes that alter the pathogenesis of the disease).
6. Faulty or insufficient data. Bimodality may be a distortion due to poor data that do not adequately conform to the naturally occurring (unimodal) distribution.
7. False conclusions based on accurate data. The second peak may be caused by valid but "noisy" data. Scientists should not assume that statistical conclusions based on a single set of data are correct. All conclusions must be constantly re-examined in light of new findings.
8. Combinations of examples 1 through 7.

Occasionally, we can determine the biological mechanism that accounts for a bimodal age distribution. For example, Kaposi sarcoma, caused by human herpesvirus-8, has two peaks in occurrence. The first peak, in young people, occurs in individuals with AIDS-related Kaposi sarcoma. The second peak occurs in older men, was a recognized disease entity prior to the AIDS epidemic (i.e., prior to 1980s), and is often referred to as "classic" Kaposi sarcoma. Classic Kaposi sarcoma is slow growing, arises on the skin, often on the leg, and does not metastasize. It tends to occur in individuals of Mediterranean descent.

Once you begin to think about diseases in terms of multimodality, there is a short leap to thinking that the common, complex diseases are composite entities, composed of small sets of separate diseases that share a clinical phenotype.

11.1.3 Rule—A disease that can be separated into biological subsets, based on a quantifiable trait, such as age, can be interpreted as an aggregate of separate diseases, each with a smaller occurrence rate than the original disease.

Brief Rationale—By definition, a disease is a pathologic condition that is biologically separable from other pathologic conditions.

In a provocative journal article entitled "The many 'small COPDs', COPD should be an orphan disease," Stephen Rennard argued that many chronic diseases are actually heterogeneous groups of diseases that we are just now learning to distinguish from one another [2] (see Glossary item, COPD; see Figure 11.3). When we begin the process of separating diseases into related but distinguishable subsets of disease, we can begin to see why the common diseases may be aggregates of less common diseases. For example, mutation in the BRCA2 gene accounts for some cases of breast cancer, but the percentage is small. In fact, all of the known breast cancer risk genes, in aggregate, account for under 10% of the incidence of breast cancer. The remaining 90% would qualify today as sporadic tumors.

FIGURE 11.3 Centrilobular emphysema, a type of COPD (chronic obstructive pulmonary disease). The cut surface of the lung shows multiple small cavities, each surrounded by black carbon deposits. These are distal airways that have extensive cavitary destruction. See color plate at the back of the book. (*Source: U.S. Centers for Disease Control and Prevention and entered into the public domain as a U.S. government work.*)

Interestingly, the same BRCA2 gene that accounts for a subset of cases of breast cancer also accounts for a miniscule subset of a rare disease: Fanconi anemia. Most cases of Fanconi anemia are caused by mutations in genes coding for protein components of the Fanconi anemia protein complex, which, along with BRCA2, helps coordinate DNA repair [3]. A small percentage of Fanconi anemia cases are caused by homozygous mutations in the BRCA2 gene.

11.1.4 Rule—Single gene mutations may account for small subsets of common diseases, but they do not account for large subsets of common diseases.

Brief Rationale—All the single gene disease mutations are rare. If this were not so, we would expect to see Mendelian inheritance, typical for monogenic diseases, among the common diseases; but we do not.

Though a rare disease hidden within a common disease accounts for only a small proportion of the total number of disease cases, the genetic cause of the rare disease subset may be much easier to find than the genetic cause of the so-called sporadic cases [4]. When one mutated gene fully accounts for a subset of cases of a disease, its statistical association with the disease can be demonstrated with a relatively small number of cases [4].

11.1.5 Rule—Rare diseases that are subsets of common diseases often occur in a younger population than the cases occurring in the larger set of individuals with so-called sporadic disease.

Brief Rationale—Rare diseases are typically germline monogenic diseases that occur in young individuals.

As discussed in Section 8.3, a rare subset of lung cancers is caused by a rearrangement in the NUT gene. As in so many other rare diseases that have a germline monogenic cause, these cancers tend to occur in a much younger age group than cancers caused by an environmental factor (i.e., smoking in this case) [5]. The same observation holds for secretory breast carcinoma, formerly known as juvenile carcinoma of breast, which occurs in a younger age group than classic ductal breast carcinoma, and which is characterized by a specific fusion gene [6]. Similarly, myelodysplastic syndrome, a preleukemic condition for which the preponderance of cases occur in elderly individuals, is known to occur in children who inherit a predisposition to losing chromosome 7 in somatic blood-forming cells [7,8].

11.1.6 Rule—In a bimodal disease wherein the disease occurs in two age groups, young and old, the strongest likelihood of finding an effective treatment resides in the younger age group.

Brief Rationale—The younger age group is more likely to have a monogenic or oligogenic cause of the disease, and this often translates into a targeted cure (see Glossary item, Oligogenic inheritance). The older age group is likely to develop disease after the accumulation of multiple epigenetic, genetic, and environmental alterations, making it difficult to find an effective treatment.

As discussed in Section 8.1, every type of cancer that is curable at an advanced stage (i.e., having multiple and widespread metastases) is a cancer of childhood. All of the cancers that typically occur late in life are incurable when they progress to an advanced stage.

11.1.2 Rare diseases usually occur with a Mendelian pattern of inheritance. Common diseases may cluster in families, but they are, with very few exceptions, non-Mendelian

11.1.7 Rule—Polygenic diseases always have a non-Mendelian pattern of inheritance.

Brief Rationale—Each gene variant in a polygenic disease was inherited independently from one another. Hence, the set of genes that together constitute the polygenic cause of disease were not present as a complete set in either parent. Hence, inheritance cannot be assigned to either parent. Hence, the inheritance pattern is non-Mendelian.

In 1919, the statistician R.A. Fisher described how polygenic variants each inherited in a Mendelian fashion can account for a trait, such as height, that is inherited in a non-Mendelian pattern [9–11]. At least 180 gene variants have been associated with variations of normal height. These 180 variants may represent only a fraction of the total number of gene variants that influence the height of individuals, as they account for only about 10% of the spread [12]. Fisher's clever insight predated the discovery of the genetic code by about 40 years.

Less is known about the non-quantitative traits: features that you either have or you have not. There seems to be at least one documented instance in which a non-quantitative trait is monogenic. Some horses have a natural ability to use a specialized gait known as the tolt. This gait is common in a certain bread of

Icelandic horse and rare among all other breeds. Most horses, no matter how hard they try, cannot tolt. Recently, the tolt was found to come from a single gene mutation [13].

In Section 3.6, we described an imaginary scenario in which a rare disease became a common disease and then became a rare disease. To refresh our memories, the example begins with a new pathogen (e.g., virus, bacteria) that emerges as a major epidemic, killing nearly every human on earth (i.e., a rare disease becomes a common disease). A few survivors have protective genes, and from these survivors, new generations of pathogen-resistant humans repopulate the planet. Over time, individuals emerge who lack resistance to the pathogen, still present in the environment, and these individuals become ill and die (i.e., common disease is now rare disease).

We have arrived at a point where we can review this imaginary scenario and form reasonable inferences about the mode of inheritance of disease in the progression of the epidemic. When the human race was confronted with a spreading virus and nearly every human died, the disease was caused by an environmental pathogen. No genetic variants were involved in this stage because just about everyone died, regardless of their genetic background. When rare survivors started to appear globally, producing some offspring that were resistant to the virus, we can infer that resistance came from pre-existing variant genes within the survivor population. The survivors came from the initially exposed general population. There was no opportunity for new gene mutations to exert an effect. Generations later, when the population of humans was partially restored, rare individuals emerged who were, once again, susceptible to the virus. These individuals may have been born with a *de novo* monogenic mutation rendering them susceptible to the virus. Alternately, these individuals may have arisen through inadvertent back-breeding for susceptibility genes that had not been completely eradicated from the resistant population gene pool.

De novo germline mutations are new (i.e., Latin, *de novo,* anew) disease-causing mutations found in the germline of organisms (i.e., passed to every somatic cell of the organism) that were not present in the germline of either parent. A *de novo* mutation may result as a new mutation in a differentiated germ cell of either parent (i.e., it was not present in all of the cells of the parent, but appeared as a mutation in the specific parental germ cell that contributed to the offspring), or it may be a new mutation in the zygote (i.e., ovum fertilized by sperm) prior to splitting into embryonic cells (see Glossary item, *De novo* germline mutation).

11.1.8 Rule—Most dominantly inherited mutations that cause early death are caused by *de novo* mutations.

Brief Rationale—Non-lethal mutations occurring in germline cells have the opportunity of entering the population germ pool if the offspring have reproductive success [14]. Not so for lethal mutations. If a mutation causes death during development (*in utero*), or during early life, prior to sexual maturation,

then there is no way that the mutation arose through inheritance from the prior generation. The mutation must have arisen *de novo*, in the affected individual.

11.1.9 Rule—Genetic diseases caused by *de novo* mutations tend to produce severe clinical consequences.

Brief Rationale—Two reasons apply. First, *de novo* mutations have never gone through the process of natural selection by which lethal inherited mutations are eliminated from the general population. Second, *de novo* mutations that produce disease are likely to be dominant genes, because the likelihood of inheriting bi-allelic *de novo* mutations is exceedingly unlikely. Dominant genes that produce disease tend to involve structural proteins, because non-structural (e.g., enzymatic) proteins are likely to be compensated for by the normal allele. Mutated structural genes can play havoc among the building blocks of tissue. Hence, *de novo* disease mutations tend to produce serious pathology.

De novo germline mutations that cause disease will occur in individuals with no family history of the disease. A variety of genetic disorders occur both as Mendelian inherited disease and as *de novo* disease. For example, Von Hippel–Lindau disease is characterized by a predisposition to develop rare hemangioblastomas of the central nervous system, small cell renal carcinoma, and several other rare cancers. It is caused by a germline mutation of the VHL tumor suppressor gene. About 20% of cases of Von Hippel–Lindau disease lack family history of the disease and are presumed to be *de novo* occurrences.

When we start to list the genetic diseases in which *de novo* mutations account for a significant proportion of cases, they all seem to be serious conditions. Here are a few examples: Hutchinson–Gilford progeria syndrome, rapid-onset dystonia-Parkinsonism, cardiofaciocutaneous syndrome, PIK3CA-related segmental overgrowth, multiple endocrine neoplasia type 1, lamin A/C-related congenital muscular dystrophy, and Phelan–McDermid syndrome (see Glossary item, *De novo* germline mutation, for additional examples).

11.1.10 Rule—Life-threatening rare monogenic diseases never become common diseases.

Brief Rationale—If a lethal monogenic disease were to become common, then evolutionary pressure would tend to exclude the gene from the population over time. In the process, we would expect the disease to become less and less common, and more and more rare.

At present, all of the common diseases known to cause high morbidity or mortality are polygenic. There are, however, several common conditions that have low morbidity and mortality that are monogenic. Two of these exceptions are diseases with no pathological consequences under typical living conditions. Pathology ensues when triggered by an environmental factor.

Consider the peculiar story of lactose intolerance. Infants are blessed with an active lactase gene that digests lactose in their mother's milk. About the time that infants are weaned from milk, the lactase gene turns off. Until about 10,000 years ago, lactose intolerance was a constitutive condition of adult humans. At that time, a new gene mutation of the MCM6 gene entered the human gene pool. The mutated gene kept the gene coding for lactase in the "on" position permanently, even after weaning. The mutated MCM6 gene apparently conferred a survival advantage, perhaps related to a newfound ability to produce cheese and other dairy products that were consumed by adults. Today, the majority of adult humans lack this new gene variant, and have some degree of difficulty digesting lactose in foods [15].

Glucose-6-phosphate dehydrogenase deficiency is characterized by hemolysis (lysis of red blood cells) and is triggered by certain drugs and chemicals. Favism, a closely related disorder, is characterized by hemolysis triggered by eating fava beans. Together, these conditions occur in nearly half a billion individuals, most of whom never suffer any clinical symptoms. **As a general rule, common genetic conditions are relatively benign.**

11.1.3 Rare diseases often occur as syndromes, involving several organs or physiologic systems, often in surprising ways; most common diseases are non-syndromic (i.e., they involve one organ or system)

A germline mutation is passed to every cell of the body, and exerts its effects during fetal development and throughout life. Inherited diseases have every opportunity to alter multiple tissues and cellular systems along the way, producing multi-organ syndromes. Contrasted with monogenic diseases, the polygenic and multi-factorial diseases can be conceptualized as a collection of conditional requirements, with each condition placing one or more restrictions on cellular expression. For example, if a polygenic disease requires an environmental agent as a disease trigger, it would be unlikely that the agent would affect all cell types equally. Exposure to the agent may be limited to a subset of cell types in the body, possibly at a particular time of exposure (e.g., *in utero*). With multiple genetic and environmental conditions applied to the common, polygenic diseases, it is a miracle that anybody ever gets sick; but we do.

In addition to comparing rare syndromes with common non-syndromes, we can also compare rare syndromes against rare non-syndromic conditions. As an example, cleft lip can occur as part of a syndrome or as an isolated congenital abnormality. The van der Woude syndrome is caused by any one of several mutations in the IRF6 gene and is characterized by cleft lip or cleft palate or cleft lip plus cleft palate, a pitted lower lip, and hypodontia (i.e., reduced number of teeth). Isolated cleft lip and isolated cleft palate are polygenic disorders and much more common in the general population than is van der Woude syndrome.

As it happens, some cases of isolated cleft lip and isolated cleft palate, and even isolated hypodontia, are occasionally found to have IRF6 mutations [1].

This is another example wherein a polygenic disease contains an allele of the same gene that causes a monogenic disease with a related phenotype. In the polygenic forms of these abnormalities, the IRF6 gene is estimated to contribute about 12% of the genetic variation leading to cleft lip or cleft palate in populations that carry IRF6 polymorphisms [16–18].

The distinction between inherited and sporadic retinoblastoma is based on a multi-event hypothesis proposed by Knudson and coworkers in 1971 [19,20]. As discussed in Section 8.3, when retinoblastomas occur in early childhood, and in both eyes, the disease is presumed to be hereditary, for which all of the mutational events causing the development of retinoblastoma are passed through the germline. Target cells in both retinas have inherited the full complement of retinoblastoma-causing mutations; hence the occurrence of bilateral tumors early in life. Sporadic cases of retinoblastoma occur in adults, and in these cases, the acquisition of a set of mutations that lead to the development of retinoblastoma occur as mutational events in somatic cells. Sporadic tumors, when they occur, are very rare, requiring more than one mutational event in somatic cells, and occurring unilaterally in one eye.

Hirschsprung disease is a rare, congenital condition characterized by the absence of enteric ganglia, preventing normal peristalsis in the affected segments of bowel (usually the distal colon), and leading to bowel obstruction. Hirschsprung disease is polygenic, and like most polygenic diseases, it produces a non-syndromic condition isolated to one organ. Nonetheless, Hirschsprung disease is a notable exception to two general rules: (1) rare diseases are monogenic; and (2) polygenic diseases occur in adults. It is worth noting that although Hirschsprung disease qualifies as a rare disease, it is one of the more common of the rare diseases. In Japan, where its incidence is among the highest reported, it occurs once in every 5000 births [21].

Among the multiple genes involved in Hirschsprung disease, a RET gene mutation is commonly found [22]. Hirschsprung disease can occur as familial or non-familial (i.e., sporadic) forms. In addition, the Hirschsprung disease phenotype (i.e., absence of enteric ganglia) occurs as one component of several rare monogenic syndromes (i.e., Shah–Waardenburg syndrome, Mowat–Wilson syndrome, congenital central hypoventilation syndrome, and Goldberg–Shprintzen syndrome) each caused by a disease gene other than RET. It has been suggested that in these syndromes, the addition of a Hirschsprung disease component occurs as the result of a polymorphism in RET or some other modifier gene that influences the development of intestinal ganglia [23,24].

11.1.4 Environmental factors play a major role in the cause of common diseases; less so in the rare diseases

The history of humans and of animals is replete with examples of environmental elements wiping out huge numbers of animals and entire species. The bubonic plague is credited with killing about 200 million individuals throughout history.

Of course, many of the environmental causes of death can be attributed directly to humans exercising free will. According to the U.S. Centers for Disease Control and Prevention, smoking takes 5.4 million lives worldwide annually [25]. If we accept that this rate applies back 50 years (i.e., to 1963), then smoking has taken 270 million lives in the past half century. As discussed in Section 3.6, the intentional release of myxoma virus killed 3 billion Australian rabbits in under a decade, reducing the rabbit population by 95%.

Environmental factors can cause rare diseases as well. There are hundreds of rare infectious diseases caused by pathogenic organisms lurking in the environment or inside our bodies (see Chapter 7). Many environmental agents produce toxic conditions that mimic the rare diseases (see Section 9.5). Nonetheless, collected evidence on thousands of rare diseases indicates that the vast majority are caused by monogenic mutations passed in the germline.

11.1.11 Rule—Common diseases have contributing causes other than gene alterations.

Brief Rationale—There is discordance in the disease occurrences among monozygotic twins. As both twins have the same genes, differences in disease occurrences must have a non-genetic cause.

We have previously described discordance in disease occurrences among close relatives and monozygotic twins (see Section 3.5) [26,27]. The discordance among monozygotic twins is not observed in the rare congenital diseases and in rare diseases occurring in children. Discordances arise later, when the twins are older and the diseases are common, complex, or have environmental etiologies. As discussed in Section 10.2, monozygotic twins look alike at birth, often growing into early adulthood as a pair of strikingly similar individuals. But as the decades go by, identical twins begin to diverge in appearance, and they develop different diseases. Fraga and coworkers found that monozygotic twins are born with nearly identical epigenetic patterns of DNA methylation and histone acetylation. This near-identity persisted in the early years, but in later years, monozygotic twins had widely divergent patterns of DNA methylation and histone acetylation. These divergent patterns were accompanied by discordances in gene expression.

11.1.5 The difference in rates of occurrence of the rare diseases compared with the common diseases is profound, often on the order of a thousand-fold, and sometimes on the order of a million-fold

11.1.12 Rule—Genetic diseases with multiple genetic variants (e.g., retinitis pigmentosa) are among the most common of the rare genetic diseases.

Brief Rationale—As previously discussed, common diseases are common because many different causes and pathways lead to the same clinical phenotype

(e.g., many paths lead to a heart attack). Many rare diseases are rare because only one particular mutation can cause the rare disease. If we have a disease that is monogenic, for which many different alleles of a single gene may lead to the disease (i.e., allelic heterogeneity), or for which mutations in any of several different genes may cause the disease (i.e., locus heterogeneity), then we might expect to see disease incidences that are higher than most rare diseases but lower than any common disease.

The collagenopathies, the inherited blistering conditions, and the inherited causes of blindness and vision impairment all have numerous genetic causes, and all are among the more common of the rare diseases. There seems to be no exception to the rule. If a rare disease lacks both locus heterogeneity and allelic heterogeneity, it will be a very rare disease (see Glossary items, Allelic heterogeneity, Locus heterogeneity).

11.1.13 Rule—When a mutation is deleterious and widespread, it is likely that the mutation also has some useful purpose.

Brief Rationale—Otherwise the mutation would have been selected against, thus lowering the number of affected individuals.

The carrier states of several autosomal recessive diseases seem to provide some protection against pathogens in the environment. For example, carriers of one cystic fibrosis gene seem to have heightened resistance to cholera.

Carriers of thalassemia and sickle cell genes seem to have heightened protection against malaria (see Glossary item, Thalassemia). Individuals with glucose-6-phosphate dehydrogenase deficiency also have some protection from malaria. These three genetic diseases that provide some protection against malaria all produce hemolysis (e.g., lysis of red blood cells). Carriers who do not develop hemolytic disease may have more red cell fragility than is seen in non-carriers. Theory suggests that when a carrier is infected by a variant of malaria that lives within blood cells, the genetically altered red blood cells, and their contained parasites, are rapidly destroyed in the spleen.

11.1.14 Rule—When you have a disease with many different possible causes, it is probably not a rare disease.

Brief Rationale—Rare diseases usually result from a single genetic alteration (e.g., inherited condition), or from a single exposure to a specific agent (e.g., *in utero* exposure to thalidomide). As the number of possible causes of a disease rises, so does the incidence of the disease. Eventually, the disease ceases to be rare.

We have all heard one or another variation of a common lament: "Everything I eat causes heart disease," or "I can't do anything without increasing my risk for cancer," or, the favorite catch-all, "Everything I like to do is bad for me!" Atherosclerosis and cancer, the two greatest killers in the wealthiest societies, are both characterized by a multitude of causations. Eggs, sugars, trans-fats, red meats, and so on, are implicated in one way or another with heart disease. Asbestos, radon, cigarettes, pollution, sunlight, and many other environmental constituents are generally believed to cause cancers in humans. When there are a great number of causes for a common disease, it usually becomes impossible to attribute any particular instance of a disease to any particular cause. The term "sporadic" is applied to a case occurrence of a disease that has no selectable cause.

In the case of cancer, we may know a lot about carcinogens, oncogenes, and heritable cancer syndromes, but, with a few exceptions, such as lung cancer, we cannot attribute any particular tumor to any particular cause (see Glossary item, Cancer-causing syndrome). Hence, most cancers are "sporadic." Occasionally, the occurrence of a tumor that is thought to be sporadic may turn out to be non-sporadic when a specific cause is identified. Likewise, a tumor thought to be non-sporadic may turn out to be sporadic when it is shown that the presumed cause did not apply in that particular instance (see Section 9.1).

11.1.6 There are many more rare diseases than there are common diseases

As discussed in Chapter 3, there are about 7000 listed rare diseases, and these do not take into account all the diseases caused by rare infectious agents. The common diseases can be counted in the dozens, or the low hundreds, depending on the incidence or prevalence data you choose to trust. This enormous difference in the number of the rare diseases and the common diseases must tell us something about their biology. If not, then at the very least, it should tell us something about the way we choose to name and categorize diseases.

To some extent, the truth of this observation depends on how you choose to count your diseases. There are well over 100 different forms of retinitis pigmentosa occurring as an isolated disease of photoreceptor cells, or as part of a syndrome (e.g., Usher syndrome, characterized by retinitis pigmentosa plus deafness). Should these be counted as one with multiple subtypes, or as multiple diseases belonging to one general class? Likewise, there are hundreds of arrhythmia-causing conduction defects. Could we not lump these diseases together?

When we consider the common diseases, we need to ask ourselves whether we have been taking the easy way out by counting all the genotypic variants that happen to have a single broad phenotype under one name. By doing so, we ignore variants that might benefit from a specific treatment suited to a particular biological subtype of disease. When we say that there are about 4.5 million individuals in the U.S. who suffer from emphysema, are we just showing our ignorance by lumping hundreds or even thousands of biologically separable diseases under one name [2]?

Perhaps the reason that there are so many rare diseases is that we know enough about them to distinguish one rare disease from another in most instances. The common diseases, which we think we understand because we encounter them frequently, are actually enigmas that cannot be classified with any scientific rigor.

11.2 A TRIP TO MONTE CARLO: HOW NORMAL VARIANTS EXPRESS A DISEASE PHENOTYPE

"One of the marks of a good model—it is sometimes smarter than you are."

—Paul Krugman

Let us try to solve a puzzle. Imagine the following hypothetical scenario. You are a geneticist studying a new rare disease. You find that in half the cases, the disease is monogenic, caused by a mutation in any one of 10 different genes. Each gene that causes the disease has one specific mutation that is associated with the disease. The diseases caused by mutations in these 10 genes all have a Mendelian pattern of inheritance. The remaining half of the cases of the disease have a familial, but non-Mendelian pattern of inheritance. You have examined all 10 disease-associated genes in every case of the non-Mendelian cases, and all were negative for every disease-associated mutation. Furthermore, all of the examined genes in the non-Mendelian disease samples contained common polymorphisms found in the general population. You conclude that the diseases occurring in the non-Mendelian cases must have a completely different etiology than the diseases occurring in the Mendelian, monogenic cases. Later on, you start to think about your findings, and you wonder if you might have been too hasty in your conclusion. Is it likely that all of the genes known to cause the Mendelian form of the disease will have no role whatsoever in the occurrences of the non-Mendelian form? You cannot help thinking that you have missed an important clue that would help you find some genetic relationship that connects the non-Mendelian cases with the Mendelian cases.

Let's imagine that we're measuring the metabolic efficiency of a pathway in 1,000,000 individuals. For the sake of discussion, individuals whose pathway operates at less than 8% of normal are considered diseased. Let us say that the pathway consists of 10 enzymes, and each enzyme is known to be polymorphic in the general population (i.e., the gene coding the enzyme varies in sequence among different individuals in the general population). Each common form of the gene alters the efficiency of the encoded enzyme, with functional activities ranging from 70 to 130% of the efficiency of standard enzymatic activity. The standard pathway efficiency occurs when all 10 enzymes in the pathway have a relative activity of 1.0, giving the overall pathway standard activity of 1.0 to the tenth power, or 1.0. The lowest possible pathway efficiency occurs when every enzyme has 0.7 of normal activity, producing an overall pathway efficiency of 0.7 to the tenth power, or 0.028 (i.e., about 3% of normal). The highest efficiency would occur when every enzyme in the pathway has an activity level of 1.3, producing an overall pathway efficiency of 1.3 to the tenth power, or about 12.7 (i.e., nearly 13 times normal).

Let's use Monte Carlo techniques to randomly assign enzyme activities for each of the 10 pathway enzymes for 1,000,000 simulated individuals, and we will see how many simulations produce a pathway that works at an efficiency under 8% of normal (see Glossary item, Monte Carlo simulation).

Here is how the simulation experiment works:

1. A random number generator assigns a value of 0.7 through 1.3 to each of 10 imaginary enzymes that comprise the imaginary pathway of interest. A randomly assigned value of 0.7 simulates an enzyme with an efficiency that is 70% of the efficiency of the normal enzyme. A randomly assigned value of 1.3 simulates an enzyme with an efficiency of 130% of normal.
2. After 10 imaginary enzymes are assigned random values within a range of 0.7 to 1.3, the simulation multiplies all of the efficiencies together, producing a product that simulates the overall efficiency of the pathway.
3. We will be looking for pathways that, by random chance, have an efficiency of less than 8% of normal. This will simulate a highly inefficient pathway produced by rather small perturbations (i.e., 0.7 normal to 1.3 normal) in the efficiencies of normal enzymes in the pathway. When the simulation produces a set of enzymes with an overall pathway efficiency less than 8% of normal, it will print out the pathway data.
4. The program repeats 1,000,000 times, roughly providing the number of individuals in a population of 1,000,000 who will have a pathway deficiency under the conditions of the simulation.

Here is the output:

1. The overall path efficiency is 6.9%
   Enzyme values: 0.89 0.74 0.72 0.71 0.85 0.70 0.76 0.88 0.70 0.75
2. The overall path efficiency is 6.9%
   Enzyme values: 0.70 0.73 0.71 0.70 0.76 0.77 0.94 0.91 0.79 0.71
3. The overall path efficiency is 7.2%
   Enzyme values: 0.71 0.87 0.71 0.97 0.72 0.72 0.70 0.71 0.94 0.72
4. The overall path efficiency is 7.8%
   Enzyme values: 0.81 0.73 0.73 0.79 0.77 0.70 0.73 0.89 0.82 0.83
5. The overall path efficiency is 7.2%
   Enzyme values: 0.80 0.74 0.77 0.70 0.71 0.79 0.77 0.73 1.05 0.70
6. The overall path efficiency is 6.8%
   Enzyme values: 0.70 0.71 0.73 0.73 0.71 0.71 0.95 0.79 0.85 0.83
7. The overall path efficiency is 7.8%
   Enzyme values: 0.73 0.72 0.73 0.77 0.78 0.78 0.87 0.73 0.95 0.74
8. The overall path efficiency is 6.7%
   Enzyme values: 0.75 0.72 0.74 0.79 0.71 0.99 0.84 0.75 0.71 0.70

9. The overall path efficiency is 6.9%
   Enzyme values: 0.73 0.74 0.82 0.78 0.74 0.88 0.81 0.78 0.71 0.70
10. The overall path efficiency is 5.9%
   Enzyme values: 0.71 0.86 0.92 0.71 0.79 0.76 0.71 0.71 0.70 0.72
11. The overall path efficiency is 7.8%
   Enzyme values: 0.74 0.78 0.84 0.70 0.71 0.74 0.78 0.88 0.93 0.71
12. The overall path efficiency is 7.8%
   Enzyme values: 0.78 0.76 0.75 0.73 0.71 0.74 0.73 0.92 0.78 0.89
13. The overall path efficiency is 6.9%
   Enzyme values: 0.94 0.79 0.71 0.72 0.90 0.70 0.75 0.71 0.78 0.71

What does it mean?

1. In the 1 million simulations, there were 13 simulations in which the overall efficiency of the pathway fell below 8% of normal. This would correspond to a rare disease occurrence (i.e., 13 in a million). The remaining 999,987 simulations yielded pathway efficiencies that were 8% or higher.
2. The 13 "disease" pathways are polygenic.
3. **The simulated pathway deficiencies all resulted from small perturbations in the normal activity of enzymes involved in the pathway.** Hence, the simulation would seem to indicate that abnormal phenotypes can result from rare combinations of common polymorphisms pre-existing within a population.

11.3 ASSOCIATING GENES WITH COMMON DISEASES

"Everything has been said before, but since nobody listens we have to keep going back and beginning all over again."

—André Gide

A GWAS (genome wide association study) is a technique for finding common SNPs (single nucleotide polymorphisms) that are statistically associated with a polygenic disease (see Glossary items, Single nucleotide polymorphism, SNP, Synonymous SNP, Silent mutation). The methodology involves hybridizing DNA from individuals with disease, as well as individuals from a control group, against an array of immobilized fragments of DNA known to contain commonly occurring SNPs (i.e., allele-specific oligonucleotides). The SNPs that hybridize against the DNA extracted from individuals with disease (i.e., the SNPs matching the case samples) are compared with the SNPs that hybridize against the controls. SNPs that show a statistical difference between case samples and control samples are said to be associated with the disease.

Of course, there are many weaknesses to this approach; one being that differences in SNPs do not necessarily imply any functional variance in the gene product (see Glossary item, Anonymous variation) [28]. In addition, differences in SNPs may lead to statistically valid results that nonetheless have no relevance to the pathogenesis of disease [29]. Aside from false-positive GWAS associations, the methodology is virtually guaranteed to miss valid SNP associations, simply because SNP arrays are not exhaustive (i.e., do not contain every SNP) [30]. For example, a rare variant of the APOE gene has been shown to be strongly correlated with longevity [31]. This variant, because it is not included among the common APOE variants included in SNP arrays, would have been missed by a GWAS study.

When strong associations are found, our ability to draw biological inferences is limited by our poor understanding of disease-related phenotypic traits and by our limited knowledge of the range and frequency of gene variants in large populations of diseased and non-diseased populations [10,32]. Useful associations are those that can be found repeatedly from laboratory to laboratory, and that can be shown to have pathogenetic relevance. To date, few disease-associated SNPs found in GWAS studies have met these criteria.

We study SNPs partly because high-throughput sequencing techniques coupled with efficient computational analyses make it relatively easy to generate lots of SNP data. We should not forget that there are many biological mechanisms that result in varieties of gene variations other than SNPs. These would include alterations in karyotype or cytogenetic alterations observable with special techniques, as well as changes too small to see with a microscope, such as small deletions, insertions, and larger insertions, inversions, and translocations. Sources of variation would also include duplications and other copy-number alterations as well as variations in non-coding regions and in coding regions of pseudogenes (see Glossary items, Genomic structural variation, GSV, Copy-number, Chromosomal disorder, Karyotype, Genetic surplus disorder, Pseudogene) [33].

Early advocates of GWAS studies assumed that common diseases were caused by commonly found gene variants; otherwise, a disease would not be common [34]. With few exceptions, this seems not to be the case. Numerous GWAS studies indicate that common gene variants account for only a small portion of the heritability of common diseases [35,36]. One notable exception is found in age-related macular degeneration, a common form of blindness. Two gene variants account for more than 70% of the genetic susceptibility to age-related macular degeneration [37]. Overall, GWAS has received harsh criticism. It has been suggested that the GWAS studies, *in toto*, have had little scientific merit and have been misleading [38–41].

A sympathetic assessment of the GWAS methodology holds that such studies have helped us find recurrent sets of pathway genes involved in the pathogenesis of common diseases. There is great value in knowing that a related set of genes are involved in a pathway operative in the development or expression of a common disease [42,43]. For example, the aforementioned SNP associations found for age-related macular degeneration led to the finding that some of the associated genes are involved in the complement activation pathway [44]. This finding led other investigators to find additional genes of the same pathway that were also associated with age-related macular degeneration [45]. Complement pathway inhibitors are currently being tested as potential agents to prevent or treat this common, disabling disease [46].

Eleven out of 30 SNPs implicated in commonly occurring hyperlipidemias are among the causal genes of monogenic rare dyslipidemias (e.g., ABCA1, PCSK9, and LDLR22) [35]. Knowing this tells us that the common dyslipidemias may respond to treatments targeted against a metabolic pathway containing the constitutive genes involved in monogenic dyslipidemias.

Likewise, five common disorders—autism spectrum disorder, attention deficit-hyperactivity disorder, bipolar disorder, major depressive disorder, and schizophrenia—are all associated with the same set of SNPs that play a role in calcium channel signaling pathway [47]. A breakthrough in understanding the role of calcium signaling in the major neuro-psychiatric disorders may lead us to effective new treatments.

By focusing attention on a pathway, scientists can start to dissect the important events in the pathogenesis of a disease. If the pathway is known to be disrupted in a monogenic disease that replicates the phenotype of a common disease, then an effective new treatment, aimed at the pathway, may be feasible. Computational methods have been developed that examine the variant genes that associate with a disease to determine whether any groups of those genes operate together in a pathway [48].

Suffice it to say that the methodology of GWAS is still in its infancy. It might be best to think of GWAS as an early foray into a host of new methodologies that will find correlations among gene variants, epigenetic variations, and expressed genes, with disease phenotypes and with treatment responses (see Glossary item, Exome sequencing) [49].

11.4 MUTATION VERSUS VARIATION

"Nothing in biology makes sense except in the light of evolution."

—Theodosius Dobzhansky

It is appropriate to return to the concept of organism speciation, because the biologic process that accounts for the acquisition of traits among members of a species also accounts for the acquisition of common diseases.

Schoolchildren are taught that new species evolve when there is some environmental condition (e.g., change in climate, scarcity of traditional food source, emergence of a new predator) that causes some individuals to die and others to live. Under this selection pressure, new mutations occur, improving the survival of individuals. At this point, natural selection steps in to preserve the helpful mutation. Eventually, a sufficient number of new mutations yield a new species, optimally adapted for its environment. Not so!

Mutation does not play an active role in speciation. The emergence of new traits is accomplished through the selection of pre-existing genetic variations that are present in the population. Variations from individual to individual within a population are a ready source of new traits, and these variations are the basis for the development of new species [50,51].

If speciation involves selecting pre-existing gene variants, it is reasonable to ask, "Where did all the variation come from?" Variations came from stable mutations that occurred over the past several billion years in the human genome, and in all of the genomes in our ancestral lineage. We have seen that the genome and the epigenome are a complex system, and there are many opportunities for DNA to be acquired (e.g., via retroviruses), expanded, duplicated, repositioned, reverse-transcribed, alternatively spliced, and so on [52]. The epigenome provides a method whereby the expression of genes in various types of cells can be expressed, or silenced, or otherwise modified. In a self-organizing system that is so complex that it seems magical, the epigenome regulates the genome, and the genome codes for the proteins that assemble the epigenome.

So, yes, mutations have a role in speciation, but the mutational process occurred in the shrouded past. When you think about it, mutation is a slow and inefficient way to move evolution forward [52]. Most mutations are deleterious or of no benefit. Developing a desirable trait through mutation is much like winning the lottery. Selection of new traits from pre-existing variations within a species is fast and easy. Animal and plant breeders use trait selection to develop new breeds within a few generations.

Selection of pre-existing traits applies to back-breeding, a process whereby long-extinct breeds of animals can be selected from traits that persisted in newer breeds. For example, a project is under way to produce aurochs (ancient predecessors of modern cattle) by back-breeding cows. If the project proceeds as planned, DNA will be extracted from museum specimens of auroch teeth and bones. Comparisons of auroch DNA with various breeds of cow DNA will determine which breeds carry auroch genes. These breeds will be interbred to produce cows with greater and greater auroch genes. Eventually, repeated back-breeding should produce an animal that is more auroch than cow.

As an interesting aside, it should be noted that many of the most significant evolutionary advances came not from gene mutations but from gene acquisitions. The primordial mitochondrion that helped to create the first eukaryotic cell was an acquisition from bacteria. The very first chloroplast in the most primitive precursor of the plant kingdom was an acquisition from cyanobacteria (see Glossary item, Cyanobacteria). The big jump in adaptive immunology came with acquisition of the RAG1 gene. This gene enabled the DNA that encodes a segment of the immunoglobulin molecule to rearrange, thus producing a vast array of protein variants [53]. The RAG1 gene, which kicked off adaptive immunity in animals, was derived from a transposon, an ancient DNA element that was acquired through horizontal gene transfer or through infection from another living organism or from a virus, not by random base mutations (see Glossary items, Adaptive Immunity, Transposon, Transposable element).

Susceptibility to common diseases is acquired through the same biological mechanisms that operate in trait selection or speciation. Pre-existing variations in multiple genes within populations determine which individuals are likely to acquire a common disease.

11.4.1 Rule—With the exception of cancer, new mutation plays no direct role in the development of common diseases.

Brief Rationale—Mutations occur infrequently, whereas genetic variation occurs universally. A common disease that occurs in billions of individuals cannot be explained on the basis of mutation.

For the most part, a solitary SNP plays no biological role whatsoever (i.e., does not account for any particular disease and does not account for any particular biological trait). Nonetheless, we know that we can find individuals who have different traits (e.g., height, hair color, weight), and different likelihoods of contracting disease (e.g., obesity, osteoarthritis, diabetes), and that these traits and susceptibilities often run in families. When we compare the DNA sequences of people who have observable biological features against the DNA sequences of people lacking these features, we find that no single SNP accounts for much of the difference, but that collections of SNPs, often in the dozens, achieve statistical significance. We presume that among all these SNP associations will be some variations that somehow account for at least a portion of the familial inheritance of those features that occur in the population.

11.4.2 Rule—Phenotype is influenced by pre-existing gene variations in the human population much more often than it is influenced by new mutations.

Brief Rationale—The rate of new mutations is low; the number of pre-existing gene variants is high.

A SNP occurs about once in every 300 nucleotides [54]! Even so, genomic structural variations (e.g., duplications, expansions, microdeletions) may account for more phenotypic variations in the human population than SNPs (see Glossary item, Genomic structural variation) [55].

Compared with pre-existing genetic variations, the contribution by new mutation is trivial. New mutations in human cells are estimated to occur at the rate of several mutations for every 100 million base pairs per cell generation [56].

11.4.3 Rule—Natural selection does not strongly apply to common diseases, except for the infectious diseases.

Brief Rationale—The most common diseases tend to be ailments that have persisted in the human population for millions of years, affecting millions or billions of individuals in the interim. If natural selection was doing its job, these common diseases would have become rare.

In the case of the infectious diseases, pathogenic organisms are the evolutionary winners of a game played out over many millions of years. These organisms, despite all of our considerable host defenses, have learned how to live inside the human body and cause disease. These successful organisms know much more about us than we know about them.

For the non-infectious common diseases, there are actually three very good reasons why natural selection plays such a small role in curbing the incidence of the common diseases.

11.4.1 Natural selection does not care much about the longevity of organisms and does not act to preserve individuals after their reproductive activities have stopped

Section 4.2 contained a discussion of long-lived and immortal species. Does it occur to the reader that there must be something fundamentally flawed with the concept of immortality, at least as it is practiced here on earth? If a species is immortal, would you not expect its population to be always increasing, until it eventually covered every inch of the planet? Instead, what we see is that some of the most short-lived species have dominated the long-lived species. You cannot go out of doors and walk more than a few feet without encountering hundreds of ants and spiders and spring-tails. Three hundred-year-old turtles, 200-year-old rougheye rockfish, and immortal *Turritopsis nutricula* jellyfish are much harder to find. Short-lived insects have won the battle for planet Earth. It just goes to show that success, in the natural world, is not determined by lifespan.

The common diseases, for the most part, are diseases of adults, who have reached an age where they have passed their disease-causing genes to a new generation. There is nothing much that natural selection can do to rid a population of genes that cause disease in individuals who have successfully procreated. Rare diseases are different; they occur primarily in infants and children. A rare disease gene that decreases reproductive fitness or that reduces the likelihood that the individual will live into late adolescence will seldom pass to the next generation.

11.4.2 Natural selection has limited ability to resist polygenic diseases

The common diseases are polygenic, but natural selection really cannot effectively deal with a disease that results from the activities of many different gene variants. How does it choose which genes reduce the fitness of an organism when a single gene works in concert with many different genes in different cells and under different physiologic conditions, often with pleiotropic effects? A gene that cooperates with other genes to enhance an organism's fitness under one set of circumstances may work with another set of genes to reduce the same organism's fitness under other circumstances. When traits are polygenic, long-term survival benefits of individual genes cannot be determined.

Though it is difficult to select for genes resistant to polygenic diseases, there is one observation that suggests that, given enough time, natural selection might be able to make some useful adjustments.

11.4.4 Rule—Most common diseases occur more often in men than in women; not so for the rare diseases.

Brief Rationale—The same rule seems to apply to all mammals. Females have lower incidences of the common diseases and live considerably longer than males. The survival of mammalian species requires that children be protected and nurtured for a prolonged period. Women, particularly mothers and grandmothers, are caregivers to children; hence their prolonged survival is a desirable trait. Nature has no special imperative to keep men living beyond their reproductive years.

In the U.S., males have a life expectancy of 73.4 years, while women live to 80.1 years, on average. Because the most common causes of death are common diseases, we can logically infer that women are protected from the common diseases to a greater extent than are men.

11.4.3 Natural selection does not play favorites. It serves predators, infectious bacteria, and animal parasites with the same indifference that it extends to their victims

In the very long run, moving back several hundred million years, natural selection may be responsible for the evolution of our innate, intrinsic, and adaptive immune systems.

Many of the common diseases are caused by pathogens that have been around for a very long time; perhaps millions of years. Natural selection works for the parasite as well as for the host organism. As humans have evolved specialized immune defenses against pathogens, so too have pathogens evolved their own methods to thwart our defenses. For many of the common infectious diseases, the best that humans can hope for is a draw. Hence, humans have failed to "evolve out" of the common infectious diseases.

Adrian Hill has noted that there may be fundamental differences between the genetics of common infectious diseases and the genetics of the common non-infectious diseases. As we will discuss in Section 12.1, based on over 400 GWAS studies, 89% of the variant genes associated with non-infectious common diseases occur in non-coding regions of the genome [36,35] (see Glossary items, Genome wide association study, GWAS). Hill has noted that in the case of the common infectious diseases, polymorphisms associated with susceptibility to infectious diseases occur in exons (i.e., coding regions). This finding may indicate that our defenses against infection may have evolved in a manner quite different from our resistance to the common, non-infectious diseases [57].

REFERENCES

1. Craig J. Complex diseases: research and applications. Nat Educ 1:1, 2008.
2. Rennard SI, Vestbo J. The many "small COPDs", COPD should be an orphan disease. Chest 134:623–627, 2008.
3. D'Andrea AD. Susceptibility pathways in Fanconi's anemia and breast cancer. N Engl J Med 362:1909–1919, 2010.
4. Li B, Leal SM. Discovery of rare variants via sequencing: implications for the design of complex trait association studies. PLoS Genet 5:e1000481, 2009.
5. French CA, Kutok JL, Faquin WC, Toretsky JA, Antonescu CR, Griffin CA, et al. Midline carcinoma of children and young adults with NUT rearrangement. J Clin Oncol 22:4135–4139, 2004.
6. Tognon C, Knezevich SR, Huntsman D, Roskelley CD, Melnyk N, Mathers JA, et al. Expression of the ETV6-NTRK3 gene fusion as a primary event in human secretory breast carcinoma. Cancer Cell 2:367–376, 2002.
7. Lizcova L, Zemanova Z, Malinova E, Jarosova M, Mejstrikova E, Smisek P, et al. A novel recurrent chromosomal aberration involving chromosome 7 in childhood myelodysplastic syndrome. Cancer Genet Cytogenet 201:52–56, 2010.
8. Shannon KM, Turhan AG, Chang SS, Bowcock AM, Rogers PC, Carroll WL, et al. Familial bone marrow monosomy 7. Evidence that the predisposing locus is not on the long arm of chromosome 7. J Clin Invest 84:984–989, 1989.
9. Fisher RA. The correlation between relatives on the supposition of Mendelian inheritance. Trans R Soc Edinb 52:399–433, 1918.
10. Ward LD, Kellis M. Interpreting noncoding genetic variation in complex traits and human disease. Nat Biotechnol 30:1095–1106, 2012.
11. Visscher PM, McEvoy B, Yang J. From Galton to GWAS: quantitative genetics of human height. Genet Res 92:371–379, 2010.
12. Zhang G, Karns R, Sun G, Indugula SR, Cheng H, Havas-Augustin D, et al. Finding missing heritability in less significant loci and allelic heterogeneity: genetic variation in human height. PLoS One 7:e51211, 2012.
13. Goldberg R. Horses' ability to pace is written in DNA. The New York Times, September 11, 2012.
14. Eyre-Walker A, Keightley PD. The distribution of fitness effects of new mutations. Nature Rev Gen 8:610–618, 2007.
15. Pribila BA, Hertzler SR, Martin BR, Weaver CM, Savaiano DA. Improved lactose digestion and intolerance among African-American adolescent girls fed a dairy-rich diet. J Am Dietetic Assoc 100:524–528, 2000.
16. Zucchero TM, Cooper ME, Maher BS, Daack-Hirsch S, Nepomuceno B, Ribeiro L, et al. Interferon regulatory factor 6 (IRF6) gene variants and the risk of isolated cleft lip or palate. N Engl J Med 351:769–780, 2004.
17. Blanton SH, Cortez A, Stal S, Mulliken JB, Finnell RH, Hecht JT. Variation in IRF6 contributes to nonsyndromic cleft lip and palate. Am J Med Genet A 137A:259–262, 2005.
18. Kondo S, Schutte BC, Richardson RJ, Bjork BC, Knight AS, Watanabe Y, et al. Mutations in IRF6 cause Van der Woude and popliteal pterygium syndromes. Nat Genet 32:285–289, 2002.
19. Knudson AG. Mutation and cancer: statistical study of retinoblastoma. Proc Natl Acad Sci USA 68:820–823, 1971.
20. Knudson AG Jr, Hethcote HW, Brown BW. Mutation and childhood cancer: a probabilistic model for the incidence of retinoblastoma. Proc Natl Acad Sci USA 72:5116–5120, 1975.

21. Suita S, Taguchi T, Ieiri S, Nakatsuji T. Hirschsprung's disease in Japan: analysis of 3852 patients based on a nationwide survey in 30 years. J Pediatr Surg 40:197–201, 2005.
22. Salomon R, Amiel J, Atti T, Pelet A, Munnich A, Lyonnet S. From monogenic to polygenic: model of Hirschsprung disease. Pathol Biol (Paris) 46:705–707, 1998.
23. de Pontual L, Pelet A, Trochet D, Jaubert F, Espinosa-Parrilla Y, Munnich A, et al. Mutations of the RET gene in isolated and syndromic Hirschsprung's disease in human disclose major and modifier alleles at a single locus. J Med Genet 43:419–423, 2006.
24. de Pontual L, Pelet A, Clement-Ziza M, Trochet D, Antonarakis SE, Attie-Bitach T, et al. Epistatic interactions with a common hypomorphic RET allele in syndromic Hirschsprung disease. Hum Mutat 28:790–796, 2007.
25. Smoking and Tobacco Use. Centers for Disease Control and Prevention. Available at: http://www.cdc.gov/tobacco/global/, viewed October 5, 2013.
26. Chatterjee A, Morison IM. Monozygotic twins: genes are not the destiny? Bioinformation 7:369–370, 2011.
27. Wong AHC, Gottesman II, Petronis A. Phenotypic differences in genetically identical organisms: the epigenetic perspective. Human Mol Gen 14:R11–R18, 2005.
28. Ikegawa S. A short history of the genome-wide association study: where we were and where we are going. Genomics Inform 10:220–225, 2012.
29. Platt A, Vilhjalmsson BJ, Nordborg M. Conditions under which genome-wide association studies will be positively misleading. Genetics 186:1045–1052, 2010.
30. Wade N. Many rare mutations may underpin diseases. The New York Times, May 17, 2012.
31. Beekman M, Blanch H, Perola M, Hervonen A, Bezrukov V, Sikora E, et al. Genome-wide linkage analysis for human longevity: genetics of healthy aging study. Aging Cell 12: 184–193, 2013.
32. Nebert DW, Zhang G, Vesell ES. From human genetics and genomics to pharmacogenetics and pharmacogenomics: past lessons, future directions. Drug Metab Rev 40:187–224, 2008.
33. Kim HL, Iwase M, Igawa T, Nishioka T, Kaneko S, Katsura Y, et al. Genomic structure and evolution of multigene families: "flowers" on the human genome. Int J Evol Biol 2012:917678, 2012.
34. Agarwal S, Moorchung N. Modifier genes and oligogenic disease. J Nippon Med Sch 72: 326–334, 2005.
35. Manolio TA, Collins FS, Cox NJ, Goldstein DB, Hindorff LA, Hunter DJ, et al. Finding the missing heritability of complex diseases. Nature 461:747–753, 2009.
36. Hindorff LA, Sethupathy P, Junkins HA, Ramos EM, Mehta JP, Collins FS, et al. Potential etiologic and functional implications of genome-wide association loci for human diseases and traits. Proc Natl Acad Sci USA 106:9362–9367, 2009.
37. Lotery A, Trump D. Progress in defining the molecular biology of age related macular degeneration. Hum Genet 122:219–236, 2007.
38. Couzin-Frankel J. Major heart disease genes prove elusive. Science 328:1220–1221, 2010.
39. Hall WD, Mathews R, Morley KI. Being more realistic about the public health impact of genomic medicine. PLoS Med 7:e1000347, 2010.
40. Guessous I, Gwinn M, Khoury MJ. Genome-wide association studies in pharmacogenomics: untapped potential for translation. Genome Med 1:46, 2009.
41. Zuk O, Hechter E, Sunyaev SR, Lander ES. The mystery of missing heritability: genetic interactions create phantom heritability. Proc Natl Acad Sci USA 109:1193–1198, 2012.
42. Field MJ, Boat T. Rare Diseases and Orphan Products: Accelerating Research and Development. Institute of Medicine (US) Committee on Accelerating Rare Diseases Research and

Orphan Product Development. The National Academies Press, Washington, DC, 2010. Available from: http://www.ncbi.nlm.nih.gov/books/NBK56189/.
43. Jakobsdottir J, Gorin MB, Conley YP, Ferrell RE, Weeks DE. Interpretation of genetic association studies: markers with replicated highly significant odds ratios may be poor classifiers. PLoS Genet 5:e1000337, 2009.
44. Edwards AO, Ritter R 3rd, Abel KJ, Manning A, Panhuysen C, Farrer LA. Complement factor H polymorphism and age-related macular degeneration. Science 308:421–424, 2005.
45. Gold B, Merriam JE, Zernant J, Hancox LS, Taiber AJ, Gehrs K, et al. Variation in factor B (BF) and complement component 2 (C2) genes is associated with age related macular degeneration. Nat Genet 38:458–462, 2006.
46. Khandhadia S, Cipriani V, Yates JR, Lotery AJ. Age-related macular degeneration and the complement system. Immunobiology 217:127–146, 2012.
47. Cross-Disorder Group of the Psychiatric Genomics Consortium, Smoller JW, Craddock N, Kendler K, Lee PH, Neale BM, et al. Identification of risk loci with shared effects on five major psychiatric disorders: a genome-wide analysis. Lancet 381:1371–1379, 2013.
48. Kim S, Xing EP. Statistical estimation of correlated genome associations to a quantitative trait network. PLoS Genet 5:e1000587, 2009.
49. Ng SB, Buckingham KJ, Lee C, Bigham AW, Tabor HK, Dent KM, et al. Exome sequencing identifies the cause of a Mendelian disorder. Nat Genet 42:30–35, 2010.
50. Mayr E. The Growth of Biological Thought: Diversity, Evolution and Inheritance. Belknap Press, Cambridge, 1982.
51. DeQueiroz K. Ernst Mayr and the modern concept of species. PNAS 102(Suppl 1): 6600–6607, 2005.
52. Lower R, Lower J, Kurth R. The viruses in all of us: characteristics and biological significance of human endogenous retrovirus sequences. Proc Natl Acad Sci USA 93:5177–5184, 1996.
53. Kapitonov VV, Jurka J. RAG1 core and V(D)J recombination signal sequences were derived from Transib transposons. PLoS Biol 3:e181, 2005.
54. Genetics Home Reference. National Library of Medicine. July 1, 2013. Available from: http://ghr.nlm.nih.gov/handbook/genomicresearch/snp, viewed July 6, 2013.
55. Korbel JO, Urban AE, Affourtit JP, Godwin B, Grubert F, Simons JF, et al. Paired-end mapping reveals extensive structural variation in the human genome. Science 318:420–426, 2007.
56. Nachman MW, Crowell SL. Estimate of the mutation rate per nucleotide in humans. Genetics 156:297–304, 2000.
57. Hill AVS. Evolution, revolution and heresy in the genetics of infectious disease susceptibility. Philos Trans R Soc Lond B Biol Sci 367:840–849, 2012.

Chapter 12

Rare Diseases and Common Diseases: Understanding their Relationships

12.1 SHARED GENES

"Because all of biology is connected, one can often make a breakthrough with an organism that exaggerates a particular phenomenon, and later explore the generality."

—Thomas R. Cech

The theme that will be developed in this section, and the next section of this chapter, is that many common diseases have a rare disease hidden within them.

12.1.1 Rule—The same genes that cause monogenic rare diseases are found in the sporadically occurring diseases for which there is phenotypic overlap.

Brief Rationale—Because there are many instances of a common disease, it would seem likely that a gene that is known to cause a particular clinical phenotype, in the case of a monogenic disease, is likely to contribute to at least some cases of a polygenic disease that has a similar phenotype.

We do not need to look very hard to find examples that demonstrate the rule:

- Germline mutations of the p53 tumor suppressor gene are present in the rare Li–Fraumeni syndrome. A somatic p53 mutation is found in about half of all human cancers [1].
- Families with germline mutations of the KIT gene develop GISTs (gastrointestinal stromal tumors). Somatic mutations of KIT occur in the majority of sporadic GISTs.
- Germline RET gene mutations occur in familial medullary carcinoma of thyroid and in sporadic cases of medullary carcinoma of thyroid [2,3].
- Germline RB1 gene mutations occur in familial retinoblastoma syndrome and in sporadic cases of retinoblastoma [4].
- Germline patched (ptc) gene mutations occur in basal cell nevus syndrome and in sporadically occurring basal cell carcinomas [5].
- Germline PTEN mutations occur in Cowden syndrome and in Bannayan–

Rare Diseases and Orphan Drugs. http://dx.doi.org/10.1016/B978-0-12-419988-0.00012-2

Riley–Ruvalcaba syndrome and both the syndromes feature occurrences of familial endometrial carcinoma. PTEN mutations are found in sporadic cases of endometrial carcinoma.

- Germline IRF6 mutations cause van der Woude syndrome, characterized by cleft lip, cleft palate, and hypodontia [6]. Sporadic cleft lip, cleft palate, and hypodontia may have variant single nucleotide polymorphisms (SNPs) in the IRF6 gene, in addition to multiple additional SNP variants associated with sporadic malformations of the types found in van der Woude syndrome [7].
- Inherited monogenic dyslipidemias can be caused by ABCA1, PCSK9, or LDLR22. These same genes are among the variant genes associated with an increased risk of occurrence of common dysplipidemias in adults [8].

It is easy to find cases wherein a rare disease accounts for a somewhat uncommon clinical presentation of a common disease.

12.1.2 Rule—Uncommon presentations of common diseases are sometimes rare diseases, camouflaged by a common clinical phenotype.

Brief Rationale—Common diseases tend to occur with a characteristic clinical phenotype and a characteristic history (e.g., risk factors, underlying causes). Deviations from the normal phenotype and history are occasionally significant. Rare diseases may produce a disease that approximates the common disease; the differences being subtle findings revealed to the most astute observers.

Here is some pithy wisdom that senior physicians love to impart to junior colleagues: "When you see hoof prints, look for horses, not zebras." The message warns young doctors that most clinical findings can be accounted for by common diseases. Nonetheless, physicians must understand that zebras, unlike unicorns and griffins, actually exist. Occasionally, a rare disease will present with the clinical phenotype of a common disease.

For example, mutations of the JAK2 gene are involved in several myeloproliferative conditions, including myelofibrosis, polycythemia vera (see Glossary item, Polycythemia), and at least one form of hereditary thrombocythemia (i.e., increased blood platelets) [9–11]. Surprisingly, somatic blood cells with JAK2 mutations are found in 10% of apparently healthy individuals [12]. The high incidence of JAK2 mutations in the general population, and the known propensity for JAK2 mutations to cause thrombocythemia and thrombosis, should alert physicians to the possibility that some cases of idiopathic thrombosis may be caused by a platelet disorder caused by undiagnosed JAK2 mutation of blood cells. As it happens, it has been shown that a JAK2 mutation can be found in 41% of patients who present with idiopathic chronic portal, splenic and mesenteric venous thrombosis [13]. Such thrombotic events are uncommon in otherwise healthy patients. The search for a zebra, in this case a cryptic myeloproliferative disorder caused by a JAK2 mutation, pays off (see Glossary item, Myeloproliferative disorder).

Zebras can hide among the horses. Consider lung cancer, the number one cause of cancer deaths in the U.S. When lung cancer occurs in a young person, you might wonder if this is a rare disease cloaked as a common disease. Midline carcinoma of children and young adults is an extremely rare type of lung cancer. It is characterized by a NUT gene mutation, not typically found in commonly occurring lung cancers of adults [14]. Hence, midline carcinoma of children and young adults is an example of a rare disease hidden in a common disease.

Secretory carcinoma, formerly known as juvenile breast cancer, is a rare form of breast cancer. It has a less aggressive clinical course than commonly occurring breast cancer, and occurs at a younger median age (i.e., about 25 years) than the median age of occurrence of common breast cancer (i.e., 61 years). In 2002, it was discovered that the expression of the ETV6-NTRK3 gene fusion is a primary event in the carcinogenesis of secretory breast carcinoma [15]. Once again, an uncommon presentation of a common tumor was found to hide a rare disease with its own characteristic genetic mutation.

Myelodysplastic syndrome, formerly known as preleukemia, is a rare blood disorder occurring almost exclusively in older individuals. The specific gene causing myelodysplastic syndrome is unknown, but recurrent cytogenetic alterations have been found in bone marrow cells, particularly losses of the long arm of chromosome 5 (i.e., 5q-) and of chromosome 7 (i.e., monosomy 7). Myelodysplastic syndrome occurs in very young children, with extreme rarity. Virtually all such childhood cases involve monosomy 7. An inherited predisposition to lose one copy of chromosome 7 in somatic cells has been reported in kindreds whose children have a high likelihood of developing myelodysplastic syndrome, or of acute leukemia. Hence, it seems that a somatic chromosomal abnormality associated with a rare disease occurring in adults is also associated with an even more rare childhood form of the disease. The childhood disease may occur when an inherited mutation predisposes children to the equivalent somatic chromosomal abnormality observed in the adult form of the disease [16,17].

As a final example, there are two recognized types of acute myelogenous leukemia (AML): AML following myelodysplasia, a preleukemia, and *de novo* AML, which develops in the absence of an observed preleukemic condition [18]. *De novo* AML can occur in children or in adults. The *de novo* AML cases in children have a different set of cytogenetic markers than those observed in adult *de novo* AML [19].

For diseases that typically occur in adults, exceptional occurrences of disease in young persons have consistently led to the discovery of a separable disease entity characterized by a distinctive genetic marker, and a distinctive biological phenotype. In all such cases, it is worth considering that biologically separable diseases, with a broadly similar phenotype, may benefit from treatments tailored to subtype.

On occasion, we find that disease entities described in old medical textbooks are not true diseases at all. They may be concoctions of several similar diseases that could not be distinguished from one another prior to the advent of advanced diagnostic technologies. A good example is malignant fibrous histiocytoma. Current thinking is that this diagnostic entity has been used as a grab-bag diagnosis for sarcomas that do not fit well into any particular category [20,21]. There is now substantial evidence, based on genetic and electron-microscopic studies, that many cases of malignant fibrous histiocytoma would have been better diagnosed as leiomyosarcomas or liposarcomas or fibrosarcomas and a host of rare sarcomas, each with its own characteristic age distribution [20–22]. By batching different tumors under a single name, it becomes impossible to conduct meaningful clinical trials; trialists cannot determine which tumor is responding to treatment.

12.1.3 Rule—Rare gene variants account for the bulk of the genetic component of common diseases.

Brief Rationale—Current genome wide association studies (GWAS) indicate that commonly occurring gene variants do not account for the bulk of the genetic component of diseases. If common genes account for a small fraction of the genetic component of disease, it seems reasonable to suspect that rare gene variants play a large role in the common diseases [23,24].

By observing the occurrences of common diseases in close relatives, fraternal twins, and monozygotic twins, compared with the occurrences of these diseases in the general population, geneticists derive an estimate of the genetic contribution of disease [8]. A great many GWAS studies have been done to find sets of common gene variants that associate with the common diseases. Based on the strength of association between common gene variants and common diseases, an estimate is obtained for the fraction of the genetic component of common diseases that are credited to common gene variants. **These studies have led to the current estimate that common gene variants account for only about 10–15% of the genetic component of common diseases [8]**.

Finding a set of common genes associated with a disease is not necessarily enlightening. For example, multiple polymorphic variants of the IL2RA gene have been found to associate with type 1 diabetes; but none of those variants have been shown to have a functional effect (see Glossary item, Anonymous variation) [25]. Furthermore, based on over 400 GWAS studies, 89% of the variant genes associated with non-infectious common diseases occur in non-coding regions of the genome [26,8]. We know almost nothing about the biological functionality of the non-coding regions of the genome [27]. When it comes down to it, we know very little about the functional consequences of any single gene polymorphism, and even less about the epistatic interactions among multiple polymorphisms.

In summary, most of what we know about the genetic basis of the common diseases seems to come from studies of the rare diseases. **Once more, we are reminded that rare diseases are not the exceptions to which the general rules need not apply; rare diseases are the exceptions upon which the general rules are based**.

12.2 SHARED PHENOTYPES

"Mille viae ducunt homines per saecula Romam" (A thousand roads lead men forever to Rome)

—Alain de Lille in *Liber Parabolarum*, circa 1175

It is almost impossible to study a rare disease without uncovering some fundamental cellular mechanism underlying a common disease [28]. The reason is simple: there are a finite number of mechanisms whereby cells can malfunction, and most of these mechanisms are encountered in pure form in one or another rare disease. Furthermore, the best way to understand a complex disease often involves understanding the rare diseases that reproduce the common disease phenotype.

12.2.1 Rule—We know more about the pathogenesis of rare diseases than we know about the pathogenesis of common diseases.

Brief Rationale—Each common disease has many causes and many pathways that contribute to the fully developed clinical phenotype. Because many cellular events are happening at once, there really is no way to design a controlled experiment that can determine the consequences of altering a single component of the system. Hence, the common diseases are all somewhat inscrutable.

For example, consider the pathologic complexity of cancer. Every measured pathway, organelle, and biochemical process is altered in cancer cells. The history of cancer research is littered by theories of carcinogenesis based on observations of malfunctioning cellular components. Here is a small sampling of paraphrased hypotheses:

"Cancer cells have unchecked proliferation, accounting for the malignant phenotype."

"Cancer cells preferentially employ anaerobic metabolism, which accounts for the malignant phenotype."

"Cancer cells have dysfunctional mitochondria, accounting for the malignant phenotype."

"Cancer cells have lost programmed senescence; hence, the non-dying cells account for the malignant phenotype."

"Cancer cells have lost cell membrane processes that control transmembrane homeostasis, giving rise to a malignant phenotype."

"The epigenome is ultimately responsible for the normal control of the genome; when the epigenome is sufficiently altered, cells cannot behave normally, and cancer results."

"Cancer cells are genetically unstable, resulting in the selection of cells with a malignant phenotype."

These theories and many others have helped fund generations of cancer researchers. All of these theories were based on valid observations. The problem has been that when everything is changed from normal in a cell, as it is in cancer, it becomes impossible to select those changes that are the underlying causes of disease [29].

What is true for cancer is true for every complex disease. We cannot determine the effects of one variable on another variable when all the variables are changing all of the time. Under such circumstances, the most we can do is to describe the phenotype of the diseases during its development, and make a reasonable guess as to what seems to be the most important events that arise as the disease progresses. The monogenic rare diseases are much easier to study; one gene changes, and one disease phenotype emerges. A monogenic disease is something that scientists can understand.

12.2.2 Rule—Common diseases are aggregates of the individual pathogenic pathways that account for the rare diseases.

Brief Rationale—Because every pathway is a product of gene expression, and because virtually every gene of functional importance is a candidate for a rare disease, it is reasonable to assume that each of the many pathways that participate in the phenotypic expression of a common disease will be expressed in one or more of the 7000+ rare diseases.

The set of rare diseases covers all the bases, so that every pathological expression of every pathway is presumably represented by a rare disease. If this is the case, you might expect similarities between the clinical phenotypes of common diseases and of rare diseases.

12.2.3 Rule—Any polygenic disease can be replicated by a monogenic disease.

Brief Rationale—The phenotype associated with a polygenic disease converges toward a physiologically permissible outcome. Because there is a monogenic disease affecting virtually every pathway available to cells, it is likely that each common disease will be replicated by at least one monogenic disease that converges to the same clinical phenotype.

We have observed that there are few common diseases, and that there are many different causes for the common diseases. If many different causes lead to a limited number of common phenotypes, can we not infer that many pathways lead to the common diseases, including the pathways found in rare diseases [30,31]?

In point of fact, there are monogenic forms of most, if not all, of the common diseases. We have encountered many of them in prior chapters:

- MODY (maturity onset diabetes of the young), also known as monogenic diabetes, refers to any of several hereditary forms of the disease. Despite its name, MODY has a childhood onset, like most other rare diseases. The "maturity onset" in its name refers to its common disease counterpart.
- Fragile X syndrome (FXS), also known as Martin–Bell syndrome, is a monogenic cause of autism.

- McKusick–Kaufman syndrome and Bardet–Biedl syndrome-6 are both diseases that include a monogenic form that causes obesity.
- Monogenic emphysema due to alpha-1-antitrypsin deficiency [32].
- Monogenic gallstone disease due to a mutation in the ABCB4 gene.
- Monogenic cardiomyopathy due to a mutation in the ABCC9 gene.
- Monogenic cardiac arrhythmia due to monogenic mutations in ion channel genes (see Section 5.3).
- Monogenic cause of migraine in familial hemiplegic migraine type 2 and familial basilar migraine due to mutations in the gene encoding the alpha-2 subunit of the sodium/potassium pump.
- Monogenic osteoarthritis, as a component of familial osteochondritis dissecans, due to mutation in the ACAN gene.
- Familial Alzheimer disease type 1 due to a mutation in the gene encoding the amyloid precursor protein.
- Monogenic, Mendelian forms of hypertension associated with proteins involved, in one way or another, with the transport of electrolytes in the renal tubules (see Section 5.4 for detailed discussion). Changes in electrolyte transport result in increased retention of sodium and to an increased volume of body fluid [33–35].
- Autoinflammatory syndromes with monogenic subtypes, including familial Mediterranean fever caused by a mutation in the MEFV gene encoding pyrin [36].

In at least one polygenic disease, Williams–Beuren syndrome, a gene associated with the disease has been assigned a specific trait, essentially establishing a monogenic disease within a polygenic disease. Williams–Beuren syndrome is a microdeletion disorder caused by a deletion of about 26 genes on the long arm of chromosome 7. It is characterized by a striking facial morphism described as "elfin," developmental delays, transient hypercalcemia, and cardiovascular abnormalities. One gene of the 26 deleted genes seems to account for all of the cardiovascular abnormalities [37]. Other features of the syndrome seem to arise collectively from the other deleted genes.

If common diseases are puzzles, then rare diseases are the pieces of the puzzle.

REFERENCES

1. Royds JA, Iacopetta B. P53 and disease: when the guardian angel fails. Cell Death Differ 13:1017–1026, 2006.
2. Marshall E. Genetic testing. Families sue hospital, scientist for control of Canavan gene. Science 290:1062, 2000.
3. Vezzosi D, Bennet A, Caron P. Recent advances in treatment of medullary thyroid carcinoma. Ann Endocrinol (Paris) 68:147–153, 2007.
4. Blanquet V, Turleau C, Gross-Morand MS, Senamaud-Beaufort C, Doz F, et al. Spectrum of germline mutations in the RB1 gene: a study of 232 patients with hereditary and nonhereditary retinoblastoma. Hum Molec Genet 4:383–388, 1995.

5. Johnson RL, Rothman AL, Xie J, Goodrich LV, Bare JW, Bonifas JM, et al. Human homolog of patched, a candidate gene for the basal cell nevus syndrome. Science 272:1668–1671, 1996.
6. Kondo S, Schutte BC, Richardson RJ, Bjork BC, Knight AS, Watanabe Y, et al. Mutations in IRF6 cause Van der Woude and popliteal pterygium syndromes. Nat Genet 32:285–289, 2002.
7. Blanton SH, Cortez A, Stal S, Mulliken JB, Finnell RH, Hecht JT. Variation in IRF6 contributes to nonsyndromic cleft lip and palate. Am J Med Genet A 137A:259–262, 2005.
8. Manolio TA, Collins FS, Cox NJ, Goldstein DB, Hindorff LA, Hunter DJ, et al. Finding the missing heritability of complex diseases. Nature 461:747–753, 2009.
9. Mead AJ, Rugless MJ, Jacobsen SEW, Schuh A. Germline JAK2 mutation in a family with hereditary thrombocytosis. New Engl J Med 366:967–969, 2012.
10. Barosi G, Bergamaschi G, Marchetti M, Vannucchi AM, Guglielmelli P, Antonioli E, et al. JAK2 V617F mutational status predicts progression to large splenomegaly and leukemic transformation in primary myelofibrosis. Blood 110:4030–4036, 2007.
11. Zhang L, Lin X. Some considerations of classification for high dimension low-sample size data. Stat Methods Med Res 2011 November 23. Available from: http://smm.sagepub.com/content/early/2011/11/22/0962280211428387.long, viewed January 26, 2013.
12. Sidon P, El Housni H, Dessars B, Heimann P. The JAK2V617F mutation is detectable at very low level in peripheral blood of healthy donors. Leukemia 20:1622, 2006.
13. Orr DW, Patel RK, Lea NC, Westbrook RH, O'Grady JG, Heaton ND, et al. The prevalence of the activating JAK2 tyrosine kinase mutation in chronic porto-splenomesenteric venous thrombosis. Aliment Pharmacol Ther 31:1330–1336, 2010.
14. French CA, Kutok JL, Faquin WC, Toretsky JA, Antonescu CR, Griffin CA, et al. Midline carcinoma of children and young adults with NUT rearrangement. J Clin Oncol 22:4135–4139, 2004.
15. Tognon C, Knezevich SR, Huntsman D, Roskelley CD, Melnyk N, Mathers JA, et al. Expression of the ETV6-NTRK3 gene fusion as a primary event in human secretory breast carcinoma. Cancer Cell 2:367–376, 2002.
16. Lizcova L, Zemanova Z, Malinova E, Jarosova M, Mejstrikova E, Smisek P, et al. A novel recurrent chromosomal aberration involving chromosome 7 in childhood myelodysplastic syndrome. Cancer Genet Cytogenet 201:52–56, 2010.
17. Shannon KM, Turhan AG, Chang SS, Bowcock AM, Rogers PC, Carroll WL, et al. Familial bone marrow monosomy 7. Evidence that the predisposing locus is not on the long arm of chromosome 7. J Clin Invest 84:984–989, 1989.
18. Head DR. Revised classification of acute myeloid leukemia. Leukemia 10:1826–1831, 1996.
19. Harrison CJ, Hills RK, Moorman AV, Grimwade DJ, Hann I, Webb DK, et al. Cytogenetics of childhood acute myeloid leukemia: United Kingdom medical research council treatment trials AML 10 and 12. J Clin Oncol 28:2674–2681, 2010.
20. Al-Agha OM, Igbokwe AA. Malignant fibrous histiocytoma: between the past and the present. Arch Pathol Lab Med 132:1030–1035, 2008.
21. Nakayama R, Nemoto T, Takahashi H, Ohta T, Kawai A, Seki K, et al. Gene expression analysis of soft tissue sarcomas: characterization and reclassification of malignant fibrous histiocytoma. Mod Pathol 20:749–759, 2007.
22. Baird K, Davis S, Antonescu CR, Harper UL, Walker RL, Chen Y, et al. Gene expression profiling of human sarcomas: insights into sarcoma biology. Cancer Res 65:9226–9235, 2005.
23. Pritchard JK. Are rare variants responsible for susceptibility to complex diseases? Am J Hum Genet 69:124–137, 2001.
24. McClellan J, King M. Genomic analysis of mental illness: a changing landscape. JAMA 303:2523–2524, 2010.

25. Belot MP, Fradin D, Mai N, Le Fur S, Zelenika D, Kerr-Conte J, et al. CpG methylation changes within the IL2RA promoter in type 1 diabetes of childhood onset. PLoS ONE 8:e68093, 2013.
26. Hindorff LA, Sethupathy P, Junkins HA, Ramos EM, Mehta JP, Collins FS, et al. Potential etiologic and functional implications of genome-wide association loci for human diseases and traits. Proc Natl Acad Sci USA 106:9362–9367, 2009.
27. Graur D, Zheng Y, Price N, Azevedo RB, Zufall RA, Elhaik E. On the immortality of television sets: "function" in the human genome according to the evolution-free gospel of ENCODE. Genome Biol Evol 5:578–590, 2013.
28. Jiang X, Liu B, Jiang J, Zhao H, Fan M, Zhang J, et al. Modularity in the genetic disease-phenotype network. FEBS Lett 582:2549–2554, 2008.
29. Berman JJ. Neoplasms: Principles of Development and Diversity. Jones & Bartlett, Sudbury, 2009.
30. Rennard SI, Vestbo J. The many "small COPDs", COPD should be an orphan disease. Chest 134:623–627, 2008.
31. Crow YJ. Lupus: how much "complexity" is really (just) genetic heterogeneity? Arthritis Rheum 63:3661–3664, 2011.
32. Stoller JK, Aboussouan LS. Alpha1-antitrypsin deficiency. Lancet 365:2225–2236, 2005.
33. Lifton RP. Molecular genetics of human blood pressure variation. Science 272:676–680, 1996.
34. Wilson FH, Kahle KT, Sabath E, Lalioti MD, Rapson AK, Hoover RS, et al. Molecular pathogenesis of inherited hypertension with hyperkalemia: the Na-Cl cotransporter is inhibited by wild-type but not mutant WNK4. Proc Natl Acad Sci USA 100:680–684, 2003.
35. Bahr V, Oelkers W, Diederich S. Monogenic hypertension. J Med Klin (Munich) 98:208–217, 2003.
36. Glaser RL, Goldbach-Mansky R. The spectrum of monogenic autoinflammatory syndromes: understanding disease mechanisms and use of targeted therapies. Curr Allergy Asthma Rep 8:288–298, 2008.
37. Pober BR. Williams-Beuren syndrome. New Engl J Med 362:239–252, 2010.

Chapter 13

Shared Benefits

13.1 SHARED PREVENTION

"Intellectuals solve problems; geniuses prevent them."

—Albert Einstein

When a toxic agent produces a small rise in the incidence of a common disease, the increase cannot be distinguished from statistical noise, and the hazard is undetected. To illustrate, imagine a disease that affects 500 million people. An increase in 5 million new cases will produce a negligible 1% increase in the affected population. If the same agent increases the incidence of a rare disease, sometimes by as few as a half-dozen cases, the rise in incidence will be noticed, and the agent can be identified. Imagine that some external toxin produces two diseases; one rare, one common. Increases in the rare disease will draw attention to both the rare and the common diseases.

13.1.1 Rule—Rare diseases are the sentinels that protect us from common diseases.

Brief Rationale—A few new cases of a rare disease will raise the suspicions of astute public health workers and can warn us that the general population has been exposed to a new or growing environmental hazard.

In most instances, a sudden increase in a rare disease has revealed totally unexpected threats to the general population, necessitating enduring improvements in industrial methods and resetting the normal mode of societal behavior. We can trace the origins of chemical carcinogenesis (the study of cancer causes) and of teratogenesis (the study of the causes of congenital malformations) to epidemiologic observations on rare diseases.

There are numerous rare diseases that have served as sentinels for environmental hazards. We listed a few of them in Section 8.4 when we were discussing cancer. Here, we extend the concept to cover all common diseases and we provide the back-stories that clarify the important role of rare diseases in disease prevention and public health.

- **Angiosarcoma of liver** → chemical carcinogenesis by carcinogens in plastics (polyvinyl chloride) [1]

Rare Diseases and Orphan Drugs. http://dx.doi.org/10.1016/B978-0-12-419988-0.00013-4

In 1974, healthcare workers noticed a handful of new cases of liver angiosarcoma, an extremely rare cancer. Some epidemiologic sleuthing led to the discovery that all of the cases of hepatic angiosarcoma occurred in individuals who were employed by the rubber and tire manufacturing industries [2,3]. Furthermore, all these individuals had worked at the site of vinyl chloride polymerization reactor vessels. Aerosolized polyvinyl chloride had previously been shown to cause tumors in rats. Here was proof enough that the agent also caused cancer in humans. Soon thereafter, the public was alerted to the hazardous effects of polyvinyl chloride [2].

- **Aplastic anemia** → bone marrow toxins (e.g., benzene and chloramphenicol)

Aplastic anemia is a condition in which there is depletion of bone marrow cells. The blood cell blasts (i.e., the stem cells that give rise to the differentiated cells that circulate in blood) are reduced in number to such a degree that the peripheral blood is seriously depleted of red cells, white cells, and platelets (see Glossary item, Blast). Healthy, red, marrow disappears, only to be replaced by adipocytes and fibrous tissue. Aplastic anemia may be the only example of a disease characterized by the disappearance of a normal tissue. Acute aplastic anemia is a medical emergency, requiring replacement blood products. Those who survive aplastic anemia, regardless of its etiology, are at increased risk of developing blood dyscrasias of clonal origin (e.g., acute leukemia, myelodysplastic syndrome) later in life.

In the late 1960s, and continuing into the 1970s, epidemiologic studies of leukemia in Istanbul linked increased cases of these rare tumors to benzene exposure among shoe industry workers [4]. In many cases, leukemias followed bouts of aplastic anemia (see Glossary item, Aplastic anemia). Benzene is a known toxin of hematopoietic tissue, and it is reasonable to assume that bone marrow toxicity secondary to occupational benzene exposure commonly preceded the development of leukemia. A reduction in benzene exposure was followed by a striking drop in the incidence of leukemia in shoe industry workers [4].

- **Cancer of the scrotum** → chemical carcinogenesis by soot and common pollutants caused by burning wood

Carcinoma of the scrotum is essentially a non-existent tumor today, but it was a common tumor of chimney sweeps back in 1775 [5]. Percivall Pott, the first cancer epidemiologist, linked hot soot rising up the pant legs of chimney sweeps to the occurrence of squamous carcinoma of the scrotum. His studies essentially eradicated a rare disease confined to a small occupational group. Pott also raised awareness to the potential dangers of air pollutants, and launched the scientific field of cancer prevention.

- **Cholangiocarcinoma** → exposure to thorotrast, a long-lived radiation transmitter contained in a radiographic contrast solution [6]

In the early days of X-ray technology, radiologists were searching for radio-opaque agents that could be swallowed or injected to provide a contrast material that outlined anatomic structures. In the 1930s, when the dangers of low-dose radioactive agents were completely unknown, radiologists seized upon thorotrast, a colloidal suspension of radioactive thorium dioxide, as an ideal contrast agent [6]. Not all of the delivered thorotrast was excreted from the body; some traveled to the reticuloendothelial cells lining the liver sinusoids (i.e., Kupffer cells). The Kupffer cells engulfed the thorotrast, storing it permanently in the liver. From its location within Kupffer cells, the captured thorotrast emitted alpha particles for years. Alpha particles have low penetrance, and the tissue cells most likely to be damaged are the cells carrying the thorotrast and the cells immediately adjacent to these cells. Tumors caused by exposure to radiological thorotrast are hepatocellular carcinoma, cholangiocarcinoma, and hepatic angiosarcoma. Hepatocellular carcinoma is a common tumor, but cholangiocarcinoma is rare and angiosarcoma of the liver is extremely rare. Increases in these two rare tumors warned workers in the fledgling field of radiology that thorotrast was a public menace.

- **Clear cell adenocarcinoma of cervix or vagina in young women→** *in utero* exposure to diethylstilbestrol (DES)

In 1971 physicians noticed that they were encountering more and more cases of young women with an extremely rare cancer: clear cell adenocarcinoma of the cervix or of the vagina [7]. Affected women had mothers who had ingested a non-steroidal synthetic estrogen (diethylstilbestrol, DES) during their pregnancies. A short, *in utero* exposure to a medication resulted in the occurrence of rare tumors many years later in the female offspring. In the original study, tumors occurred in an age range of 7 years to 27 years; 91% were in young women over the age of 14 [7,8]. Exposed males were not at risk of cancer [9]. At the time, there were no known examples of transplacental carcinogenesis occurring in humans. If the tumors caused by *in utero* exposure to DES were common cancers, such as breast carcinoma or colon carcinoma, the link connecting the drug to cancer would have been impossible to discover. Just as thalidomide kindled interest in the relatively new field of transplacental teratogenesis (see below), DES helped create the field of transplacental carcinogenesis.

- **Mesothelioma→** asbestos exposure

Asbestos is an insulating material that was used extensively in the mid-twentieth century. Asbestos inhalation is associated with an increased risk of bronchogenic lung cancer, the number one cause of cancer deaths in the U.S. [10]. In most cases of asbestos-related lung cancer, patients have a history of smoking. If asbestos caused bronchogenic lung cancer exclusively, we probably would have no idea of its carcinogenicity, because its effect among the many smoking-related lung cancers would be negligible.

In 1960, a link was established between occupational asbestos exposure in miners and rare mesotheliomas of the pleura and peritoneum, the tissues lining the lung cavity and the abdominal cavity, respectively [11]. The news came too late to help individuals who were exposed to asbestos in its heyday, during the booming construction years of World War II. In those days, naval ship-workers, eager to protect vessels from fire, lavished pipes and ceilings with asbestos insulation. In so doing, they exposed themselves to asbestos dust. Their family members, who washed their dusty uniforms, were also exposed. Single exposures to the dust could cause mesotheliomas, and these mesotheliomas tended to occur 20 to 40 years following exposure.

Today, we treat asbestos as a serious occupational and environmental hazard. At enormous cost, we have abated our exposures to asbestos found in attic and pipe insulation, brake liners, and cigarette filters. Once again, occurrences of a rare cancer warned us to take measures to reduce exposure to a hazardous substance. Asbestos carcinogenicity also taught us a new lesson; solid and nonreactive agents could cause cancer.

- **Mad hatter disease →** occupational exposure to heavy metals (mercury)

Beginning in the seventeenth century, European hatters used mercury in the preparation of felt from animal fur. Mercury vapors induced tremors in the hatters, and the disease came to be known, callously, as mad hatter disease. Occurrences of the disease among hatters continued for at least two more centuries. By the mid-nineteenth century, the link between mercury exposure and tremors was a scientific certainty. England passed laws to protect hatters from mercury exposure, and the incidence of disease declined. Though a rarity, new cases of mad hatter disease have occurred in the twentieth century. Today, mercury is used in gold extraction. As a result, mercury contamination has increased in gold mining areas, and new cases of occupational tremors are occurring among gold workers [12].

- **Phocomelia →** medication-induced teratogenesis via thalidomide

Thalidomide was first marketed in 1957 as a treatment for morning sickness. Thalidomide was used in many countries, but not in the U.S., where it was denied Food and Drug Administration (FDA) approval. Extreme cases of phocomelia characterized by congenital absences of one or more limbs were reported for the first time in Germany. Soon thereafter large numbers of congenital limb malformations were reported throughout Europe. The occurrences of limb malformations were traced to a common exposure: thalidomide. The drug was removed from the European market in 1961. By that time, over 10,000 children were affected worldwide [13]. Current theory holds that thalidomide kills cells that are needed for limb development, with embryotoxicity highest in weeks 3 and 8 following conception [13]. Thalidomide sparked scientific interest in the field of transplacental teratogenesis (see Glossary item, Teratogenesis).

- **Radium jaw and phossy jaw →** occupational exposure to toxins (phosphorus vapors)

From the mid-nineteenth century until the first decade of the twentieth century, phosphorus was used as an ingredient in so-called easy-strike matches. Many workers in this industry developed disfiguring, disabling, and life-threatening necrotic lesions extending from gingiva into underlying bone (i.e., mandible or maxilla). It took nearly a half-century to associate the jaw condition (eventually called "phossy jaw") with phosphorus exposure, and to eliminate phosphorus in the manufacture of matches [14].

Today, bisphosphonates are drugs that retard the normal process of bone resorption. Bisphosphonates are used to treat osteoporosis. In high doses, bisphosphonates treat bone that has been eroded by metastatic tumor deposits. When used at high doses, bisphosphonates can produce a condition indistinguishable from phossy jaw. It is presumed that phosphorus exposure creates chemical intermediates that are similar to the chemical structure of today's bisphosphonates, and that both toxins produce the same jaw conditions via the same chemical reactions [15]. The twenty-first century lesion is sometimes called "bisphossy jaw" [15]. Historical experiences with an extinct and rare condition, phossy jaw, have served as cautionary lessons for other toxic agents that can cause osteonecrosis of the jaw (e.g., radium, heavy elements).

- **Thyroid carcinoma in children →** radiation-induced cancers

In the wake of the 1986 Chernobyl nuclear accident, three epidemiologic alarms were tripped [16]:

1. Following the accident, there was an increase in the incidence of an uncommon form of cancer: thyroid cancer. As described earlier, it is relatively easy to demonstrate increases in the occurrence of a rare cancer; whereas a quantitatively identical increase in the occurrences of a common cancer is likely to go undetected.
2. The thyroid tumors occurred preferentially in exposed children. In the case of Chernobyl, those children most likely to develop thyroid cancer were between 0 and 4 years of age at the time of the accident [16]. A comparable increase in thyroid cancers was not observed among exposed adults.
3. Tumors arose relatively soon after the Chernobyl accident. The increase in thyroid cancers was first detected 3–4 years after the accident [16]. Most cancers take decades to develop. When cancers occur in just a few years, it is a sign that the exposure was large, or that the carcinogen was very potent. In this case, the majority of thyroid cancers occurred in Gomel, Belarus, the region exposed to the greatest concentrations of iodine-131, the chief radioactive component of the fallout, and a chemical known to concentrate in the thyroid after absorption [16].

- **Acral melanoma in African-Americans →** UV light-induced melanomas

Sometimes, observations made on rare variants of common diseases can resolve issues that could not be adequately understood with observations on common diseases alone. Malignant melanoma is a common cancer, accounting for about 75% of skin cancer deaths, with about 48,000 deaths worldwide [17]. It has long been recognized that the highest incidence of melanomas among white persons occur in regions with the highest exposures to sunlight, and that the skin areas most likely to develop melanoma correspond roughly to the skin areas that receive the most sunlight (i.e., face, neck, and exposed limbs). Furthermore, individuals with the least amount of UV-protective skin pigmentation (i.e., fair-skinned, blonde or red-headed individuals) have the highest rates of melanoma.

Though it is generally accepted that sunlight is a cause of melanoma, there is debate as to the relative importance of sunlight to melanoma formation [18]. Could other factors have equal or greater roles in the carcinogenesis of melanoma? In the U.S., about 20% of melanomas occur in Black African and Asian populations, and the causes of melanoma in highly pigmented individuals is unknown [18].

The U.S. National Cancer Institute's Surveillance, Epidemiology and End Results Program (SEER) public use data sets quantify the occurrences of cancers, by type, among a large segment of the U.S. population. Using SEER data, we can express the occurrences of the different types of melanomas as an adjusted ratio of tumors occurring in white individuals compared with tumors of the same type occurring in black individuals [19]:

11.55 Mixed epithelial and spindle cell melanoma
13.07 Malignant melanoma
16.75 Spindle cell melanoma
25.81 Nodular melanoma
30.62 Lentigo maligna melanoma
32.93 Melanoma *in situ*
39.46 Superficial spreading melanoma
40.83 Lentigo maligna
77.01 Superficial spreading melanoma *in situ*
00.90 Acral lentiginous melanoma

With the exception of the last entry on the list, the findings were as expected. In the U.S., melanoma occurs much less frequently in African-Americans than in white individuals. The most extreme difference was found in the *in situ* superficial spreading melanoma, with an occurrence rate 77 times higher in white persons than in African-American persons (see Glossary item, *In situ*). The data lends credence to the hypothesis that melanin protects skin from the short-term and long-term harmful effects of ultraviolet light: sunburn, solar elastosis, epidermal skin cancer (primarily squamous cell carcinoma and basal cell carcinoma), and melanoma. There is remarkable internal consistency in the list. Where a tumor appears, it is often closely followed by a variant of the same tumor. This indicates that closely related tumors, which have the same general cell type (in this case, tumors of melanocyte origin), most likely have the same biological causes. Otherwise, why would they aggregate in the list?

There is one glaring exception on the list; a rare type of melanoma known as acral lentiginous melanoma. Only 940 such cases were found among the SEER records [19]. The data show that white individuals and African-American individuals have about the same incidence ratio (i.e., 0.90) for this rare form of melanoma.

Acral lentiginous melanoma is a variant of melanoma that occurs in non-pigmented skin: the sole of the foot, the palm of the hand, and under fingernails or toenails. Acral parts are non-pigmented in white individuals and in African-American individuals. If melanoma were caused by exposure to sunlight, then you would expect that white persons and African-Americans would have the same low incidence of acral lentiginous melanomas. This is precisely the case. In this example, data on a rare variant of a common cancer greatly strengthened a hypothesis created for a common cancer.

13.1.2 Rule—When a new toxin is introduced to a population, individuals with a rare disease will be among the first to succumb to its ill-effects.

Brief Rationale—Many of the rare diseases have mild variants that cause minimal pathology under normal circumstances. When physiological systems are overwhelmed or otherwise stressed by a toxin, a clinical phenotype may emerge.

In most instances, rare diseases act to reduce the functionality of a metabolic process. For example, there are a variety of rare diseases that impair immunity. If a new fungal pathogen is introduced to a general population, you might expect that individuals with impaired immunity will be among the most vulnerable to infection.

There are several rare diseases that reduce the ability of cells to survive agents that break or otherwise disrupt normal chromosomal function (e.g., ataxia telangiectasia, Nijmegen breakage syndrome, Fanconi anemia) (see Glossary item, Ataxia telangiectasia). We find that patients with these rare diseases are highly sensitive to new agents or procedures that reduce chromosomal integrity, such as excessive radiation exposure from therapeutic or diagnostic radiology devices [20].

Individuals in the early stages of rare diseases, or who have mild forms of the disease, or who are carriers of the disease, may display a heightened sensitivity to agents that interfere with key disease pathways. These subclinical pathway deficiencies provide the scientific rationale for the so-called "evocative" and "suppressive" diagnostic tests.

13.1.3 Rule—When a common disease occurs in a young individual, in the absence of a diagnosed inherited cancer syndrome, then an environmental cause should be suspected.

Brief Rationale—Common diseases typically occur in middle-aged and elderly patient populations. When a common disease occurs in a young person, and it is not caused by an inherited syndrome, then a high exposure to an environmental agent is likely.

As discussed in Section 3.2, common diseases (e.g., heart disease, cancer, metabolic diseases) tend to develop over many years, affecting an older population. When we start to see an increase of common diseases in a young population, we need to start wondering whether these individuals are exposed to very high levels of the same agents that typically cause disease in older individuals, or whether they are exposed to a new agent not previously linked to the disease.

Cancers of the mouth in teenage boys who play baseball is an example wherein adolescents are exposed to high local concentrations of a presumably weak carcinogen (i.e., chewing tobacco placed on a favorite spot between the gingiva and the buccal mucosa). Type 2 diabetes in children is presumably also caused by high childhood exposure to a common set of contributing factors (e.g., inactivity, poor diet). In the case of type 2 diabetes in children, we cannot rule out the participation of newly introduced environmental agents.

It is worth remembering that when we drop our guard, a rare disease can become a common disease. In 1963, there were only 17 reported cases of malaria in Sri Lanka. Five years later, after mosquito control measures were relaxed, 440,000 cases were reported [21].

Before smoking became a national pastime, bronchogenic carcinoma was a rare tumor. If smoking had been curtailed when cigarettes were a novelty, the current worldwide epidemic of lung cancer would have been prevented. When we do not take adequate measures to limit our exposure to agents that cause rare diseases (e.g., asbestos, benzene, silicates, radiation), we can expect the incidence of the rare diseases to continuously rise.

13.2 SHARED DIAGNOSTICS

"Many patients with rare diseases today have difficulty in finding providers with the expertise and resources to diagnose and treat their conditions."

—Committee on Accelerating Rare Diseases Research and Orphan Product Development, Institute of Medicine of the National Academies (U.S.), 2010 [22]

In former times, the sole purpose of diagnostics was to apply a name of a disease to a clinical condition. If the name was known, a treatment could be applied. If no treatment was available, there was always prayer. Today, it is not sufficient to simply provide a name for a disease. Diagnosis today covers a wide range of activities, including:

- Risk prediction—Determining whether an individual is at increased risk of developing a disease at some unspecified future time.

- Screening—Determining whether an individual falls into a separable group of individuals who are likely to have the disease at the time of screening. After screening, further studies would be necessary to determine whether the individual actually has the disease.
- Early detection—Determining whether an individual has a disease at an early stage of development. Early detection is often confused with disease screening. Early detection determines whether an individual has the disease at an early, usually pre-clinical, stage. Screening (see above) determines whether an individual is likely to have the disease.
- Molecular diagnosis—Determining the presence of disease with a molecular technique performed on very small samples of tissues. Molecular diagnosis typically replaces, supplements, or confirms traditional diagnostic methods, such as surgical biopsy.
- Subtyping—Determining which biological subtype of disease applies to an individual.
- Response prediction—Determining whether an individual with a disease is likely to respond to a particular treatment.
- Staging—Determining the extent to which a disease has advanced within an individual.
- Surveillance for minimal residual disease and for recurrence—The objective of minimal residual disease surveillance is to determine whether there are any traces of disease, not observable by standard clinical examination, that persist following treatment. The objective of recurrence surveillance is to determine whether a disease has recurred after remission.

When we review the various new diagnostic activities, we see that they recapitulate the steps of disease pathogenesis: the conditions that place an individual at risk of developing disease, the earliest steps in pathogenesis, the development of precursor lesions, response pathways, and disease progression. New laboratory tests are designed to measure markers for the genes and pathways that account for the components of pathogenesis.

> **13.2.1 Rule—Any specialized diagnostic techniques applied to a common disease will likely draw on knowledge obtained from one or more rare diseases.**
>
> **Brief Rationale—Genes and pathways that lead to rare diseases are the pathogenetic building blocks of common diseases. We can expect that these same genes and pathways will serve as new diagnostic markers.**

Let us look once again at the list of modern-day diagnostic activities. This time, we will relate a diagnostic activity for a common disease to a known feature of a rare disease.

- **Risk prediction**

If we wanted to know our risk of developing a heart attack, a reasonable way to do so would be to measure blood levels of various constituents that are known to be causally related to rare causes of heart disease. For example, individuals with inherited hypercholesterolemia, or inherited hypertriglyceridemia, or inherited low levels of HDL are at increased risk for heart attacks. A battery of blood tests using these same three blood constituents that characterize rare conditions that carry elevated risks of heart attacks are routinely measured on members of the general population to predict risk of heart attacks.

- **Screening**

Individuals who carry specific mutations in the BRCA1 and BRCA2 tumor suppressor genes have an increased risk for developing breast cancer. Mutations in these two genes account for 5–10% of all the breast cancers in the general population [23]. Currently, screening for BRCA1 or BRCA2 mutations is available for members of families wherein one of the following criteria are met: (1) multiple breast and/or ovarian cancers within a family (often diagnosed at an early age); (2) two or more primary cancers in a single family member (more than one breast cancer, or breast and ovarian cancer); (3) cases of male breast cancer. In this example, the causal mutation in a rare, inherited, familial breast cancer syndrome is employed as a screening test for a small, selected set of individuals drawn from a large population.

New gene markers for rare diseases are all candidate biomarkers for common diseases (see Glossary item, Candidate gene approach). Of course, whether a candidate biomarker becomes a useful screening test for the general population will depend on many factors, including cost, reproducibility, accuracy, sensitivity, specificity, and the prevalence of the biomarker in the diseased and disease-free members of the general population.

- **Early detection**

With few, if any, exceptions, it is easier to treat a disease in its early stages when the patient is relatively healthy, than it is to treat an advanced stage disease when the patient is in poor health.

An example of early detection would include finding an immunoglobulin spike in an electrophoresis blood test. Normal blood contains many different immunoglobulin molecular species that separate individually by size and electric charge on a blood protein electrophoresis preparation. When there is a spike in the test pattern, this indicates that one species of immunoglobulin molecule appears in blood in high concentration. A spike is produced by a clonal expansion of an immunoglobulin-producing cell (i.e., a plasma cell). Multiple myeloma is a cancer composed of neoplastic plasma cells (see Figure 13.1). In the presence of advanced symptoms of multiple myeloma (e.g., characteristic bone lesions, elevated calcium levels), an immunoglobulin spike is almost always associated with malignancy. In the absence of clinical findings, a monoclonal spike might indicate a very early form of disease. Monoclonal gammopathy of undetermined significance (MGUS) is a proliferative lesion of plasma cells occurring in otherwise asymptomatic patients that sometimes progresses

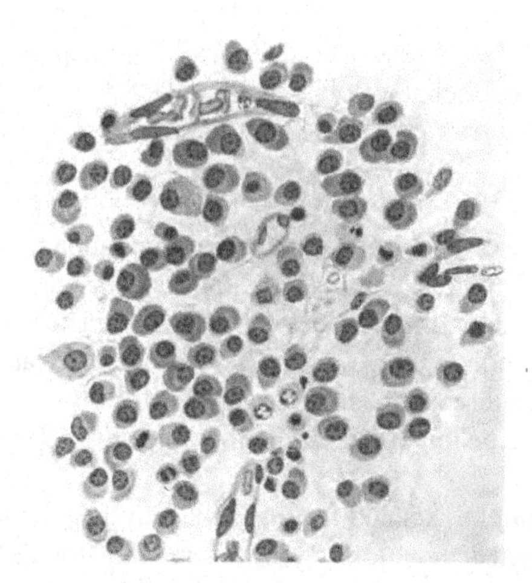

FIGURE 13.1 A thin histopathologic section of multiple myeloma, featuring a rather uniform collection of neoplastic cells that closely resemble normal plasma cells. The nuclei are round and eccentric (i.e., seemingly pushed against the cytoplasmic membrane). The cytoplasm has a uniform dark color, indicating a high protein concentration, with the exception of a light area touching the nucleus, the site of intense synthesis of immunoglobulins and other proteins. (*Source: MacCallum WG. A Textbook of Pathology* [26].)

to multiple myeloma (see Glossary item, Monoclonal gammopathy of undetermined significance). Virtually every case of multiple myeloma is preceded by MGUS [24]. An immunoglobulin spike found in an asymptomatic individual is an example of a laboratory observation that detects multiple myeloma in an early stage of development (i.e., MGUS). Because most cases of MGUS do not progress to multiple myeloma within the lifetime of the individual (i.e., patients typically die of some other cause before the MGUS progresses into multiple myeloma), a strategy of watchful waiting with bone lesion surveillance may be appropriate [25].

- **Molecular diagnosis**

The same mutations found in inherited rare diseases are likely to be found in one or more common diseases; hence, the methods developed to detect a mutation in a rare disease can often be applied to a common disease. For example, LEOPARD syndrome type 3, Noonan syndrome type 7, and cardiofaciocutaneous syndrome are all caused by mutations in the BRAF gene [27]. The methods that were developed to detect BRAF mutations in rare diseases can be applied to diagnosing a common type of thyroid cancer (i.e., papillary carcinoma of the thyroid) that also carries BRAF mutations [28].

As discussed previously, germline PTEN mutations occur in Cowden syndrome and Bannayan–Riley–Ruvalcaba syndrome, two inherited disorders associated with a high rate of endometrial carcinomas. PTEN mutations are found in 93% of sporadically occurring endometrial carcinomas [29]. PTEN is a diagnostic marker for precancerous endometrial intraepithelial neoplasia (see Glossary item, Intraepithelial neoplasia) [30]. PTEN is also used as an indicator of poor prognosis in oligodendroglioma [31].

- **Subtyping**

Distinguishing a rare disease from a subtype of a common disease is an exercise in hair-splitting. As previously discussed, lung cancers occurring in young individuals often occur along the midline and have a characteristic mutation in the NUT gene [32]. Is midline lung carcinoma of the young a rare disease, or is it a variant of a common disease? Distinguishing the different subsets of a common disease may prove to be important if the different subsets respond differently to treatment protocols.

As another example, isolated hearing loss may be sporadic, or it may be caused by an inherited mutation. Genetic screening for the different known inherited causes of sensorineural hearing loss may uncover clinically distinct subtypes of commonly occurring deafness.

Every phenotypic subtype of a common disease is fair game for rare disease researchers who will search for a genetic marker that characterizes the subtype, and that may have been acquired through inheritance or through a new germline mutation. When subtypes of a common disease are characterized by a genetic marker, new diagnostic tests can be developed that target the mutation or its pathway.

Furthermore, every seemingly sporadic occurrence of a common disease may include formes frustes presentations of inherited syndromes limited to an isolated organ (see Glossary item, Forme fruste). For example, of the so-called sporadic colon cancers, up to 6% are due to one form or another of inherited monogenic conditions that raise the risk of developing colorectal cancer and other tumors (e.g., Lynch syndrome) [33]. Finding the genetic marker for a rare cancer syndrome in individuals with a common colon cancer can serve as a warning that the patient may develop additional cancers of the colon or other organs, and that the patient's relatives may also be at risk.

- **Response prediction**

For the patient, there are few words more heartbreaking than, "I'm sorry, but you're not responding to treatment."

Decades ago, when a cancer patient received a course of treatment with a chemotherapeutic agent and the tumor did not shrink, it was common for physicians to summarize the situation by remarking, "the patient failed the treatment." Nowadays, physicians are taught not to blame the patient for a therapeutic failure. It is better to admit, "the treatment failed the patient."

Many of the drugs used today are targeted against a specific molecule or pathway involved in the pathogenesis of disease. Because common diseases are complex, a pathway that is operative in one patient may play a minor role in other patients that have the same disease. It is important to know which patients will respond to which targeted treatments. Knowing that a patient is unlikely to respond to a particular treatment is always disappointing news; yet, such knowledge allows the physician to search for an alternate treatment protocol most suited to the patient.

13.2.2 Rule—Every diagnostic gene or pathway is a potential drug target.

Brief Rationale—Pathways, and the genes that code for pathway proteins, that are crucial to the pathogenesis of disease are the logical targets of therapeutic agents.

Today, it is possible to tie a treatment to a test that predicts responsiveness. For example, trastuzumab (trade name, Herceptin) is a monoclonal antibody that interferes with the Her2 receptor. Certain breast cancers have high levels of the Her2 receptor, and the receptor is thought to play an important role in driving tumor cell proliferation. Interfering with the receptor can reduce the rate of tumor cell growth; but only in the subtype of breast cancers that have high levels of Her2. Thus, the diagnostic test (i.e., tumors with high levels of Her2 receptors) is tied to treatment (i.e., with trastuzumab, a drug that targets the Her2 receptor). **By tying a diagnostic test to a drug target, treatment is reserved for those individuals who are most likely to have a satisfactory therapeutic response.**

- **Staging**

Staging is important for every type of disease. Physicians need to know whether they are dealing with a localized disease, or whether the disease has spread to distant organs.

Among the common diseases, staging is most important for cancer. If a cancer is localized to its site of origin, then surgical removal of the cancer is curative. This is true even if the tumor is highly malignant, locally invasive, and with a predisposition for distant metastasis. If the tumor is removed before it has metastasized, then the tumor is cured. Contrariwise, if a tumor has metastasized, then complete excision of the tumor at its primary site is non-curative. Even if the tumor is generally an indolent neoplasm with a low likelihood of distant metastasis, it may eventually kill the patient if it bucked the odds and metastasized prior to its primary site excision. For these reasons, physicians try to stage cancers to determine whether the tumor has metastasized from its origin.

When metastases are very small, it can be difficult or impossible to observe them using currently available scanning techniques (e.g., bone scans, computed tomography). As a surrogate technique, oncologists have been trying to find a battery of molecular markers, found in DNA extracted from a sample of the primary tumor (i.e., the tumor at its site of origin), that can predict the tumor stage [34,35].

Predicting tumor stage by examining a profile of genes is a tricky business. Aggressive tumors that are likely to metastasize are composed of heterogeneous subclones, each with its own set of up-regulated and down-regulated genes that contribute to a malignant phenotype. A sampling of tumor from the center of the tumor mass, where blood flow is lowest and where cell death rates are highest, is likely to contain cells that are well suited for survival in an ischemic environment. Such cells may not exhibit a gene profile that is predictive of metastasis or invasion, which use another set of genes and pathways.

There is scant published evidence at this time to suggest that molecular profiling on primary tumors will provide staging information that is more accurate than currently available staging techniques (e.g., radiologic imaging studies). Still, if gene profiling on a primary tumor can tell us something about the likely stage of the tumor, it may well be worth the effort to explore this technology. Many of the rare cancer syndromes are caused by genes that confer biological properties associated with an aggressive malignant phenotype (e.g., genetic instability, cell proliferation, resistance to apoptosis, cell junction abnormalities, respiratory defects). Hence, the same genes that cause rare, inherited cancers are likely to be predictive biomarkers.

- **Surveillance for minimal residual disease and recurrence**

The issue of finding residual disease and recurrent disease is most relevant to the infectious diseases and to cancer. For the infectious diseases, residual and recurrent disease often involves measuring titers of infectious organisms (e.g., viruses, bacteria, fungi) in blood samples. For the most part, searches for residual infectious organisms or response-to-infection markers are most useful for infectious diseases that have chronic, subclinical, or latent phases. For example, IRF5 is an interferon regulatory factor that influences susceptibility to systemic lupus erythematosus, an autoimmune disease. IRF7 is a related factor, and both IRF5 and IRF7 participate in systemic responses to infection. These markers have been used to detect inflammatory diseases, including recurrent systemic lupus erythematosus [36] and latent Epstein–Barr virus infections [37].

For the cancers, minimal residual disease and recurrence are monitored for altered genes and proteins that characterize the primary tumor (e.g., alpha fetoprotein blood levels in recurrent liver cancer, high levels of prostate-specific antigen in recurrent prostate cancers, human chorionic gonadotropin in recurrent choriocarcinoma and germ cell tumors, carcinoembryonic antigen in recurrent colon cancers). Because the rare cancers tend to have specific monogenic markers (e.g., bcr/abl fusion gene in chronic myelogenous leukemia, CD117 gene mutations in gastrointestinal stromal tumors, myc translocation in Burkitt lymphoma), the most promising tests for minimal disease and for recurrent disease will come from the rare cancers [38].

13.3 SHARED CURES

"Thus, we are poised to make rapid advances in the understanding and, in an increasing number of cases, the treatment of rare diseases. As past research has demonstrated, some of these advances will undoubtedly illuminate disease mechanisms and treatment avenues for more common conditions."

—Committee on Accelerating Rare Diseases Research and Orphan Product Development, Institute of Medicine, National Academies [22]

"What's sauce for the goose is sauce for the gander."

—Proverb

When I speak to colleagues about "shared cures" for the rare diseases and the common diseases, they often ask the following question: "Why don't we cure the common diseases first? We'll get the most benefit from our research if we use the drugs under development for the common diseases to treat the rare diseases a bit later."

It is very difficult to steer scientists away from their belief that common diseases are more important than rare diseases. It is a simple fact that a few dozen common diseases account for the vast majority of the morbidity and mortality suffered by humans. A breakthrough in treating any of the common diseases will benefit many more people than an advance in any of the rare diseases. The reasoning is flawless, but the results have been disappointing. **In the past 50 years, most of the major advances in medicine have involved the rare diseases. Advances in the common diseases have come consequent to discoveries made on rare diseases**.

As it happens, the rare diseases are much easier to understand and treat than the common diseases. If we waited for medical scientists to cure the common diseases, we would miss our currently available opportunity to cure diseases, either rare or common.

13.3.1 Rule—Rare diseases are easier to treat than common diseases.

Brief Rationale—Rare diseases have simple genetic defects, have little heterogeneity, and have few metabolic options with which they can evade targeted treatments.

How many life-threatening common diseases can be successfully treated by avoiding certain foods? None, of course, but a considerable number of inherited metabolic diseases are treated with special diets (e.g., phenylketonuria, galactosemia, tyrosinemia, homocystinuria, maple syrup urine disease). In each, a deficiency in one protein disrupts a pathway whose pathologic consequences can be ameliorated by the avoidance of certain nutritional components.

When we look at the cancers that can be cured, even when diagnosed at an advanced stage of disease, they are exclusively rare cancers [39]. As described in Chapter 8, rare cancers tend to have a single genetic cause, and the phenotype of the resulting tumors is simple, with minimal heterogeneity. A drug that effectively kills one of the cells in a rare cancer is likely to kill all the other cells of the same tumor mass. A simple pathogenesis leads to a simple cure.

13.3.2 Rule—Every chemical with a known biological activity (i.e., an effect on some biological function) is useful in the treatment of one or more diseases.

Brief Rationale—If a drug has a biological action on a biological pathway, and if the biological pathway is involved in one disease or another, the drug will likely have some effect on diseases that depend on those pathways. Because there are many thousands of diseases, the odds are that an agent that modifies the activity of any pathway will have some value in one or more diseases. As you might expect, chemicals that modify cellular pathways are seldom "non-toxic." The most we can hope for is that at the doses prescribed, their beneficial actions will surpass their toxic actions.

Ryanodine receptor 2 mutations are responsible for several rare arrhythmia syndromes in humans (e.g., forms of catecholaminergic polymorphic ventricular tachycardia and arrhythmogenic right ventricular dysplasia, as discussed in Section 5.3). Individuals with these disorders can be treated with drugs that stabilize the receptor. Damage to ryanodine receptor 2 seems to occur as a component of common heart failure, leading to calcium leak and arrhythmia. Preliminary studies indicate that drugs that stabilize the receptor may ameliorate all types of heart failure and the lethal arrhythmias that ensue [40].

Alexion is a pharmaceutical company that specializes in developing drugs intended to treat rare diseases. For example, Alexion discovered and developed eculizumab (trade name Soliris), a first-in-class terminal complement inhibitor. Eculizumab was approved by the FDA in 2007 for the treatment of paroxysmal nocturnal hematuria; and in 2011 for the treatment of atypical hemolytic uremic syndrome. Subsequently, eculizumab was tested for its effectiveness for several common diseases. Eculizumab was a candidate treatment for so-called dry age-related macular degeneration, a common disease, though it was not shown to be effective [41]. On the brighter side, eculizumab has been shown to prevent acute and chronic rejection in certain subsets of patients who received renal transplants [42]. When you have a drug that is known to target a particular member of an active physiologic pathway, it is likely to have some benefit in one or more common diseases whose clinical phenotype is due, in part, to aberrations of the same pathway.

13.3.3 Rule—All drugs that are safe and effective against rare diseases will be used in the treatment of one or more common diseases.

Brief Rationale—The rare diseases, as an aggregate group, comprise every possible pathogenic pathway available to cells. Hence, pathogenic pathways that are active in the common diseases will be active in one or more rare diseases. Agents that target pathways in the rare diseases are candidate treatments for the common diseases with which they share active pathways.

Wrinkling is one of the most common physical conditions. Every man and woman who lives long enough will wrinkle a bit. For some individuals, wrinkling is a problem that merits medical attention. Botox (botulism toxin) is the drug *du jour* for treating wrinkles. Botox is also one of the most powerful poisons known. How did it come about that Botox emerged as a popular wrinkle treatment? What pharmaceutical company would dare enter clinical trials designed to test a highly toxic poison on a set of individuals that have a trivial condition that does not cause functional impairment?

Botox was originally developed, tested, and approved to treat several rare diseases characterized by uncontrolled blinking. After approval was awarded, Botox was found to be extremely effective for rare spasmodic conditions, including spasmodic torticollis (i.e., wry neck). In the course of treating rare diseases, it was noticed that Botox injections could temporarily erase wrinkles. The rest is history. The Botox story exemplifies how an effective treatment developed for rare diseases can gain popularity as a treatment for a common condition.

13.3.4 Rule—It is much more useful to treat a disease pathway than it is to treat the individual gene mutation or its expressed protein.

Brief Rationale—Many different diseases may respond to a drug that targets a pathogenic pathway, while only one genetic variant of one rare disease is likely to respond to a drug that targets the disease-causing gene or its expressed protein.

There is a very important lesson to be learned: **Treat the pathway, not the gene**. This lesson is somewhat counter-intuitive and is received with some skepticism from experienced medical researchers. Nonetheless, it is a core principle that diseases are caused by perturbed pathways, and that the successful treatment of diseases has always involved compensating, in one way or another, for pathway disturbances. Let us review some examples that demonstrate the point.

As discussed in Section 10.3, imatinib (trade name Gleevec) inhibits tyrosine kinase, an enzyme involved in a pathway that drives the growth of various rare tumors and proliferative diseases (e.g., chronic myelogenous leukemia, gastrointestinal stromal tumor, hypereosinophilic syndrome) [43–47]. Pathways with increased tyrosine kinase activity, and pathways whose tyrosine kinase activity is particularly sensitive to the inhibiting action of imatinib, would make the best drug targets. Because imatinib is targeted to a key protein in a general pathway that contributes to a proliferative phenotype, it has potential benefit in several different diseases.

Bevacizumab (trade name Avastin) is an angiogenesis (i.e., vessel-forming) inhibitor (see Glossary item, Angiogenesis). All cancers require vessel growth. **In theory, bevacizumab is a universal tumor growth inhibitor because its target is the non-neoplastic mesenchymal cells that form the vessels that feed growing tumor cells**. Bevacizumab is employed in the treatment of common cancers, including cancers of the colon, lung, breast, kidney, ovaries, and brain (i.e., glioblastoma). Bevacizumab produces tumor shrinkage in more than half of vestibular schwannomas occurring in neurofibromatosis 2 [48]. As you might expect, bevacizumab has its greatest value in diseases for which neovascularization has a required role in pathogenesis. Two non-cancerous diseases of vascularization treated with angiogenesis inhibitors are hereditary hemorrhagic telangiectasia [49] and various forms of ocular neovascularization, including common age-related macular degeneration [50].

Because pathways are interconnected, a drug that is effective against a component of a pleiotropic pathway may be effective against multiple diseases. For example, Janus kinase genes (e.g., AK1, JAK2, JAK3, TYK2) influence the growth and immune responsiveness in various blood cells through their activity on cytokines. In Section 12.1, we learned that mutations of the JAK2 gene are involved in several myeloproliferative conditions, including myelofibrosis, polycythemia vera, and at least one form of hereditary thrombocythemia [51–53].

Inhibitors of JAK genes have been approved for the treatment of various diseases that involve heightened proliferation of lymphocytes in immune reactions, or blood cells in myeloproliferative disorders. Ruxolitinib has been approved in the U.S. for use in psoriasis, myelofibrosis, and rheumatoid arthritis [54]. A host of JAK pathway inhibitors are either approved or under clinical trials for the treatment of allergic diseases, rheumatoid arthritis, psoriasis, myelofibrosis, myeloproliferative disorders, acute myeloid leukemia, and relapsed lymphoma [55].

13.3.5 Rule—Common diseases and rare diseases that share a pathway are likely to respond to the same pathway-targeted drug.

Brief Rationale—Pathogenesis (i.e., the biological steps that lead to disease) and clinical phenotype (i.e., the biological features that characterize a disease) are determined by cellular pathways. If a pathway has a crucial role in the development of disease, then you would have reason to hope that drugs that disrupt the pathway will alter the progression and the expression of the disease, whether the disease is common or rare.

Individuals with a rare resistance to HIV infection have a specific deletion in the gene that codes for the CCR5 co-receptor. The gene plays a role in the entry of HIV into cells; no entry, no infection. As it happens, both HIV virus and smallpox virus enhance their infectivity by exploiting a receptor, CCR5, on the surface of white blood cells. This shared mode of infection may contribute to the cross-protection against HIV that seems to come from smallpox vaccine. It has been suggested that the emergence of HIV in the 1980s may have resulted, in part, from the cessation of smallpox vaccinations in the late 1970s [56]. The same, rare CCR5 gene deletion that protects against HIV infection may very well protect against smallpox infection. We may never know with certainty whether this is true because smallpox has been eradicated, along with smallpox experiments. Nonetheless, knowledge of the role of CCR5 in HIV infection has inspired the development of a new class of HIV drugs targeted against entry receptors [57].

Individuals with genetic absence of Duffy antigen receptor for chemokines (i.e., DARC, formerly known as Duffy blood group antigen) are protected from malaria caused by *Plasmodium vivax*. It turns out that entry of the parasite requires participation by DARC [58,59]. A new vaccine candidate for *P. vivax* malaria targeted against the Duffy binding protein was developed based on observations of naturally occurring resistance in individuals lacking DARC [60,58].

Osteoporosis–pseudoglioma syndrome is a rare disease characterized clinically by multiple bone fractures and various eye and neurologic abnormalities. It is caused by loss-of-function mutations in the low-density lipoprotein receptor-related protein-5 (LRP5). LRP5, under normal conditions, reduces the production of serotonin in the gut. Based on rare disease research directed towards understanding the role of LRP5, agents that compensate for the reduction in LRP5 by reducing gut serotonin are candidate drugs for the treatment of both rare osteoporosis–pseudoglioma syndrome and common osteoporosis [61,22,62].

Of course, advances in the common diseases may have value in treating rare diseases. Losartan is an effective drug against one of the most common diseases of humans: hypertension. Losartan blocks the angiotensin II type 1 receptor, and it also blocks TGF-alpha (transforming growth factor-alpha). In Marfan syndrome, a rare disease of connective tissue, growth of the aortic root may lead to life-threatening aortic aneurysm. A reduction in TGF-alpha activity following losartan treatment reduces growth of the aortic root, and slows the progression of aortic root distension in Marfan syndrome [63].

Shared cures for the rare diseases and the common diseases do not occur as low-probability events in an unpredictable world. Knowledge of disease biology leads us to conclude that whenever a cure for a rare disease is found, there is a high likelihood that this same cure will have practical application in the treatment of a common disease. Pharmaceutical companies understand that rare disease research is often the prelude to common disease research [64].

It is crucially important to appreciate the role of rare diseases in drug development. **Ultimately, if funding agencies do not appreciate how cures for the rare diseases will lead to cures for the common diseases, the field of rare disease research will continue to be underfunded and generally ignored by the medical research community**.

REFERENCES

1. Falk H, Creech JL Jr, Heath CW Jr, Johnson MN, Key MM. Hepatic disease among workers at a vinyl chloride polymerization plant. JAMA 230:59–68, 1974.
2. Wagoner JK. Toxicity of vinyl chloride and poly(vinyl chloride): a critical review. Environ Health Perspect 52:61–66, 1983.
3. Hill RB, Anderson RE. Pathologists and the autopsy. Am J Clin Pathol 95(Suppl):42–49, 1991.
4. Aksoy M. Hematotoxicity and carcinogenicity of benzene. Environ Health Perspect 82:193–197, 1989.

5. Gerber C, von Hochstetter AR, Schuler G, Hofmann V, Rosenthal C. Penis carcinoma in a young chimney sweep. Case report 200 years following the description of the first occupational disease. Schweiz Med Wochenschr 125:1201–1205, 1995.
6. Sharp GB. The relationship between internally deposited alpha-particle radiation and subsite-specific liver cancer and liver cirrhosis: an analysis of published data. J Radiat Res 43:371–380, 2002.
7. Herbst AL, Ulfelder H, Poskanzer DC. Association of maternal stilbestrol therapy and tumor appearance in young women. New Engl J Med 284:878–881, 1971.
8. Herbst AL, Scully RE, Robboy SJ. The significance of adenosis and clear-cell adenocarcinoma of the genital tract in young females. J Reprod Med 15:5–11, 1975.
9. Strohsnitter WC, Noller KL, Hoover RN, Robboy SJ, Palmer JR, et al. Cancer risk in men exposed in utero to diethylstilbestrol. J Natl Cancer Inst 93:545–551, 2001.
10. Kamp DW. Asbestos-induced lung diseases: an update. Transl Res 153:143–521, 2009.
11. Wagner JC, Sleggs CA, Marchand P. Diffuse pleural mesothelioma and asbestos exposure in the North Western Cape Province. Br J Ind Med 17:260–271, 1960.
12. Eisler R. Mercury hazards from gold mining to humans, plants, and animals. Rev Environ Contam Toxicol 181:139–198, 2004.
13. Knobloch J, Ruther U. Shedding light on an old mystery: thalidomide suppresses survival pathways to induce limb defects. Cell Cycle 7:1121–1127, 2008.
14. Marx RE. Uncovering the cause of "phossy jaw" circa 1858 to 1906: oral and maxillofacial surgery closed case files—case closed. J Oral Maxillofac Surg 66:2356–2363, 2008.
15. Jacobsen C, Zemann W, Obwegeser JA, Grätz KW, Metzler P. The phosphorous necrosis of the jaws and what can we learn from the past: a comparison of "phossy" and "bisphossy" jaw. Oral Maxillofac Surg, December 28, 2012.
16. Cardis E, Hatch M. The Chernobyl accident—an epidemiological perspective. Clin Oncol (R Coll Radiol) 23:251–260, 2011.
17. Lucas R, McMichael T, Smith W, Armstrong B. Solar ultraviolet radiation: global burden of disease from solar ultraviolet radiation. Environmental Burden of Disease, Series 13, Springer, 2006.
18. Armstrong BK, Kricker A. How much melanoma is caused by sun exposure? Melanoma Res 3:395–401, 1993.
19. Berman JJ. Methods in Medical Informatics: Fundamentals of Healthcare Programming in Perl, Python, and Ruby. Chapman and Hall, Boca Raton, 2010.
20. Pollard JM, Gatti RA. Clinical radiation sensitivity with DNA repair disorders: an overview. Int J Radiat Oncol Biol Phys 74:1323–1331, 2009.
21. Lemon SM, Sparling PF, Hamburg MA, Relman DA, Choffnes ER, Mack A. Vector-Borne Diseases: Understanding the Environmental, Human Health, and Ecological Connections, Workshop Summary. Institute of Medicine (US) Forum on Microbial Threats. Washington (DC). National Academies Press, (US), 2008.
22. Field MJ, Boat T. Rare diseases and orphan products: accelerating research and development. Institute of Medicine (US) Committee on Accelerating Rare Diseases Research and Orphan Product Development. The National Academies Press, Washington, DC, 2010. Available from: http://www.ncbi.nlm.nih.gov/books/NBK56189/
23. Campeau PM, Foulkes WD, Tischkowitz MD. Hereditary breast cancer: new genetic developments, new therapeutic avenues. Human Gen 124:31–42, 2008.
24. Landgren O, Kyle RA, Pfeiffer RM. Monoclonal gammopathy of undetermined significance (MGUS) consistently precedes multiple myeloma: a prospective study. Blood 113:5412–5417, 2009.

25. Hillengass J, Weber MA, Kilk K, Listl K, Wagner-Gund B, Hillengass M, et al. Prognostic significance of whole-body MRI in patients with monoclonal gammopathy of undetermined significance. Leukemia 28:174–178, 2014.
26. MacCallum WG. A Textbook of Pathology, 2nd edition. WB Saunders Company, Philadelphia and London, 1921.
27. Sarkozy A, Carta C, Moretti S, Zampino G, Digilio MC, Pantaleoni F, et al. Germline BRAF mutations in Noonan, LEOPARD, and cardiofaciocutaneous syndromes: molecular diversity and associated phenotypic spectrum. Hum Mutat 30:695–702, 2009.
28. Xing M, Tufano RP, Tufaro AP, Basaria S, Ewertz M, Rosenbaum E, et al. Detection of BRAF mutation on fine needle aspiration biopsy specimens: a new diagnostic tool for papillary thyroid cancer. J Clin Endocr Metab 89:2867–2872, 2004.
29. Omim. Online Mendelian Inheritance in Man. Available from: http://omim.org/downloads, viewed June 20, 2013.
30. Jarboe EA, Mutter GL. Endometrial intraepithelial neoplasia. Semin Diagn Pathol 27:215–225, 2010.
31. Sasaki H, Zlatescu MC, Betensky RA, Ino Y, Cairncross JG, Louis DN. PTEN is a target of chromosome 10q loss in anaplastic oligodendrogliomas and PTEN alterations are associated with poor prognosis. Am J Pathol 159:359–367, 2001.
32. French CA, Kutok JL, Faquin WC, Toretsky JA, Antonescu CR, Griffin CA, et al. Midline carcinoma of children and young adults with NUT rearrangement. J Clin Oncol 22:4135–4139, 2004.
33. Kastrinos F, Stoffel EM. The history, genetics, and strategies for cancer prevention in Lynch syndrome. Clin Gastroenterol Hepatol 2013. pii: S1542–S3565(13)01047-1.
34. Gokmen-Polar Y, Cook RW, Goswami CP, Wilkinson J, Maetzold D, Stone JF, et al. A gene signature to determine metastatic behavior in thymomas. PLoS One 8:e66047, 2013.
35. Fang ZQ, Zang WD, Chen R, Ye BW, Wang XW, Yi SH, et al. Gene expression profile and enrichment pathways in different stages of bladder cancer. Genet Mol Res 12:1479–1489, 2013.
36. Sweeney SE. Hematopoietic stem cell transplant for systemic lupus erythematosus: interferon regulatory factor 7 activation correlates with the IFN signature and recurrent disease. Lupus 20:975–980, 2011.
37. Zhang L, Pagano JS. IRF-7, a new interferon regulatory factor associated with Epstein-Barr virus latency. Mol Cell Biol 17:5748–5757, 1997.
38. Waterhouse M, Bertz H, Finke J. A fast and simple approach for the simultaneous detection of hematopoietic chimerism, NPM1, and FLT3-ITD mutations after allogeneic stem cell transplantation. Ann Hematol 93:293–298, 2014.
39. Holland Frei Cancer Medicine. Kufe D, Pollock R, Weichselbaum R, Bast R, Gansler T, Holland J, Frei E, eds. BC Decker, Ontario, Canada, 2003.
40. Yamamoto T, Yano M, Xu X, Uchinoumi H, Tateishi H, Mochizuki M, et al. Identification of target domains of the cardiac ryanodine receptor to correct channel disorder in failing hearts. Circulation 117:762–772, 2008.
41. Leung E, Landa G. Update on current and future novel therapies for dry age-related macular degeneration. Expert Rev Clin Pharmacol 6:565–579, 2013.
42. Legendre C, Sberro-Soussan R, Zuber J, Rabant M, Loupy A, Timsit MO, et al. Eculizumab in renal transplantation. Transplant Rev (Orlando) 27:90–92, 2013.
43. Berman J, O'Leary TJ. Gastrointestinal stromal tumor workshop. Hum Pathol 32:578–582, 2001.
44. Heinrich MC, Joensuu H, Demetri GD, Corless CL, Apperley J, Fletcher JA, et al. Phase II, open-label study evaluating the activity of imatinib in treating life-threatening malignancies known to be associated with imatinib-sensitive tyrosine kinases. Clin Cancer Res 14:2717–2725, 2008.

45. Heinrich MC, Corless CL, Demetri GD, Blanke CD, von Mehren M, Joensuu H, et al. Kinase mutations and imatinib response in patients with metastatic gastrointestinal stromal tumor. J Clin Oncol 21:4342–4349, 2003.
46. Selvi N, Kaymaz BT, Sahin HH, Pehlivan M, Aktan C, Dalmizrak A, et al. Two cases with hypereosinophilic syndrome shown with real-time PCR and responding well to imatinib treatment. Mol Biol Rep 40:1591–1597, 2013.
47. Cools J, DeAngelo DJ, Gotlib J, Stover EH, Legare RD, Cortes J, et al. A tyrosine kinase created by fusion of the PDGFRA and FIP1L1 genes as a therapeutic target of imatinib in idiopathic hypereosinophilic syndrome. New Eng J Med 348:1201–1214, 2003.
48. Plotkin SR, Merker VL, Halpin C, Jennings D, McKenna MJ, Harris GJ, et al. Bevacizumab for progressive vestibular schwannoma in neurofibromatosis type 2: a retrospective review of 31 patients. Otol Neurotol 33:1046–1052, 2012.
49. Bose P, Holter JL, Selby GB. Bevacizumab in hereditary hemorrhagic telangiectasia. N Engl J Med 360:2143–2144, 2009.
50. Eyetech Study Group. Anti-vascular endothelial growth factor therapy for subfoveal choroidal neovascularization secondary to age-related macular degeneration: Phase II study results. Ophthalmology 110:979–986, 2003.
51. Mead AJ, Rugless MJ, Jacobsen SEW, Schuh A. Germline JAK2 mutation in a family with hereditary thrombocytosis. New Eng J Med 366:967–969, 2012.
52. Barosi G, Bergamaschi G, Marchetti M, Vannucchi AM, Guglielmelli P, Antonioli E, et al. JAK2 V617F mutational status predicts progression to large splenomegaly and leukemic transformation in primary myelofibrosis. Blood 110:4030–4036, 2007.
53. Zhang L, Lin X. Some considerations of classification for high dimension low-sample size data. Stat Methods Med Res, November 23, 2011. Available from: http://smm.sagepub.com/content/early/2011/11/22/0962280211428387.long, viewed January 26, 2013.
54. Mesa RA, Yasothan U, Kirkpatrick P. Ruxolitinib. Nat Rev Drug Discov 11:103–104, 2012.
55. Pesu M, Laurence A, Kishore N, Zwillich SH, Chan G, O'Shea JJ. Therapeutic targeting of Janus kinases. Immunol Rev 223:132–142, 2008.
56. Smallpox Demise Linked to Spread of HIV Infection. BBC News, May 17, 2010.
57. Huang Y, Paxton WA, Wolinsky SM, Neumann AU, Zhang L, He T, et al. The role of a mutant CCR5 allele in HIV-1 transmission and disease progression. Nat Med 2:1240–1243, 1996.
58. Arevalo-Herrera M, Castellanos A, Yazdani SS, Shakri AR, Chitnis CE, Dominik R, et al. Immunogenicity and protective efficacy of recombinant vaccine based on the receptor-binding domain of the Plasmodium vivax Duffy binding protein in Aotus monkeys. Am J Trop Med Hyg 73:25–31, 2005.
59. Miller LH, Mason SJ, Clyde DF, McGinniss MH. The resistance factor to Plasmodium vivax in blacks. The Duffy-blood-group genotype, FyFy. N Engl J Med 295:302–304, 1976.
60. Hill AVS. Evolution, revolution and heresy in the genetics of infectious disease susceptibility. Philos Trans R Soc Lond B Biol Sci 367:840–849, 2012.
61. Long F. When the gut talks to bone. Cell 135:795–796, 2008.
62. Zhang W, Drake MT. Potential role for therapies targeting DKK1, LRP5, and serotonin in the treatment of osteoporosis. Curr Osteoporos Rep 10:93–100, 2012.
63. Chiu HH, Wu MH, Wang JK, Lu CW, Chiu SN, Chen CA, et al. Losartan added to beta-blockade therapy for aortic root dilation in Marfan syndrome: a randomized, open-label pilot study. Mayo Clin Proc 88:271–276, 2013.
64. Report on Rare Disease Research, its Determinants in Europe and the Way Forward. Ayme S, Hivert V, eds. INSERM, May 2011. Available from: http://asso.orpha.net/RDPlatform/upload/file/RDPlatform_final_report.pdf, viewed February 26, 2013.

Chapter 14

Conclusion

14.1 PROGRESS IN THE RARE DISEASES: SOCIAL AND POLITICAL ISSUES

"Research policies should note the specific benefits of research into rare diseases for gaining information, such as the cause of more common and multi-factorial diseases. This may justify weighting of research funds towards rare diseases."

—International Conference for Rare Diseases and Orphan Drugs [1]

Society today encourages the discussion of illness. Children with rare diseases are not, as a matter of social policy, shunned, cloistered within their homes, excluded from educational opportunities, or denied the customary and necessary experiences of a normal and happy childhood. No longer are people forbidden to mention cancer in polite company. Women with breast cancer are often eager to discuss their situation with their friends and with other women who are breast cancer survivors.

The Internet, smartphones, Facebook, and Twitter are outlets for individuals who seek guidance, fellowship, and support services. Individuals and families coping with the medical and emotional challenges incident to rare diseases will find companions around the globe if they have Internet access.

There are now hundreds of rare disease organizations, representing over 30 million Americans; they all maintain an active, supportive Web presence [2]. In addition to the organizations that support awareness for individual diseases, there are groups that consolidate support under a single banner: NORD (National Organization for Rare Diseases), ZebraWatch (rare diseases are sometimes called zebra diseases, because zebras are less common than horses), and EuroDis (Rare Diseases in Europe). The Genetic Alliance is an advocacy organization composed of a network of member organizations. This includes more than 1200 disease-specific groups and various universities, government agencies, and interested profit and non-profit corporations and organizations.

February 29, 2008, a rarely occurring date in a leap year, marked the first observance of "Rare Disease Day," an international initiative designed to draw attention to rare diseases as an important public health issue. In 2012, about 60 countries actively participated in the event.

Rare Diseases and Orphan Drugs. http://dx.doi.org/10.1016/B978-0-12-419988-0.00014-6

Many countries have passed legislation ensuring that the rare diseases receive research funding, that pharmaceutical companies are encouraged to produce medications for the rare diseases, and that individuals and families receive necessary medical and emotional support. In the U.S., some of the most important political milestones have been the following:

- **Public Law 97-414, the Orphan Drug Act of 1983** defines rare diseases and provides sponsors of drugs intended to treat rare diseases with protection from competition (i.e., 7 years of market exclusivity), tax credits, and various other incentives.
- **Public Law 101-629, the Safe Medical Devices Act of 1990** provides incentives and exemptions for devices, much as prior legislation covered drugs. The Act applies to devices to treat or diagnose diseases affecting fewer than 4000 individuals [3].
- **Public Law 105-115, the FDA Modernization Act of 1997** grants an exemption for orphan drugs from drug approval application fees that would otherwise apply [3]. Amendments to the Act in 2007 include the Best Pharmaceuticals for Children Act (Public Law 110-85), which encourages the recruitment of children into clinical trials.
- **Public Law 107-280, the Rare Diseases Act of 2002** directed the National Institutes of Health (NIH) to establish an Office of Rare Diseases and, through this office, to support regional centers of excellence or clinical research into the rare diseases [4]. The Act also increased funding for the development of diagnostics and treatments for the rare diseases [5].
- **Public Law 108-155, the Pediatric Research Equity Act of 2003** is a somewhat ambivalent law that requires applications for new drugs to test for safety and effectiveness in relevant pediatric populations, while providing full and partial waivers from the law when such testing is considered impractical.
- **Public Law 110-233, the Genetic Information Nondiscrimination Act of 2008** makes it illegal to discriminate against employees or applicants for employment based on their personal genetic information (i.e., whether individuals or family members have a genetic disease or condition, or whether individuals are at risk of developing a disease or condition based on genetic testing).
- **Public Law 111-80, the Agriculture, Rural Development, Food and Drug Administration, and Related Agencies Appropriations Act of 2010** authorized the FDA to appoint a review group to recommend design improvements for preclinical and clinical trials aimed at preventing, diagnosing, and treating rare diseases [3].
- **Public Law 111-148, the Patient Protection and Affordable Care Act of 2010, known widely as ObamaCare** requires insurance companies to cover all applicants regardless of pre-existing conditions. As generally interpreted, the Act will eliminate lifetime caps on benefits.

Current social, legal, and scientific realities provide a nurturing environment for grass roots rare disease movements. Despite these advances, primitive ideas about the rare diseases remain entrenched in the human psyche. Not long ago, the medical profession referred to children with congenital malformations as "monsters." The idea that a physician, ostensibly devoted to the well-being of his patients, would convey to a new mother that she gave birth to a monster seems cruel and stupid. Yet, even today, medical journals use the term "monster" when it suits them [6].

Rare disease organizations must be aware that there are many ways to subvert the best of intentions. The case of *Greenberg v. Miami Children's Hospital* provides a stark example. The Greenbergs, and about 150 other families, provided funds, tissues, and a range of services in support of Dr. Reuben Matalon's efforts to find the gene responsible for Canavan disease. He succeeded, and promptly patented the gene for his employer, Miami Children's Hospital (U.S. Patent 5,679,635, October 21, 1997). The Miami Children's Hospital charged a royalty fee for the test. The families, thinking that their donations of time, materials, and money had supported an altruistic effort, were shocked that Miami Children's Hospital was trying to profit from the misfortune of children with Canavan disease; hence, the lawsuit. In the end, Miami Children's Hospital won, and was permitted to continue to charge royalties for the use of their diagnostic test (see Glossary item, Intellectual property).

Rare disease advocates have a tendency to assume that everyone shares their priorities. This is definitely not the case. It can be a hard sell to convince the average voter that our limited healthcare funding for the common diseases should be diverted to the rare diseases. Funding for the common diseases has always prevailed over funding for the rare diseases. If there were a reversal, with the rare diseases suddenly taking the bulk of healthcare expenditures, the public would ask whether it is reasonable to treat every individual with a rare disease, regardless of cost, regardless of proven benefit, and regardless of the effects on the national economy.

Though recent progress in the rare diseases has been rapid and profound, there are effective treatments for only a few hundred rare diseases out of a total of about 7000. Many of these treatments are incredibly expensive. Will we, as a society, be able to pay for the development of new drugs and related treatments for all of the rare diseases, and will we be able to pay for these new drugs and treatments when they become available? When it comes to healthcare research and treatment, rare and common, it comes down to two crude but inescapable questions: "How badly do we want all this healthcare?" and "Who is going to pay for it?"

Families coping with rare diseases must somehow endure the profit-driven pricing policies of pharmaceutical companies. In the U.S., the cost of drugs is controlled by the pharmaceutical companies, and the pharmaceutical companies are restrained by the market (i.e., how much people are willing to pay). In the case of life-threatening diseases, individuals and insurance companies are willing to pay quite a lot. Procarbazine is effective against Hodgkin lymphoma, and was approved for that purpose back in 1969. Inexpensive to synthesize, procarbazine sold for decades at under a dollar per pill. In 2005, the price jumped to $56 per pill. In 2005, only one company was marketing the drug. Without competitors, the manufacturer could ask almost any price; and they did.

There is a class of rare diseases known as lysosomal storage diseases. Gaucher, Pompe, and Fabry diseases are examples. In each disease, cells fail to break down a lipid molecule, resulting in intracellular accumulations that interfere with normal cellular functions. Such diseases can be successfully treated with drugs that replace the specific enzymes lacking in affected individuals. The cost of a year's supply of medicine ranges from $200,000 to $400,000.

For a variety of reasons discussed earlier, individuals with rare diseases often receive "off-label" drug treatments (see Glossary item, Off-label). Physicians are permitted to prescribe drugs for purposes other than the approved purposes, as listed on the manufacturer's label; hence, the term "off-label." It is impossible to know with any precision or confidence the prevalence of off-label treatments in the rare diseases. Still, it is commonly held that about 90% of all treatments for rare diseases lack FDA approval. Furthermore, most off-label treatments are conducted without the benefit of a series of well-designed, statistically valid clinical trials establishing the safety and efficacy of the drug for its intended patient population [7]. Medicare and private insurers pay for a variety of off-label uses of drugs, particularly when these drugs are used to treat rare diseases.

By lifting the lifetime cap on benefits, as provided by the Patient Protection and Affordable Care Act of 2010, the spiraling long-term costs associated with off-label treatments will effectively shift to public and private insurers. The cost of care for individuals with rare diseases is certain to become a point of national debate. It is in the interest of rare disease advocates to ensure that the treatments used for rare diseases are safe, effective, based on good scientific evidence, and fairly priced.

14.2 SMARTER CLINICAL TRIALS

"The subphenotyping of COPD [chronic obstructive pulmonary disease] into separate groups based on mechanism sets the stage for the rational development of therapeutics."

—Stephen Rennard [8]

Modern clinical trials had great success in the late 1960s and early 1970s, when highly effective chemotherapeutic agents were found to be effective against a wide range of rare, childhood cancers. The prospective randomized control trial, performed on children with cancer, was so successful that it served as a requirement for drug testing for the past half-century.

Today, large, randomized prospective clinical trials are the standard for common diseases such as cancer. The problem has been that none of the drugs tested on adults with cancer has had the kind of curative successes that we saw with the childhood tumors. Larger, longer, and increasingly expensive studies were conducted to demonstrate incremental improvements in chemotherapeutic regimens. Though there have been successes in clinical trials for the common cancers occurring in adults, no trial on common cancers has yielded the spectacular successes witnessed for the rare childhood cancers.

14.2.1 Rule—Clinical trials are the best method ever developed to determine whether a drug is safe and effective for a particular purpose in a particular target population. Nonetheless, clinical trials cannot provide the clinical guidance we need to develop all of the new medications that will be needed to conquer the common diseases.

Brief Rationale—We simply do not have the money, time, and talent to perform all the anticipated clinical trials for the common diseases.

Modern clinical trials are long and expensive. It takes about 10 to 15 years for an experimental drug to be developed [9]. Only five in 5000 compounds that have preclinical testing will enter clinical trials [9]. The cost of developing a drug and bringing it to market is about $1 billion [10].

Clinical trials can be very large. In the realm of cancer trials, the Prostate, Lung, Colorectal and Ovarian Cancer Screening Trial (PLCO, NIH/NCI trial NO1 CN25512) serves as an example. The PLCO is a randomized controlled cancer trial. Between 1992, when the trial opened, and 2001, when enrollment ended, 155,000 participants were recruited [11]. The study will end in 2016.

It can be difficult or impossible to enroll all the patients required for a clinical trial. In an analysis of 500 planned cancer trials, 40% of trials failed to accrue the minimum necessary number of patients. Of cancer trials that have passed through preclinical, phase I clinical, and phase II clinical trials, three out of five failed to achieve the necessary patient enrollment to move into the final phase III clinical trial [12]. Most clinical trials for cardiovascular disease, diabetes, or depression are designed to be even larger than cancer trials [12].

Overall, about 95% of drugs that move through the clinical trial gauntlet will fail [13]. Of the 5% of drugs that pass, their value may be minimal. To pass a clinical trial, a drug must have proven efficacy. It need not be curative; only effective. Of the drugs that pass clinical trials, some will have negligible or incremental benefits. After a drug has reached market, its value to the general population might be less than anyone had anticipated. Clinical trials, like any human endeavor, are subject to error [14–16]. Like any human endeavor, clinical trials need to be validated in clinical practice [10]. It may take years or decades to determine whether a treatment that demonstrated a small but statistically significant effect in a clinical trial will have equivalent value in everyday practice.

Funders of medical research are slowly learning that there simply is not enough money or time to conduct all of the clinical trials that are needed to advance medical science at a pace that is remotely comparable to the pace of medical progress in the first half of the twentieth century.

14.2.2 Rule—Clinical trials for common diseases have limited value if the test population is heterogeneous; as is often the case.

Brief Rationale—Abundant evidence suggests that most common diseases are heterogeneous, composed of genotypically and phenotypically distinct disease populations, with each population responding differently with the clinical trial.

The population affected by a common disease often consists of many distinct genetic and phenotypic subtypes of the disease; essentially many different diseases. A successful clinical trial for a common disease would require a drug that is effective against different diseases that happen to have a somewhat similar phenotype. One-size-fits-all therapies seldom work as well as anticipated, and more than 95% of the clinical trials for common diseases fail [13].

14.2.3 Rule—Clinical trials for the rare diseases are less expensive, can be performed with less money, and provide more definitive results than clinical trials on common diseases.

Brief Rationale—Common diseases are heterogeneous and produce a mixed set of results on subpopulations. This in turn dilutes the effect of a treatment and enlarges the required number of trial participants. Rare diseases are homogeneous, thus producing a uniform effect in the trial population, and thus lowering the number of trial participants required to produce a statistically convincing result.

Rare diseases often have a single genetic aberration, driving a single metabolic pathway that results in the expression of a rather uniform clinical phenotype. This means that a drug that succeeds in one patient will likely succeed in every patient who has the same disease. Likewise, a drug that fails in one patient will fail in all the other patients. This phenomenon has enormous consequences for the design of clinical trials. When the effects of drugs are consistent, the number of patients enrolled in clinical trials can be reduced, compared with the size of clinical trials wherein the effects of drugs are highly variable among the treated population. In general, clinical trials targeted on rare diseases or on genotypically distinct subsets of common diseases require fewer enrolled participants than trials conducted on heterogeneous populations that have a common disease [13].

It is easy to assume that because rare diseases affect fewer individuals than do the common diseases, it would be difficult to recruit a sufficient number of patients into an orphan drug trial. Due to the energetic and successful activities of rare disease organizations, registries of patients have been collected for hundreds of different conditions. For the most part, patients with rare diseases are eager to enroll in clinical trials. The rare disease registries, made available to clinical trialists, eliminate the hit-or-miss accrual activities that characterize clinical trials for common diseases.

In an effort to increase the scientific and clinical value of clinical trials, trialists often include ancillary studies in their trial designs. These ancillary studies may consist of molecular studies on tissue biopsies obtained from trial subjects. Using biopsy samples, different responses to a treatment can be correlated with a genetic marker or a genetic profile measured on tissues. In instances for which rare disease organizations collect and store biopsies obtained from their registered members, ancillary studies for orphan drug trials can be performed quickly, and with less expense than comparable studies on common diseases.

In the U.S., the FDA is poised to provide guidance to organizations and corporations conducting clinical trials on orphan drugs [17]. It is crucial that trial sponsors stay in close touch with FDA staff during the planning stages of drug trials. A little advice from a regulator can avoid the heartbreak that comes when an effective drug fails approval due to poor trial design.

Trials on orphan drugs commonly accrue human subjects from vulnerable populations (e.g., children, mentally impaired subjects, subjects with multiple life-threatening conditions). In such cases, human subjects may not be able to provide informed consent, and a parent or guardian will need to be consulted (see Glossary item, Informed consent). Trialists must be sensitive to the special needs of their subjects and their families. When recruiting members for a clinical safety board or an institutional review board, it is important to select individuals who are sensitive to the social issues raised by rare disease clinical trials, and who have no financial ties to the trialists or their sponsors [17].

14.2.4 Rule—Clinical trials on a common disease can be reduced to one or more trials of a subtype of the disease.

Brief Rationale—The heterogeneity of populations with a common disease allows trialists the freedom to design small trials for subsets of individuals who have a particular genotype (i.e., a gene marker or a gene expression profile), a particular mode of inheritance (i.e., Mendelian), or a distinguishing clinical phenotype (e.g., early onset disease).

A recurring theme in this book is that common diseases are collections of genotypically distinct diseases that share a common phenotype and common disease pathways [8,18,19]. If there is some reason to expect a drug to be particularly effective against a defined subset of individuals with a common disease, it may be worthwhile to design the trial for these individuals. The pharmaceutical company Genentech employed this strategy when it developed the breast cancer drug trastuzumab (trade name Herceptin). Trastuzumab is a monoclonal antibody against the HER2 receptor (see Section 13.2). In this case, preclinical evidence indicated that trastuzumab might be effective against breast cancers that had high levels of HER2. By limiting their study to individuals with HER2-positive breast cancers, the company achieved success with a relatively small number of trial participants [13].

As previously discussed, it is easy to find rare diseases that pose as variant subsets of common diseases (e.g., B-K mole syndrome patients composing a subset of individuals at high risk of developing melanoma; BRCA gene-positive individuals as a subset of individuals at high risk of breast cancers; patients with alpha-1-antitrypsin deficiency as a subset of emphysema cases). A clinical trial specifically aimed at a rare subset of a common disease might facilitate later trials directed at other subsets of the same disease.

Such clinical trials are in progress. The I-SPY 2 trial matches treatments against subgroups of breast cancer patients whose tumor cells match particular molecular profiles [13]. In the I-SPY 2 trial, multiple drugs are tested on relatively small, selected subgroups of cancer patients. As results are collected, unsuccessful drugs are phased out and replaced by other drug candidates, all within the same trial [13].

14.3 FOR THE COMMON DISEASES, ANIMALS ARE POOR SUBSTITUTES FOR HUMANS

"The proper study of Mankind is Man."

—Alexander Pope in "An Essay on Man," 1734.

Common diseases are complex, as is the response of humans to treatments for the common diseases. Are we likely to find adequate animal models for common diseases?

14.3.1 Rule—For the common diseases of humans, there are no adequate animal models.

Brief Rationale—The common diseases are complex, the end result of many genetic and environmental factors. There is no reason to expect that a complex set of factors interacting in humans could be replicated in an animal.

Rodents, especially mice and rats, are often used in disease research. Historically, the drug development process employs mouse models to identify candidate drugs for clinical trials in humans [20]. Few such mouse-inspired trials have shown success [21–24]. In a review of human clinical trials based on research data collected from mouse models, every one of 150 clinical trials of inflammatory responses in humans was a failure [20].

In the vascular field, there are animal models for stroke. Based on animal models, about 500 candidate drugs were proposed as neuroprotective agents in human stroke. Of the 500 candidate drugs, only two were shown to be of value for humans [23].

In the field of cancer research, carcinogens induce cancers in rodents, and the cancers that occur in rodents and humans share a set of fundamental properties: continuous growth, autonomous growth, invasiveness, metastasis (see Glossary item, Autonomous growth). Beyond these features, most animals models deviate from their human counterparts. Here are a few examples:

- Rodent tumors develop over a very short period of time, limited by the short life expectancy of the mouse or rat. A strong carcinogen can produce

palpable mouse tumors in mere weeks. The commonly occurring tumors in humans require years to develop.
- In most strains of rodent, tumors lack molecular markers commonly found in human tumors (e.g., p53). The cytogenetic markers for rodent tumors are different from the cytogenetic markers for human tumors. In fact, the karyotype, physical mappings of genes, causal genes, and gene polymorphisms of rodent tumors are all quite different from human tumors (see Glossary items, Synteny, Haplotype).
- Animals metabolize drugs differently compared to humans.
- Viruses, bacteria, and other organisms that cause human cancer are different from the organisms causing cancer in animals.
- The diet of animals is different from the diet of humans.
- The host factors of animals, including immune status, are different from those of humans.

In a remarkable paper by Shachaf and coworkers, a malignant tumor in mice was cured by inactivating its oncogene [25]. The authors used a transgenic mouse that overexpressed the MYC oncogene in liver cells (see Glossary item, Transgenic). Such mice develop liver tumors in about 3 months. When MYC gene overexpression was experimentally inactivated, the liver tumors regressed. Furthermore, the liver underwent a restorative process, recruiting normal hepatocytes and bile duct cells from the formerly cancerous cell population. This experiment suggests that regardless of the many steps involved in the pathogenesis of cancer, the cancer can be cured at advanced stages of growth when a key gene driving the growth of the tumor is suppressed. In this mouse model, carcinogenesis did not involve the many steps in tumor development observed for the commonly occurring cancers of humans. With a short, 3-month development time, the mouse tumor did not accommodate the long period of tumor development in humans, wherein multiple mutations accumulate.

Giurato and coworkers used a similar MYC inactivation model [26]. In their mouse model, MYC overexpression in lymphoid cells produced aggressive lymphomas in transgenic mice. MYC suppression in tumor cells caused regression of the primary lymphomas but did not cause regression of all the transplanted lymphomas (see Glossary items, Regression, Spontaneous regression). One out of four transplanted lymphomas persisted when its MYC expression was suppressed. The authors demonstrated that the persistently growing tumors had acquired another abnormality: p53 inactivation. When a second genetic abnormality was acquired, MYC inactivation alone could no longer regress the malignant phenotype.

In the common tumors of humans, genetic instability leads to the accumulation of many different genetic abnormalities. Only the rare tumors of humans are characterized by simple genetic abnormalities. The transgenic mouse models demonstrating tumor regression following oncogene suppression were not modeling common cancers of humans; they were modeling rare, monogenic cancers of humans.

The closer we look at the pathogenesis of human diseases, the less we can seriously entertain the notion that animal diseases can model common diseases. In the case of rare diseases, which are often monogenic and simple, it may be possible to find animal models.

14.3.2 Rule—Animals can model some of the rare human diseases.

Brief Rationale—Because many of the rare diseases are caused by mutations in one gene, a defect in an orthologous gene may produce a disease in an animal model that resembles the human phenotype.

An orthologue is a gene found in different organisms that evolved from a common ancestor's gene. When an ancestral gene is passed to descendant species, the function of the orthologous gene is likely to be similar in all of the species that contain the gene. Hence, a rare disease caused by an alteration in a single gene may produce a similar phenotype if it occurs in the orthologous gene in another species. Such occurrences are called orthodiseases. For example, *Drosophila* contains homologues of the genes that cause tuberous sclerosis, a hamartoma-cancer syndrome in humans. The brain tubers (hamartomas of the neuroectoderm, also called phakomas), for which tuberous sclerosis takes its name, contain large, multinucleate neurons. Loss of function of these genes in *Drosophila* produces enlarged cells with many times the normal amount of DNA [27]. The tuberous sclerosis orthodisease in *Drosophila* is being studied to help us understand cell growth control mechanisms in humans.

A form of Cornelia de Lange syndrome is caused by mutations in NIPBL (see Section 10.1). Human NIPBL is homologous to Nipped-B in *Drosophila melanogaster*, a type of fly, and is presumed to be an orthologue. Cornelia de Lange syndrome is characterized by structural abnormalities of the face, limb reduction, growth delay, and mental retardation, a phenotype that cannot be replicated in *Drosophila melanogaster*.

There are good examples of mouse model orthodiseases (see Glossary item, Orthodisease). Cisd2, a candidate aging-associated gene in humans, causes premature aging and a shortened lifespan in Cisd2-null mice [28]. Xeroderma pigmentosum, complementation group f, which causes photosensitivity and a heightened risk of early skin cancer in humans, can be simulated in mice with an XPF-dependent loss in telomeres. Ligneous conjunctivitis due to plasminogen deficiency in humans can be modeled by a similar conjunctivitis in mice lacking the plasminogen gene [29]. Rare rhabdoid tumors in humans are modeled by INI1-negative rhabdoid tumors in mice [30].

Having an orthodisease does not guarantee experimental success. Zebrafish, an organism popular among developmental biologists, can be infected with mycobacteria. A gene in the zebrafish was shown to modulate its susceptibility to mycobacterial infection [31]. Naturally, there was hope that the orthologous gene in humans would be associated with human susceptibility to tuberculosis. Despite a large study involving 9115 subjects, no such association was found [32]. As we learned in Chapter 10, the genetic cause of a disease is different from its pathogenesis. **Though a gene mutation may be found in a human disease and its orthodisease, the pathogenesis of the disease (i.e., the cellular events that lead to the clinical expression of disease) may diverge considerably in animals and humans.**

Animal models sometimes bypass pathogenesis entirely. In an animal model for Parkinson disease, researchers may resort to directly injuring brain tissue to produce a nigrostriatal dopamine deficiency to mimic the end result of progressive Parkinsonism [23]. In such a model, the complex pathogenesis of human Parkinson disease, phenotypically expressed as a slow, progressive degeneration and loss of target neurons, is simply omitted.

When pathogenesis in different species is conserved, orthodiseases may teach us something about the pathogenesis of human disease [33]. So far, experience would suggest that this is seldom the case. The same organisms used with great success by developmental biologists may fail miserably as disease response predictors in humans [34].

Assuming for a moment that a valid animal model could be obtained for a common human disease, there would be no special reason to expect that a treatment effective in an animal model would be similarly effective in humans; there are just too many response variables. In recent years, a substantial literature has emerged questioning whether there is scientific evidence to support the use of animals as models for human disease [21–24,35,36].

14.3.3 Rule—Animals cannot serve as models for human responses to treatment.

Brief Rationale—The responses of an organism to a drug are species specific and complex, and are determined by traits that have evolved over time to maximize the survival of the species.

Gram-negative bacteria can produce shock in animals via lipopolysaccharide, a molecule found in their cell walls. Mice have high resistance to the shock-inducing effect of lipopolysaccharide. The dose of lipopolysaccharide causing death in mice happens to be 1 million times the dose that causes fever in humans and about 1000–10,000 times more than the dose that causes shock in humans [37].

The mechanisms through which different organisms respond to the same toxin may be wildly different. Human responses cannot be reliably extrapolated from animal responses. Simon LeVay recounts a clinical trial that went terribly wrong [38]. The drug being tested, TGN1412, was a monoclonal antibody developed by a biotechnology company. Following preliminary safety tests in laboratory animals, a safe dose was selected for humans. Eight paid healthy volunteers were assembled. These subjects would be the first humans to receive the test drug under any conditions. In a single session, six of the eight volunteers were infused with TGN1412, and two volunteers were infused with a placebo. In about an hour, all six of the subjects receiving TGN1412 developed cytokine storm, a life-threatening condition in which an immune-response precipitates shock and a wide range of extreme system-wide responses, including multi-organ failure. Prompt treatment saved all their lives. Two of the six had prolonged hospital courses. The six subjects who received the drug, deemed safe on the basis of animal tests, must now deal with long-term medical consequences of the event [38,39].

Each species reacts to stimuli in typical, species-specific ways. Humans with unrelated types of injuries will react with a systemic response that varies little from person to person [20]. Whether the injury is trauma, burns, or endotoxemia, the physiologic response in humans is reflected in a shared gene expression profile. The stereotypical human response to injury is not replicated in mice; neither is the gene expression profile. It is of no surprise, then, that of about 150 candidate anti-inflammatory agents developed from mouse inflammation models, none have passed trials in humans [20]. You would not expect fine-tuned protective mechanisms, developing over millions of years, to apply to mice and to men.

If we were to abandon using animals in preclinical trials, how would we know that a drug is safe to test in humans (see Glossary item, Preclinical trial)? When a new drug is administered in humans for the very first time (a so-called first-in-man trial), extraordinary measures must be taken to minimize risk. Guidelines for first-in-man trials are available to researchers and should be followed closely [40]. Realistically, though, whenever a human is given a drug for the first time, there is no way to guarantee safety. If the rare diseases are to serve as the vanguard of clinical trials, as suggested in the previous section of this chapter, society will need to balance risks against benefits. It is likely that rare disease organizations, patients, and their advocates will have much to say on the subject as the research community gears up to design small clinical trials targeted against rare diseases and rare variants of common diseases.

14.4 HUBRIS

"It is likely that the complexity of complex diseases may ultimately limit the opportunities for accurate prediction of disease in asymptomatic individuals as unraveling their complete causal pathways may be impossible."

Cecile Janssens and Cornelia van Duijn [41]

In 1949, Linus Pauling and his coworkers showed that sickle cell anemia is a disease produced by an inherited alteration in hemoglobin, yielding a molecule that is separable from normal hemoglobin by electrophoresis [42]. In 1956, Vernon Ingram and J.A. Hunt sequenced the hemoglobin protein molecule (normal and sickle cell) and showed that the inherited alteration in sickle cell hemoglobin is due to a single amino acid substitution in the protein sequence [43].

The sickle cell variant is a relatively common genetic trait. In the U.S., it occurs most commonly in African-Americans. When the sickle cell trait is inherited from both parents, the offspring develops sickle cell anemia. In the 1950s, it was widely assumed that the discovery of the molecular basis of sickle cell disease, followed by the development of a simple screening test for the sickle cell trait, would lead inevitably to the eradication of sickle cell anemia. Individuals who carried the sickle cell trait would refrain from mating with other individuals who carried the trait. Consequently, no new cases of sickle cell disease would arise.

Here we are, nearly 60 years later. Has sickle cell disease been eradicated? Not at all. Using publicly available mortality records provided by the CDC (U.S. Centers for Disease Control and Prevention) we can determine the reported incidence of deaths occurring in individuals with sickle cell anemia [44]:

In 1996, the U.S. rate of sickle cell disease in death certificates was 30.54 per 100,000
In 1999, the U.S. rate of sickle cell disease in death certificates was 33.36 per 100,000
In 2002, the U.S. rate of sickle cell disease in death certificates was 33.79 per 100,000
In 2004, the U.S. rate of sickle cell disease in death certificates was 36.47 per 100,000

Sickle cell anemia is still with us, and there seems to be no drop at all in its incidence in the U.S. If anything, the death rate has been increasing. This trend is not what anyone, back in the 1950s, expected or wanted.

Historically, it has been difficult or impossible to eradicate diseases that are locked into the social fabric of our lives. Alcoholism and other substance addictions, obesity, and sexually transmitted diseases have persisted despite medical advances. Air and water pollution continue despite endless warnings from public health officials and concerned individuals.

There are sound, scientific reasons for pursuing research into the rare diseases, but there are a large number of unresolved societal issues that need to be addressed before the full benefit of rare disease research will have much impact.

We do not know how society will react to advances in the genetics of rare diseases and common diseases, but we can imagine the kinds of decisions that we will be facing in the next decade. For the rare diseases, there will be tests for determining adult carriers of disease genes, and tests for *in utero* diagnosis of rare diseases (e.g., genetic analysis of chorionic villus samples, or fetal ultrasound). For the common diseases, there will be profiles available for determining whether an individual has a set of gene variants conferring a high risk of developing a disease. For those individuals with diagnosed rare diseases and common diseases, there will be molecular profiles that determine which patients are most likely to benefit from expensive drug treatments.

In the very near future, individuals will decide whether they want to have their complete genomes sequenced and analyzed. Currently, it is not clear whether predictive testing will be of much value for the common diseases [41], but many of the mutations that cause rare diseases are well described and whole genome surveys of such genes are feasible (see Glossary item, Predictive test). Are we, as a society, prepared to cope with detailed knowledge about our disease-carrying genes? Will individuals with "bad" genes be shunned by potential mates? Will we be choosing our spouses to avoid homozygosity for undesirable gene traits? Will our future doctors adopt the principles of eugenics?

As the physicist Niels Bohr said, "Prediction is very difficult, especially if it's about the future." **One thing is certain: the rare diseases will play a very large role in the future of medicine.**

REFERENCES

1. Forman J, Taruscio D, Llera VA, Barrera LA, Cote TR, Edfjäll C, et al. The need for worldwide policy and action plans for rare diseases. Acta Paediatr 101:805–807, 2012.
2. National Organization for Rare Diseases. Available from: http://www.rarediseases.org/
3. Field MJ, Boat T. Rare Diseases and Orphan Products: Accelerating Research and Development. Institute of Medicine (US) Committee on Accelerating Rare Diseases Research and Orphan Product Development. The National Academies Press, Washington, DC, 2010. Available from: http://www.ncbi.nlm.nih.gov/books/NBK56189/
4. Rare Diseases Clinical Research Consortia (RDCRC) for the Rare Diseases Clinical Research Network (U54). NIH Request for Applications Number: RFA-OD-08-001, issued February 8, 2008.
5. Rare Diseases Act of 2002, Public Law 107-280, 107th U.S. Congress, November 6, 2002.
6. Dhall U, Kayalvizhi I, Magu S. Acardius acephalus monster—a case report. J Anat Soc India 54:26–28, 2005.
7. Radley DC, Finkelstein SN, Stafford RS. Off-label prescribing among office-based physicians. Arch Intern Med 166:1021–1026, 2006.
8. Rennard SI, Vestbo J. The many "small COPDs", COPD should be an orphan disease. Chest 134:623–627, 2008.
9. Orphan Drugs in Development for Rare Diseases; 2011 Report. America's Biopharmaceutical Research Companies. Available at: http://www.phrma.org/sites/default/files/pdf/rarediseases2011.pdf, viewed July 14, 2013.
10. Berman JJ. Principles of Big Data: Preparing, Sharing, and Analyzing Complex Information. Morgan Kaufmann, 2013.
11. Prostate, Lung, Colorectal & Ovarian Cancer Screening Trial (PLCO). Available from: http://prevention.cancer.gov/plco, viewed August 22, 2013.
12. English R, Lebovitz Y, Griffin R. Forum on Drug Discovery, Development, and Translation. Institute of Medicine, 2010.
13. Leaf C. Do clinical trials work? The New York Times, July 13, 2013.
14. Bossuyt PM, Reitsma JB, Bruns DE, Gatsonis CA, Glasziou PP, Irwig LM, et al. The STARD statement for reporting studies of diagnostic accuracy: explanation and elaboration. Clin Chem 49:7–18, 2003.
15. Ioannidis JP. Why most published research findings are false. PLoS Med 2:e124, 2005.
16. Ioannidis JP. Some main problems eroding the credibility and relevance of randomized trials. Bull NYU Hosp Jt Dis 66:135–139, 2008.

17. Wizemann T, Robinson S, Giffin R. Breakthrough Business Models: Drug Development for Rare and Neglected Diseases and Individualized Therapies Workshop Summary. National Academy of Sciences, 2009.
18. Crow YJ. Lupus: how much "complexity" is really (just) genetic heterogeneity? Arthritis Rheum 63:3661–3664, 2011.
19. Wade N. Many rare mutations may underpin diseases. The New York Times, May 17, 2012.
20. Seok J, Warren HS, Cuenca AG, Mindrinos MN, Baker HV, Xu W, et al. Genomic responses in mouse models poorly mimic human inflammatory diseases. Proc Natl Acad Sci USA 110:3507–3512, 2013.
21. Pound P, Ebrahim S, Sandercock P, Bracken MB, Roberts I. Reviewing Animal Trials Systematically (RATS) Group. Where is the evidence that animal research benefits humans? BMJ 328:514–517, 2004.
22. Hackam DG, Redelmeier DA. Translation of research evidence from animals to humans. JAMA 296:1731–1732, 2006.
23. van der Worp HB, Howells DW, Sena ES, Porritt MJ, Rewell S, O'Collins V, et al. Can animal models of disease reliably inform human studies? PLoS Med 7:e1000245, 2010.
24. Rice J. Animal models: not close enough. Nature 484:S9, 2012.
25. Shachaf CM, Kopelman AM, Arvanitis C, Karlsson A, Beer S, Mandl S, et al. MYC inactivation uncovers pluripotent differentiation and tumour dormancy in hepatocellular cancer. Nature 431:1112–1117, 2004.
26. Giuriato S, Ryeom S, Fan AC, Bachireddy P, Lynch RC, Rioth MJ, et al. Sustained regression of tumors upon MYC inactivation requires p53 or thrombospondin-1 to reverse the angiogenic switch. Proc Natl Acad Sci USA 103:16266–16271, 2006.
27. Tuberous Sclerosis Complex in Flies Too? A Fly Homolog to TSC2, Called Gigas, Plays a Role in Cell Cycle Regulation, No attributed author. Available from: http://www.ncbi.nlm.nih.gov/books/bv.fcgi?rid=coffeebrk.chapter.25, July 27, 2000.
28. Chen YF, Kao CH, Chen YT, Wang CH, Wu CY, Tsai CY, et al. Cisd2 deficiency drives premature aging and causes mitochondria-mediated defects in mice. Genes Dev 23:1183–1194, 2009.
29. Drew AF, Kaufman AH, Kombrinck KW, Danton MJ, Daugherty CC, Degen JL, et al. Ligneous conjunctivitis in plasminogen-deficient mice. Blood 91:1616–1624, 1998.
30. Roberts CW, Galusha SA, McMenamin ME, Fletcher CD, Orkin SH. Haploinsufficiency of Snf5 (integrase interactor 1) predisposes to malignant rhabdoid tumors in mice. Proc Natl Acad Sci USA 97:13796–13800, 2000.
31. Tobin DM, Vary JC Jr, Ray JP, Walsh GS, Dunstan SJ, Bang ND, et al. The lta4h locus modulates susceptibility to mycobacterial infection in zebrafish and humans. Cell 140:717–730, 2010.
32. Curtis J, Kopanitsa L, Stebbings E, Speirs A, Ignatyeva O, Balabanova Y, et al. Association analysis of the LTA4H gene polymorphisms and pulmonary tuberculosis in 9115 subjects. Tuberculosis (Edinb) 91:22–25, 2011.
33. McGary KL, Parka TJ, Woodsa JO, Chaa HJ, Wallingford JB, Marcottea EM. Systematic discovery of nonobvious human disease models through orthologous phenotypes. Proc Natl Acad Sci USA 107:6544–6549, 2010.
34. Saey TH. Rare genetic tweaks may not be behind common diseases: variants thought to be behind inherited conditions prove difficult to pin down. Science News 183:11, January 26, 2013.
35. Mitka M. Drug for severe sepsis is withdrawn from market, fails to reduce mortality. JAMA 306:2439–2440, 2011.

36. Davis MM. A prescription for human immunology. Immunity 29:835–838, 2008.
37. Warren HS, Fitting C, Hoff E, Adib-Conquy M, Beasley-Topliffe L, Tesini B, et al. Resilience to bacterial infection: difference between species could be due to proteins in serum. J Infect Dis 201:223–232, 2010.
38. LeVay S. When Science Goes Wrong. Twelve Tales from the Dark Side of Discovery. Plume, New York, pp 160–180, 2008.
39. Berman JJ. Machiavelli's Laboratory. 2010. Available from: http://www.julesberman.info/integ/machfree.htm, viewed August 22, 2013.
40. Expert Scientific Group on Phase One Clinical Trials: Final Report. Her Majesty's Stationery Office, November 30, 2006. Available from: http://www.dh.gov.uk/prod_consum_dh/groups/dh_digitalassets/@dh/@en/documents/digitalasset/dh_073165.pdf, viewed October 14, 2009.
41. Cecile A, Janssens JW, van Duijn CM. Genome-based prediction of common diseases: advances and prospects. Human Mol Gen 17:166–173, 2008.
42. Pauling L, Itano HA, Singer SJ, Wells IC. Sickle cell anemia, a molecular disease. Science 110:543–548, 1949.
43. Ingram VM. A specific chemical difference between globins of normal and sickle-cell anemia hemoglobins. Nature 178:792–794, 1956.
44. Berman JJ. Methods in Medical Informatics: Fundamentals of Healthcare Programming in Perl, Python, and Ruby. Chapman and Hall, Boca Raton, 2010.

Appendix I

List of Genes Causing More than One Disease

There are numerous examples wherein mutations in one gene may result in more than one diseases. In some cases, each of the diseases caused by the altered gene are fundamentally similar (e.g., spherocytosis and elliptocytosis, caused by mutations in the alpha-spectrin gene; Usher syndrome type IIIA and retinitis pigmentosa-61 caused by mutations in the CLRN1 gene). In other cases, diseases caused by the same gene may have no apparent relation to one another (Stickler syndrome type III and fibrochondrogenesis-2 and a form of non-syndromic hearing loss all caused by mutations in the COL11A2 gene).

In the following list, each disease-causing gene is followed by the different diseases caused by gene alterations. By reading this list closely, you may discover heretofore unknown pathogenetic relationships among diseases.

ABCB6 gene
The Lan(-) blood group phenotype
Microphthalmia, isolated, with coloboma 7

ACTA2 gene
Moyamoya disease-5
Form of thoracic aortic aneurysm

ACYLTRANSFERASE GENE
Fish-eye disease
Norum disease

ALPHA-SPECTRIN GENE
Hereditary spherocytosis-3
Elliptocytosis-2

ALPHA-SYNUCLEIN GENE
Parkinson disease-1
Autosomal dominant Parkinson disease-4

Rare Diseases and Orphan Drugs. http://dx.doi.org/10.1016/B978-0-12-419988-0.00023-7

ALX4 gene
Frontonasal dysplasia-2
Parietal foramina-2

ANO5 gene
Gnathodiaphyseal dysplasia; gdd, or osteogenesis imperfecta with unusual skeletal lesions
Limb-girdle muscular dystrophy-2L
Miyoshi muscular dystrophy-3

ARX gene
Proud syndrome
Form of non-specific X-linked mental retardation

ATN1 gene
Dentatorubral-pallidoluysian atrophy
Haw River syndrome

ATR gene
Seckel syndrome-1
Form of ataxia telangiectasia

BAG3 gene
Autosomal dominant myofibrillar myopathy
Dilated cardiomyopathy-1HH

BAP1 gene
Susceptibility to uveal melanoma
Predisposition to malignant mesothelioma upon asbestos exposure

BCS1L gene
Bjornstad syndrome
GRACILE syndrome

BUB1B gene
Mosaic variegated aneuploidy syndrome-1 (see Glossary item, Aneuploidy)
Form of premature chromatid separation

C20ORF54 gene
Brown–Vialetto–Van Laere syndrome, a ponto-bulbar palsy with deafness
Fazio-Londe disease

CACNA1A gene
Familial hemiplegic migraine
Spinocerebellar ataxia 6

CACNA1F gene
X-linked cone-rod dystrophy-3
Aland Island eye disease

CARD15 gene
Early-onset sarcoidosis
Blau syndrome

CASK gene
FG syndrome-4 ("FG" are the initials of the first proband)
Mental retardation, X-linked, with or without nystagmus
Mental retardation and microcephaly with pontine and cerebellar hypoplasia

CAVEOLIN-3 GENE
Limb-girdle muscular dystrophy type 1C
Tateyama type of distal myopathy

CEP152 gene
Autosomal recessive primary microcephaly-4
Seckel syndrome-5

CEP290 gene
Bardet–Biedl syndrome 14
Joubert syndrome 5
Leber congenital amaurosis 10
Meckel syndrome 4
Senior–Loken syndrome 6

CHAT (Choline acetyltransferase) gene
Presynaptic congenital myasthenia syndrome with episodic ataxia
Familial infantile myasthenia gravis

CHX10 gene
Microphthalmia, isolated-2
Microphthalmia with coloboma-3
Isolated colobomatous microphthalmia-3

CLCN5 gene
X-linked recessive hypophosphatemic rickets
X-linked recessive nephrolithiasis with renal failure
Dent disease-1

CLN8 gene
Neuronal ceroid lipofuscinosis-8
Progressive epilepsy with mental retardation

CLRN1 gene
Usher syndrome type IIIA
Retinitis pigmentosa-61

COL11A2 gene
Stickler syndrome type III
Fibrochondrogenesis-2
Form of non-syndromic hearing loss

COL2A1 gene
Stickler syndrome type I, sometimes called membranous vitreous type
Osteoarthritis with mild chondrodysplasia
Achondrogenesis type II
Czech dysplasia

COL7A1 gene
Classic dystrophic epidermolysis bullosa pruriginosa
Non-syndromic congenital nail disorder-8

COL9A1 gene
Form of autosomal recessive form of Stickler syndrome
Multiple epiphyseal dysplasia-6

COL9A2 gene
Multiple epiphyseal dysplasia-2
Stickler syndrome type V

COLLAGEN GENE
Autosomal dominant epidermolysis bullosa dystrophica
Pretibial dystrophic epidermolysis bullosa
Stickler syndrome
Strudwick type of spondyloepimetaphyseal dysplasia
Spondyloperipheral dysplasia
Ehlers–Danlos syndrome type IV

CONNEXIN-26 GENE
Keratitis-ichthyosis-deafness syndrome
Deafness, autosomal dominant-3A

CRYAB gene
Posterior polar cataract-2
Fatal infantile hypertonic myofibrillar myopathy

CYLD gene
Familial cylindromatosis
Multiple familial trichoepithelioma-1
Brooke-Spiegler syndrome

DOCK8 gene
Hyper-IgE recurrent infection syndrome, also known as Job syndrome
Autosomal dominant mental retardation-2

DYM gene
Dyggve–Melchior–Clausen disease
Smith–McCort dysplasia

DYNC1H1 gene
Autosomal dominant axonal Charcot–Marie–Tooth disease type 2O
Autosomal dominant mental retardation-13

ENPP1 gene
Generalized arterial calcification of infancy-1
Autosomal recessive hypophosphatemic rickets-2

ESCO2 gene
SC phocomelia syndrome, also known as SC pseudothalidomide syndrome
Roberts syndrome

FBLN5 gene
Autosomal recessive cutis laxa type IA
Macular degeneration, age-related-3

FBN1 gene
Acromicric dysplasia
Stiff skin syndrome
Autosomal dominant form of isolated ectopia lentis
Weill–Marchesani syndrome-1
Weill–Marchesani syndrome-2
Geleophysic dysplasia-2

FGFR1 gene
Trigonocephaly-1
8p11 myeloproliferative disorder

FGFR2 gene
Beare–Stevenson cutis gyrata syndrome
Form of craniosynostosis
Classic Crouzon syndrome

FGFR3 gene
Muenke craniosynostosis syndrome
Hypochondroplasia

CATSHL syndrome
Crouzon syndrome with acanthosis nigricans

FIG4 gene
Charcot–Marie–Tooth type 4J
Form of autosomal dominant ALS
Amyotrophic lateral sclerosis 11

FLNA gene
Terminal osseous dysplasia
FG syndrome-2
X-linked cardiac valvular dysplasia

FLNC gene
Filamin C-related myofibrillar myopathy
Distal myopathy-4 (MPD4), also known as Williams distal myopathy

FMR1 gene
Fragile X tremor/ataxia syndrome
Fragile X mental retardation syndrome

FOXL2 gene
Blepharophimosis, ptosis, and epicanthus inversus syndrome, with premature ovarian failure (BPES type I)
Blepharophimosis, ptosis, and epicanthus inversus syndrome, without premature ovarian failure (BPES type II)

FREM1 gene
Bifid nose with or without anorectal and renal anomalies
Trigonocephaly-2

GATA2 gene
Primary lymphedema with myelodysplasia
Dendritic cell, monocyte, B lymphocyte, and natural killer lymphocyte deficiency

GDAP1 gene
Autosomal recessive axonal CMT with vocal cord paresis
Autosomal recessive demyelinating CMT4A
Autosomal recessive axonal Charcot–Marie–Tooth disease type 2K

GDF3 gene
Klippel–Feil syndrome-3
Isolated microphthalmia with coloboma-6
Isolated microphthalmia-7

GDF6 gene
Klippel–Feil syndrome-1
Isolated microphthalmia-4

GJA1 gene
Syndactyly type III
Oculodentodigital dysplasia
Atrioventricular septal defect 3

GJB2 gene
Autosomal recessive deafness-1A
Hystrix-like ichthyosis-deafnesss syndrome

GJC2 gene (encodes gap junction protein, gamma 2)
Autosomal recessive spastic paraplegia-44
Hereditary lymphedema type IC
Form of Pelizaeus–Merzbacher disease

GLUCOKINASE GENE
Familial hyperinsulinemic hypoglycemia-3
Maturity onset diabetes of the young-2

GNAS gene
Progressive osseous heteroplasia
Pseudopseudohypoparathyroidism
Pseudohypoparathyroidism type Ia

GPR143 gene
Ocular albinism type I
X-linked congenital nystagmus-6
Nystagmus 6, congenital, X-linked

HCN4 gene
Brugada syndrome-8
Autosomal dominant form of sick sinus syndrome

HEDGEHOG GENE
Holoprosencephaly-3
Isolated microphthalmia with coloboma-5

HPRT gene
Lesch–Nyhan syndrome
Kelley–Seegmiller syndrome

HRG gene
Histidine-rich glycoprotein deficiency
Thrombocythemia-11

HSPB8 gene
Axonal Charcot–Marie–Tooth disease type 2L
HMN2A

IGHMBP2 gene
Distal hereditary motor neuronopathy type VI (dHMN6 or HMN6)
Spinal muscular atrophy, with respiratory distress-1

INF2 gene
FSGS5
Focal segmental glomerulosclerosis-5
Charcot–Marie–Tooth disease E with focal segmental glomerulonephritis

JAK2 gene
Thrombocythemia-3
Polycythemia vera, the most common form of primary polycythemia

KCNE2 gene
ATFB4
Form of atrial fibrillation
Long QT syndrome-6

KCNH2 gene
Long QT syndrome-2
Short QT syndrome-1

KCNJ11 gene
Hyperinsulinemic hypoglycemia-2 (HHF2)
TNDM3

KCNJ5 gene
Familial hyperaldosteronism type III
Long QT syndrome-13

KCNQ1 gene
Form of Jervell and Lange–Nielsen syndrome (JLNS1)
Form of autosomal dominant atrial fibrillation
ATFB3 (607554)
Short QT syndrome-2

KIF1A gene
Hereditary sensory neuropathy type IIC
Form of mental retardation

KLF1 gene
Congenital dyserythropoietic anemia type IV (see Glossary item, Dyserythropoiesis)
Form of hereditary persistence of fetal hemoglobin

KRT74 gene
Hypotrichosis simplex of the scalp-2
Autosomal dominant form of woolly hair
Hypotrichosis simplex of the scalp-2

LDB3 gene
Left ventricular non-compaction-3
Form of dilated cardiomyopathy with or without left ventricular non-compaction

LMNA gene
Form of autosomal recessive axonal CMT
Slovenian type heart-hand syndrome

LRP4 gene
Cenani–Lenz syndactyly syndrome
Sclerosteosis-2

LRP5 gene
Familial exudative vitreoretinopathy-4
Autosomal dominant osteopetrosis type I

MATRILIN-3 GENE
Form of multiple epiphyseal dysplasia
Form of autosomal recessive spondyloepimetaphyseal dysplasia

MECP2 gene
Form of neonatal severe encephalopathy
Classic Rett syndrome

MED12 gene
Lujan–Fryns syndrome
Opitz–Kaveggia syndrome, also known as FG syndrome-1

MLL2 gene
Kabuki syndrome-1
Otitis media in infancy

MSX1 gene
Form of selective tooth agenesis
Orofacial cleft 5
Witkop syndrome

MYH6 gene
Familial hypertrophic cardiomyopathy-14
Form of dilated cardiomyopathy

MYH7 gene
Form of scapuloperoneal myopathy
Hypertrophic cardiomyopathy-1
Cardiomyopathy, dilated, 1S

MYH9 gene
Fechtner syndrome
May–Hegglin anomaly
Sebastian syndrome

NEMO gene
Anhidrotic ectodermal dysplasia with immunodeficiency, osteopetrosis, and lymphedema
Atypical mycobacteriosis, familial
Familial incontinentia pigmenti
Invasive pneumococcal disease, recurrent isolated, type 2

NF1 gene
Neurofibromatosis-1
Watson syndrome
Neurofibromatosis-Noonan syndrome variant of neurofibromatosis-1

NHS gene
Nance–Horan syndrome
X-linked congenital cataract

NKX2-5 gene
Atrial septal defect of the secundum type, with or without atrioventricular conduction defects
Congenital non-goitrous hypothyroidism-5
Hypoplastic left heart syndrome-2

NOTCH2 gene
Hajdu–Cheney syndrome
Alagille syndrome-2

NPHP1 gene
Senior–Loken syndrome-1
Form of Joubert syndrome plus nephronophthisis

NPHP3 gene
Meckel syndrome, type 7
Nephronophthisis-3

NPHP4 gene
Form of Senior–Loken syndrome that maps to 1p36
Type 4 nephronophthisis

NPHP6 gene
Form of Senior–Loken syndrome that maps to 12q21-32
Joubert syndrome-5

NR0B1 gene
X-linked congenital adrenal hypoplasia with hypogonadotropic hypogonadism
46,XY sex reversal-2

NR5A1 gene
Premature ovarian failure-7
Form of 46,XY sex reversal

NRAS gene
Form of Noonan syndrome (NS6)
Form of autoimmune lymphoproliferative syndrome, designated type IV (ALPS4)

NSD1 gene
Familial Sotos syndrome
Weaver syndrome-1
Classic Sotos syndrome

OPTN gene
Amyotrophic lateral sclerosis-12
Form of adult-onset primary open angle glaucoma (POAG), designated GLC1E

P63 GENE
Ectodactyly, ectodermal dysplasia, and cleft lip/palate syndrome-3
Split-hand/split-foot malformation

PAX3 gene
Craniofacial-deafness-hand syndrome
Waardenburg syndrome type-3
Waardenburg syndrome type-1

PDE6B gene
Autosomal dominant congenital stationary night blindness-2
Form of retinitis pigmentosa

PDE8B gene
Autosomal dominant striatal degeneration
Primary pigmented nodular adrenocortical disease-3

PDX1 gene
Congenital pancreatic agenesis
Maturity onset diabetes of the young-4

PIGA gene
Paroxysmal nocturnal hemoglobinuria
Multiple congenital anomalies-hypotonia-seizures syndrome-2

PLA2G6 gene
Neurodegeneration with brain iron accumulation-2A
Neurodegeneration with brain iron accumulation-2B
Adult-onset dystonia-Parkinsonism, also known as Parkinson disease-14

PLEC1 gene
Epidermolysis bullosa simplex with pyloric atresia
Epidermolysis bullosa simplex
Autosomal recessive limb-girdle muscular dystrophy type 2Q

POLG gene
Alpers syndrome
Neurogastrointestinal encephalopathy

POLYMERASE-GAMMA GENE
Autosomal recessive progressive external ophthalmoplegia (PEOB)
Sensory ataxic neuropathy, dysarthria, and ophthalmoparesis

POMGNT1 gene
Walker–Warburg syndrome (WWS) or muscle-eye-brain disease
Muscular dystrophy-dystroglycanopathy-B3
Muscular dystrophy-dystroglycanopathy-C3

PRKAR1A gene
Acrodysostosis with hormone resistance
Carney complex, type 1

PROMININ-1 GENE
Stargardt disease-4

Retinal macular dystrophy-2
Cone-rod dystrophy-12

PRPS1 gene
Arts syndrome
X-linked deafness-1

PRRT2 gene
Familial infantile convulsions with paroxysmal choreoathetosis
Benign familial infantile seizures-2
Paroxysmal kinesigenic dyskinesia

PSEN1 gene
Dilated cardiomyopathy-1U
Familial acne inversa-3
Form of early onset Alzheimer disease

PTPN11 gene
Noonan syndrome-1
Metachondromatosis

PYCR1 gene
Autosomal recessive cutis laxa type IIIB
Autosomal recessive cutis laxa type IIB

RAB27A gene
Melanosis with immunologic abnormalities with or without neurologic impairment
Griscelli syndrome type 2

RAF1 gene
Form of Noonan syndrome
LEOPARD syndrome-2

RDS gene
Retinitis pigmentosa-7
Adult-onset vitelliform macular dystrophy (AVMD)

RET gene
Susceptibility to Hirschsprung disease-1
Multiple endocrine neoplasia-2B
Familial medullary thyroid carcinoma MTC

ROR2 gene
Brachydactyly type B1
Autosomal recessive Robinow syndrome

RPE65 gene
Leber congenital amaurosis-2
Form of autosomal recessive retinitis pigmentosa

RPGR gene
Retinitis pigmentosa-3
X-linked cone-rod dystrophy
X-linked retinitis pigmentosa with recurrent respiratory infections

RPGRIP1 gene
Autosomal recessive cone-rod dystrophy-13
Leber congenital amaurosis-6

SAMHD1 gene
Aicardi–Goutieres syndrome-5
Chilblain lupus-2

SCN1A gene
Febrile seizures, familial, type 3A
Familial hemiplegic migraine-3

SCN1B gene
Generalized epilepsy with febrile seizures plus, type 1
Brugada syndrome-5

SCN2A gene
Benign familial neonatal-infantile seizures-3
Early infantile epileptic encephalopathy-11

SCN4A gene
Hypokalemic periodic paralysis type 2
Form of congenital myasthenic syndrome

SCN5A gene
Brugada syndrome-1
Long QT syndrome-3
Sick sinus syndrome (some cases)
Atrial fibrillation (some cases)
Dilated cardiomyopathy (some cases)

SEMA4A gene
Form of retinitis pigmentosa
Cone-rod dystrophy-10

SH3TC2 gene
Charcot–Marie–Tooth disease type 4C
Mild mononeuropathy of the median nerve

SHH gene
Holoprosencephaly-3
Microphthalmia with coloboma 5

SLC16A1 gene
Erythrocyte lactate transporter defect
Form of hyperinsulinemic hypoglycemia

SLC25A19 gene
Amish lethal microcephaly
Thiamine metabolism dysfunction syndrome-3
Bilateral striatal degeneration and progressive polyneuropathy

SLC26A4 gene
Enlarged vestibular aqueduct
Pendred syndrome

SLC2A1 gene
Dystonia 18 (DYT18)
Autosomal recessive primary hypertrophic osteoarthropathy-2

SLC33A1 gene
Spastic paraplegia-42
Congenital cataracts, hearing loss, and neurodegeneration

SLC34A1 gene
Autosomal recessive form of Fanconi renotubular syndrome
Hypophosphatemic nephrolithiasis/osteoporosis-1
Fanconi renotubular syndrome-2

SLC4A1 gene
Band 3 Coimbra
Waldner blood group expression
Autosomal recessive distal renal tubular acidosis with hemolytic anemia

SLC4A11 gene
Corneal endothelial dystrophy-2
Fuchs endothelial corneal dystrophy-4

SMAD4 gene
Myhre syndrome
Juvenile polyposis syndrome

SOS1 gene
Gingival fibromatosis-1
Form of Noonan syndrome

SOST gene
Craniodiaphyseal dysplasia, autosomal dominant
Sclerosteosis
Van Buchem disease

STAT1 gene
Autosomal recessive susceptibility to mycobacterial and viral infections
Familial chronic mucocutaneous candidiasis-7

SYCP3 gene
Spermatogenic failure 4
Recurrent pregnancy loss 4

TGFBR2 gene
Loeys–Dietz syndrome type 2B
Hereditary non-polyposis colorectal cancer-6

TITIN GENE
Autosomal dominant dilated cardiomyopathy-1G
Limb-girdle muscular dystrophy type 2J
Tardive tibial muscular dystrophy

TMEM216 gene
Meckel syndrome type 2
Joubert syndrome-2

TNFRSF13B gene
Immunoglobulin A (IgA) deficiency-2
Common variable immunodeficiency-2

TREX1 gene
Aicardi–Goutieres syndrome-1 (can also be caused by mutations in the SAMHD1, TREX1, or ribonuclease H2 genes)
Chilblain lupus-1

TRPV4 gene
Brachyolmia type 3
Metatropic dysplasia
Parastremmatic dwarfism
Form of scapuloperoneal spinal muscular atrophy
Maroteaux type of spondyloepiphyseal dysplasia
Kozlowski type of spondylometaphyseal dysplasia
Congenital distal spinal muscular atrophy
Hereditary motor and sensory neuropathy type IIC

TTR gene
Form of hereditary amyloidosis
Euthyroidal hyperthyroxinemia

TULP1 gene
Retinitis pigmentosa-14
Leber congenital amaurosis-15

VHL gene
Von Hippel–Lindau syndrome
Familial erythrocytosis-2

VSX1 gene
Posterior polymorphous corneal dystrophy-1
Craniofacial anomalies and anterior segment dysgenesis syndrome

WAS gene
Wiskott–Aldrich syndrome
X-linked thrombocytopenia
X-linked neutropenia

WDR35 gene
Cranioectodermal dysplasia-2
Short rib-polydactyly syndrome type V

WNK1 gene
Hereditary sensory and autonomic neuropathy type IIA
Form of pseudohypoaldosteronism type II

Appendix II

Rules, Some of Which are Always True, and All of Which are Sometimes True

"Believe those who are seeking the truth. Doubt those who find it."

—André Gide

It is almost impossible to learn something new about a rare disease without gaining some insight into the general biology of human diseases. To prove the point, general assertions on the biology of diseases, based on observations of the rare diseases, were inserted throughout the text. The so-called rules lack the rigor of experimental proof. Some of the assumptions upon which these rules are based may prove false over time. Right or wrong, rules have educational value. They teach us how to find relationships among facts, and they give us the opportunity to think critically about received wisdom. Most importantly, these rules demonstrate that Medicine is not purely experimental; it is also inferential.

What follows is a listing composed of all the rules that appeared earlier in the text. Readers are invited to peruse the list, and, if any assertion is sufficiently provocative, return to the text for a full discussion.

1.1.1 Rule—Rare diseases are easily misdiagnosed, and are often mistaken for a common disease or for some other rare disease.

Brief Rationale—It is impossible for any physician to attain clinical experience with more than a small fraction of the total number of rare diseases. When it comes to rare diseases, every doctor is a dilettante.

1.2.1 Rule—Rare diseases are not the exceptions to the general rules of disease biology; they are the exceptions upon which the general rules are based.

Brief Rationale—All biological systems must follow the same rules. If a rare disease is the basis for a general assertion about the biology of disease, then the rule must apply to the common diseases.

1.2.2 Rule—Every common disease is a collection of different diseases that happen to have the same clinical phenotype (see Glossary item, Phenotype).

Brief Rationale—Numerous causes and pathways may lead to the same biological outcome.

1.2.3 Rule—Rare diseases inform us how to treat common diseases.

Brief Rationale—When we encounter a common disease, we look to see what pathways are dysfunctional, and we develop a rational approach to

Rare Diseases and Orphan Drugs. http://dx.doi.org/10.1016/B978-0-12-419988-0.00024-9

prevention, diagnosis, and treatment based on experiences drawn from the rare diseases that are driven by the same dysfunctional pathways.

2.1.1 Rule—A small number of diseases account for most instances of morbidity or mortality.

Brief Rationale—Pareto's principle applies to biological systems.

2.1.2 Rule—Funding for disease research adheres to Pareto's principle.

Brief Rationale—The diseases that kill the greatest number of individuals receive the highest levels of funding, in the simple-minded expectation that advances against common diseases will provide the greatest benefit to society.

2.1.3 Rule—The cancers that account for the majority of cancer deaths occur in elderly individuals.

Brief Rationale—Common diseases are caused by cellular events that accumulate over time or that arise over time. Hence, the chance of developing a common disease increases steadily as individuals age.

2.1.4 Rule—The most common causes of death, if eliminated entirely, will not greatly increase human life expectancy.

Brief Rationale—Elderly individuals who do not die from one common disease will likely die from some other common disease.

2.3.1 Rule—We may have reached the limit by which we can understand the common diseases through direct genetic studies.

Brief Rationale—The common diseases of humans are complex, and biological complexity cannot be calculated, predicted or solved, even with supercomputers.

2.3.2 Rule—Biological systems are much more complex than naturally occurring non-biological systems (i.e., galaxies, mountains, volcanoes) and man-made physical systems (e.g., jet airplanes, computers).

Brief Rationale—The components of biological systems, unlike the components of non-biological systems, have multiple functions, dependencies, and regulatory systems. We cannot predict how any single component of a biological system will react under changing physiologic conditions.

3.2.1 Rule—There is almost no overlap in the types of tumors that occur in children, all of which are rare, and the common tumors that occur later in life.

Brief Rationale—The tumors of adults are different from the tumors of children because these two sets of tumors have different causes and different pathogeneses.

3.3.1 Rule—No common disease is monogenic.

Brief Rationale—In the past several decades, medical scientists have found thousands of rare diseases, each with a monogenic cause. Scientists have not found a single instance wherein a monogenic cause accounts for all the cases of a common disease.

3.4.1 Rule—When a rare disease is non-syndromic, some particular combination of conditions must apply.

Brief Rationale—Additional conditions, beyond the single genetic defect underlying the rare disease, constrain the expression of disease to a specific organ.

3.4.2 Rule—Single gene disorders tend to be syndromic; polygenic/multifactorial disorders tend to be non-syndromic.

Brief Rationale—Single gene disorders are caused by a gene alteration that is present in every cell in the body; hence, any tissue has a chance of suffering a functional or anatomic abnormality due to the gene alteration. Polygenic disorders are caused by a combination of gene variants that occur in the normal human population (i.e., the variant genes are not defective). The expression of disease follows a collection of events and environmental influences occurring over time. The likelihood that these occur in many different tissues is remote; hence, most polygenic diseases are non-syndromic.

3.4.3 Rule—Eponymic disorders (i.e., diseases with a name of a person) are usually syndromic.

Brief Rationale—It can be too taxing to name a syndromic disease by listing the various organs and abnormalities that comprise the syndrome. It is much easier to apply a person's name to the disease, and be done with it.

3.4.4 Rule—A high proportion of diseases caused by regulators of transcription are syndromic.

Brief Rationale—Regulators of transcription have many functions, effecting many genes, and may produce changes in more than one organ, at more than one moment in development [1,2].

3.4.5 Rule—Common diseases can be conceptualized many different ways, all of which are objectively correct.

Brief Rationale—Because many conditions and factors can produce a common disease, it is impossible to exclude any single mechanism as a valid cause.

3.4.6 Rule—Common diseases have many causes; that is why they are common. Rare diseases have a small number of causes; that is why they are rare.

Brief Rationale—Common diseases have many contributing causes. It is impossible to think that all of these causes will activate the same pathways, in the same sequence, and in the same timeframe, for each instance of disease. It is much more likely that an assortment of pathways lead eventually to a collection of pathologic conditions that share a similar phenotype. In the case of rare diseases, many of which are caused by a specific mutation in a specific gene, the pathways follow the same course, over a similar timeframe, to produce very similar phenotypic outcomes in an age-restricted population (e.g., young children).

3.6.1 Rule—Every common disease was, at some point, a rare disease.

Brief Rationale—Every epidemic begins with a solitary case. Common diseases are equivalent to epidemics that settle in to stay.

3.6.2 Rule—Some of yesterday's common diseases are today's rare diseases.

Brief Rationale—The fundamental theory underlying all medical research is that we can eliminate diseases that we fully understand.

3.7.1 Rule—In common diseases, different pathways lead to a somewhat constrained set of clinical phenotypes. In rare diseases, single gene mutations activate a specific pathway producing a characteristic phenotype.

Brief Rationale—Common diseases have many contributing causes. It is impossible to think that all of these causes will activate the same pathways, in the same sequence, and in the same timeframe, for each instance of disease. It is much more likely that an assortment of pathways all lead eventually to a similar phenotype. In the case of rare diseases, many of which are caused by a specific mutation in a specific gene, the pathways follow the same course, over a similar timeframe, to produce very similar phenotypic outcomes in an age-restricted population (e.g., young children).

4.1.1 Rule—We do not have a scientifically meaningful definition for the diseases of aging.

Brief Rationale—We do not know the cellular basis of aging, hence we cannot determine whether a disease qualifies as a disease of aging on a cellular basis. The majority of the so-called diseases of aging are conditions that make individuals look like old persons, or they are conditions that happen to occur more often in elderly individuals than in young individuals.

4.1.2 Rule—Aging is not caused by a single gene.

Brief Rationale—If aging were caused by a single gene, you would expect rare occurrences of loss-of-function mutations of the gene, leading to instances of human immortality. Outside of science fiction, immortal humans do not exist.

4.4.1 Rule—The epidermis, the gut, and the bone marrow do not age.

Brief Rationale—These three tissues are constructed to continuously regenerate. Continuously regenerating tissues, like continuously regenerating animals and plants, do not senesce. It is not unusual to find elderly individuals with no histopathological signs of degeneration in these three tissues.

4.4.2 Rule—Long-lived or immortal organisms have continual cell growth.

Brief Rationale—Aging is a degenerative process that occurs in cells that have lost the ability to divide. Organisms that maintain a population of cells that grow continuously or that maintain a permanent source of stem cells (i.e., cells that renew themselves and that renew other cells in the organism) can only experience aging in the non-dividing subpopulation.

4.4.3 Rule—On a cellular basis, aging is a process confined to non-renewable cell populations.

Brief Rationale—Long-lived cells that cannot replace themselves, such as fully differentiated neurons, muscle cells, cartilage cells, have no biological destiny other than degeneration and death.

5.4.1 Rule—Common diseases that have phenotypic overlap with a rare disease will often have genotypic overlap as well.

Brief Rationale—Polymorphisms are common in the population. If a rare disease gene is known to cause a particular phenotype, it is reasonable to expect that functional variations of the rare disease gene will contribute to the clinical phenotype expressed in a polygenic disease.

6.2.1 Rule—Our knowledge of the relationships between rare infectious diseases and common infectious diseases is dependent upon our access to an accurate and comprehensive taxonomy of living organisms.

Brief Rationale—There are over 1400 known pathogenic organisms, and it would be impossible to develop individual methods to prevent, diagnose, and treat each of these organisms. Taxonomies drive down the complexity of infectious diseases, and permit us to find the biological and clinical relationships among rare infections and common infections.

6.3.1 Rule—Common infectious diseases are spread, directly or indirectly, from one infected human to another human.

Brief Rationale—If an organism has succeeded to thwart human defense systems, and if it can spread from person to person, then it is likely to infect lots of persons.

6.3.2 Rule—Rare infectious diseases are seldom transmissible from human to human.

Brief Rationale—Infections that fail to move from person to person cannot effectively spread through a population.

6.3.3 Rule—The list of rare infectious diseases is growing rapidly; the list of common diseases is more or less static

Brief Rationale—Improvements in the taxonomic designations of infectious organisms, the availability of highly advanced reference laboratories capable of accurately identifying infectious organisms, increases in the number of immune-compromised patients susceptible to infections by organisms that are not otherwise pathogenic, the increased usage of indwelling therapeutic devices, the emergence of new pathogens, and the ease with which infections can be transported from place to place throughout the world have all contributed to the increase in newly encountered rare infectious diseases.

6.4.1 Rule—A large portion of human diseases of unknown etiology will eventually be shown to have an infectious etiology.

Brief Rationale—It is difficult to satisfy Koch's postulates for every type of infectious disease (see Glossary item, Koch's postulates). Nonetheless, if efforts to find a non-infectious cause of a disease fail, and if the temporal and geographic pattern of disease occurrences resembles the typical pattern of an infectious epidemic, then an infectious etiology is likely.

7.1.1 Rule—Any infection that occurs in a healthy individual can manifest as a more serious infection if the individual becomes immune-compromised.

Brief Rationale—The immune system keeps infections in check. When the immune status is compromised, the clinical expression of an infection worsens.

7.1.2 Rule—The most common site of presentation of infectious disease in individuals who are immune-compromised is the mouth.

Brief Rationale—The mouth is the dirtiest place in the body, with a greater variety of potentially pathogenic commensals than any other site, many of which live exclusively in periodontal tissues. When an immune-deficient state provides opportunistic pathogens with an occasion to grow and invade, the mouth is first site of attack.

7.1.3 Rule—The types of cancers that are known to arise soon after immunosuppression (i.e., weeks or months) are all caused by oncogenic viruses.

Brief Rationale—Viruses are capable of inducing tumors rapidly when not kept in check by immune systems. No other cause of cancer produces tumors in adult humans in a short timeframe with no apparent latency period.

7.1.4 Rule—Immune deficits are usually polygenic or have an environmental cause.

Brief Rationale—We are constantly being reminded of Darwin's cruel game; monogenic causes of immunosufficiency are rare because they reduce fitness. Infectious diseases are extremely common, and every infectious disease marks a defeat in the human body's battle against invasive organisms. When we study families with increased susceptibility to certain types of infection, we seldom observe Mendelian patterns of inheritance; instead, we observe non-Mendelian patterns indicative of polygenic inheritance [3]. Though there are dozens or monogenic immunodeficiency syndromes, they account for a very small fraction of the instances of immune deficiency in the general population.

7.2.1 Rule—The autoimmune diseases as a group are pathogenetically related to one another.

Brief Rationale—All autoimmune diseases involve some of the same components of a complex pathway that leads to the development of antibodies.

7.2.2 Rule—The common autoimmune diseases involve one or two organs; seldom more.

Brief Rationale—The adaptive immune system is designed to produce antigen-specific antibodies. Assuming that the defective pathway is limited to the adaptive immune system, the likelihood that a disorder will yield an antibody that cross-reacts with many different tissues is small.

7.2.3 Rule—The common autoimmune diseases have a polygenic origin.

Brief Rationale—Though the common autoimmune diseases (i.e., autoimmune thyroid diseases, type 1 diabetes mellitus, pernicious anemia, rheumatoid arthritis, and vitiligo) tend to run in families, they seldom display a simple Mendelian inheritance pattern. Inheritance that is non-Mendelian usually has a polygenic origin.

7.2.4 Rule—Autoimmune disorders that result from a dysfunction of the innate immune system are rare and tend to produce systemic disease involving multiple tissues [4].

Brief Rationale—The distinction between "self" and "non-self" proteins is a function of the innate immune system [5]. When the immune system cannot ignore "self" antigens, the effects tend to be systemic.

7.2.5 Rule—There is an environmental component to the autoimmune diseases.

Brief Rationale—Autoimmune diseases, like any other diseases with a non-genetic component, are not encountered in neonates, and the overall incidence of the autoimmune diseases increases with age.

7.2.6 Rule—Physiologic systems influence the development of autoimmune diseases.

Brief Rationale—Autoimmune diseases can occur in men or women, but most occur preferentially in women. The preferential occurrence of virtually every autoimmune disease in individuals of a particular gender suggests that some intrinsic physiologic condition contributes to diseases susceptibility.

8.2.1 Rule—Most common cancers are caused by environmental agents.

Brief Rationale—The vast majority of cancers occur at body sites that are directly exposed to chemical, physical, or biological agents delivered by food, water, and air. The tissues that receive the highest levels of exposure are the same tissues that yield the highest number of tumors. Tissues of the body that are not directly exposed to outside agents (e.g., muscle, connective tissues) are not sites at which common cancers develop.

8.2.2 Rule—In adults, diseases of cells derived from ectoderm or from endoderm typically have an environmental cause.

Brief Rationale—Tissues deriving from ectoderm and endoderm are exposed to toxins at higher levels than are the tissues that derive from mesoderm. When a disease targets ectodermal- or endodermal-derived cells in adults, it is likely to have a toxic etiology. Cells of mesodermal origin (i.e., the inside cells) are typically spared, because they are less exposed to the environment.

8.2.3 Rule—Most of the metabolism of foreign compounds entering the human body is handled by cells derived from endoderm or ectoderm.

Brief Rationale—It stands to reason that the cells that receive the brunt of environmental toxins will be the cells that are adapted to detoxify exogenous chemicals.

8.2.4 Rule—Most chemical carcinogens need to be metabolized before they are converted to an active (i.e., mutagenic) molecular form.

Brief Rationale—Activated carcinogens are highly reactive molecules that can bind to just about any kind of molecule. Naturally active carcinogens would react with, and be neutralized by, non-genetic molecules before they could reach DNA. Highly carcinogenic molecules exist as stable, inactive molecular species that are metabolized within cells to active molecules that react with DNA.

8.3.1 Rule—Virtually all cancers of childhood have a germline genetic component to their pathogenesis.

Brief Rationale—The common cancers have multi-step etiologies, requiring many years to develop, and occurring in adults. Children simply do not have the opportunity to express diseases that involve repeated exposures to commonly occurring environmental agents. Hence, cancers in children develop from inborn mutations. Cancer-causing germline mutations are rare; hence, childhood cancers are rare.

8.3.2 Rule—Rare tumors are much more likely to have a single cause, a single carcinogenic pathway, a single inherited gene, or a single acquired marker, than are any of the common tumors.

Brief Rationale—Many different factors can lead to a common cancer; that is why the cancer is common. Only very specific and highly unlikely factors (e.g., genetic mutation) lead to rare cancers; that is why they are rare.

8.3.3 Rule—In a tumor that can occur as a rare, inherited form, or as a common, sporadic form, we always learn the most by studying the rare, inherited form and later extending our gained knowledge to the common, sporadic form.

Brief Rationale—Only the subset of cases arising from an inherited germline mutation can be studied in affected and unaffected relatives.

8.3.4 Rule—If you look hard enough, you can usually find examples of syndromic disorders accounting for what might otherwise be considered to be a sporadic or non-syndromic childhood cancer.

Brief Rationale—A germline mutation having the biological power to cause cancer might be expected to produce some additional phenotypic effects in the organism.

8.3.5 Rule—There is no such thing as a mutation that is necessary and sufficient, by itself, to cause cancer.

Brief Rationale—In the worst of the inherited cancer syndromes, tumors do not occur in every organ, or even in every individual who carries the cancer-causing mutation. The empiric absence of a 100% penetrant cancer mutation (i.e., one that always causes cancer) suggests that more than one event or condition must prevail during carcinogenesis.

8.3.6 Rule—In contrast to rare cancers, common cancers are characterized by many different mutations in many different genes, and the affected genes will vary from patient to patient and from tumor sample to tumor sample within the same patient.

Brief Rationale—Common cancers are genetically unstable.

8.4.1 Rule—Carcinogenesis, the pathogenesis of tumors, is a multi-step process.

Brief Rationale—Interventions can stop the process of carcinogenesis at various points in tumor development (e.g., the precancer stage), indicating the presence of multiple biological steps, each with characteristic properties and vulnerabilities.

8.4.2 Rule—Each step in carcinogenesis is a potential target of cancer prevention.

Brief Rationale—The key thing to know about carcinogenesis is that it occurs in steps. Because there are multiple steps in carcinogenesis, there are multiple opportunities for blocking the progression of cancer [6,7].

8.4.3 Rule—Rare cancers and rare cancer syndromes have helped us to dissect the various steps of carcinogenesis.

Brief Rationale—We see rare cancers and rare cancer syndromes that target various cellular processes occurring throughout carcinogenesis. These would include polymorphisms in genes that metabolize carcinogens at the time of initiation, that repair DNA (e.g., xeroderma pigmentosum), that preserve the integrity of DNA replication, that control microsatellite stability (e.g., hereditary non-polyposis colon cancer syndrome), that control apoptosis, that activate tumor suppressor genes (e.g., Li–Fraumeni syndrome) and tumor oncogenes (BCR/

ABL fusion gene in chronic myelogenous leukemia), that drive hyperplasia of particular cell types (e.g., c-KIT gastrointestinal stromal tumors), and so on.

8.4.4 Rule—Rare cancers are easier to cure than common cancers.

Brief Rationale—The malignant phenotypes of rare cancers are often driven by a single genetic alteration or a single cellular pathway. It is feasible to target and inhibit a single pathway with a single drug. Common cancers are driven by hundreds or thousands of aberrant pathways. We currently have no way of inhibiting all of the possible pathways that drive the malignant phenotype in common cancers.

9.1.1 Rule—Sporadic diseases are non-sporadic diseases that we do not understand.

Brief Rationale—A sporadic disease, by definition, occurs randomly with no known cause. Diseases do not occur at random and without cause. Once the cause is understood, the sporadic disease becomes non-sporadic.

9.2.1 Rule—A single pleiotropic gene is likely to be associated with several phenotypically unrelated diseases.

Brief Rationale—Genes with pleiotropic pathological effects, and genes that alter a pathway that operates in many different types of cells, are likely to play a role in the pathogenesis of more than one disease, simply because they perturb many different cellular processes.

9.2.2 Rule—Monogenic rare diseases that express in late adolescence, or in adulthood, are likely to require additional events (i.e., somatic genetic mutations, toxic exposures, or the accumulation of molecular species or cellular alterations caused by the original genetic defect) that occur over time.

Brief Rationale—If this were not the case, every inherited genetic defect would be expected to express itself clinically at birth or in early childhood.

9.4.1 Rule—At least for the near future, personalized medicine will be impossible to attain, except for a small and select set of diseases and genotypes.

Brief Rationale—The genotypes of the simplest diseases are highly complex. Factors that modify genotype and phenotype (e.g., genome, epigenome, cellular physiology, and environmental agents) are even more complex. Because we cannot cope with these complexities, we cannot reliably diagnose every individual with a genetic disease, nor can we rationally predict how therapeutic responses will differ from individual to individual.

9.5.1 Rule—Phenocopy diseases are typically mimics of rare diseases; not common diseases.

Brief Rationale—The prototypical phenocopy disease involves a single agent having a specific effect on a single pathway in a limited number of cell types. In theory, any pathway can be altered by a drug to produce a phenotype that mimics a monogenic disease. A simple interruption of normal cellular function of a gene or a pathway is consistent with what we see in rare diseases and in phenocopy diseases, and lacks the cumulative acquisition of multiple genetic or cellular aberrations that typically characterize the common diseases.

10.1.1 Rule—Mutations of pathways in unspecialized cells tend to produce disorders that are found in multiple organs.

Brief Rationale—If a cellular function is general (i.e., occurring in many different types of cells), a defect in the cellular function is likely to cause dysfunction in many different organs.

10.1.2 Rule—Regardless of the path taken, many pathologic processes will converge to the same pathologic condition.

Brief Rationale—There are a limited number of ways that the body can respond to malfunctions.

10.1.3 Rule—A large set of cellular defects account for a relatively small number of possible pathologic conditions.

Brief Rationale—In any complex system, there are a limited number of functional parts, but each functional part can break down due to a vast number of possible defects.

10.1.4 Rule—Regardless of the complexity of a system, the outcomes are typically repeatable and stable.

Brief Rationale—All existing biological systems, despite their complexity, converge toward stability. If a biological system were unstable, it would cease to exist.

10.2.1 Rule—The genome establishes the identity of an organism; the epigenome establishes the behavior of the individual cells within the organism.

Brief Rationale—Each terrestrial organism can be distinguished from every other terrestrial organism by its unique sequence of DNA. Furthermore, each organism can be identified as a member of a particular species based on DNA sequences inherited from a common ancestor. Within the organism, there are many different cell types, each with a specific behavior (e.g., hepatocyte, neuron, keratinocyte). Because each cell of an organism contains the same DNA, cell-type behavior cannot be determined by the genome. Because cell type is inherited (i.e., a hepatocyte produces another hepatocyte, never a keratinocyte), there must be a heritable component of somatic cells that acts on the genome to direct its activities. We call this the epigenome.

10.2.2 Rule—The epigenome influences which tissues will be affected by a disease-causing gene mutation.

Brief Rationale—In Section 10.1, we described how the expression of genes in different types of cells accounts for much of the tissue specificity of genetic diseases. We can now recognize that the difference of gene expression in different types of cells is determined by the epigenome.

10.2.3 Rule—Conditions of the epigenome can determine which disease, among several, may result from a gene mutation.

Brief Rationale—The epigenome determines gene expression; an unexpressed protein-coding gene, regardless of the mutations it may contain, cannot cause disease. The epigenome determines which genes are expressed, and in which tissues.

10.2.4 Rule—In general, epigenetic hypermethylation leads to gene suppression. Hypomethylation promotes gene expression.

Brief Rationale—As an empirical observation, hypermethylated chromosomal regions, such as the X-chromosome Barr body, are genetically hypoactive.

10.2.5 Rule—The disease phenotype (i.e., the clinical presentation) of rare diseases caused by epigenetic mutations is impossible to predict at present.

Brief Rationale—Cell regulation via the epigenome is extremely complex. A simple epigenome modification will have pleiotypic effects that are beyond our ability to measure. Hence, the best we can do at present is to observe the changes in trait that occur when the epigenome is altered, and try to make some sense of our observations.

10.2.6 Rule—The epigenome, unlike the genome, is constantly changing in every type of cell throughout development and throughout adult life.

Brief Rationale—Cells have elaborate DNA repair systems that do a fairly good job at maintaining an unchanging sequence of nucleotides throughout life. The epigenome can be easily altered. We see the effects of changes in the epigenome when we look at the process of cellular differentiation, which is based on epigenomic modification.

10.2.7 Rule—Children are afflicted by diseases with a strong genetic influence, while the elderly are afflicted by diseases with a strong epigenetic influence.

Brief Rationale—Children have not lived long enough to accumulate epigenetic changes. The elderly have lived so long that any genetic disease would have manifested decades earlier.

10.2.8 Rule—Acquired epigenetic alterations play a much greater role in the common diseases than in the rare diseases.

Brief Rationale—The rare diseases are typically diseases of children, which are driven by genetics. The common diseases happen to be diseases of adults and the elderly, who have lived long enough to accumulate epigenetic alterations.

10.3.1 Rule—Diseases with the same clinical phenotype tend to exhibit similar aberrant metabolic pathways regardless of any differences in the underlying defects that caused the diseases.

Brief Rationale—Earlier in this chapter, we discussed the following rule: "Regardless of the path taken, many pathologic processes will converge to the same pathologic condition." Here, we approach the phenomenon of disease convergence from the opposite direction. If we have two conditions that have converged to the same clinical phenotype, can we assume that both conditions employ the same cellular pathways? Probably so, because diseases are manifestations of pathologic conditions in specific types of cells. Cells of a specific type are highly restrained to express a limited set of cell-type-specific pathways.

10.3.2 Rule—When a disease phenotype is fully accounted for by the function of one pathway, then we can expect every disease having the same phenotype to have alterations in the same pathway.

Brief Rationale—If alternate single pathways or alternate sets of pathways could account for the equivalent phenotype produced by a single pathway, then we would expect to have encountered some instances of this phenomenon. No such instances are known. In the absence of such encounters, a single pathway for all instances of the clinical phenotype seems likely.

10.3.3 Rule—An agent that causes a common disease will also cause multiple rare diseases.

Brief Rationale—Common diseases are complex, and an agent that produces common disease must exert biological effects on many different pathways. Some of these pathways are likely to produce rare clinical phenotypes.

10.4.1 Rule—Complex physiological pathways that are unique to humans (e.g., coagulation pathways, immune pathways, neural pathways, metabolism pathways) can only be understood by studying rare diseases affecting steps in the pathway.

Brief Rationale—Researchers have had great success collecting cases of rare deficiencies, and then piecing together a plausible and testable complex pathway.

10.5.1 Rule—There are monogenic precursors of common diseases.

Brief Rationale—Every disease begins with a single error.

10.5.2 Rule—Precursor lesions are more common than the common diseases that they produce.

Brief Rationale—Precursor lesions are dependent upon additional events and processes that push pathogenesis forward. If these additional events and processes fail to occur, then the precursor cannot progress. Because every fully developed disease has a precursor, while not every precursor develops into a disease, we can infer that there must be more precursor lesions than developed diseases.

10.5.3 Rule—Monogenic precursors are easier to treat than fully expressed common diseases.

Brief Rationale—A monogenic precursor lesion has one gene that is the underlying cause of its expressed phenotype, and the one gene may influence as few as one pathway. If the causal pathway can be restored to a normal level of activity, the precursor lesion may regress or otherwise fail to develop into a fully expressed disease phenotype.

10.5.4 Rule—More than one type of disease may have the same precursor lesion.

Brief Rationale—There are a limited number of pathologic pathways in cells, and these pathways tend to converge to a smaller number of pathways during pathogenesis. Hence, it seems likely that there will be some precursor lesions that are common to more than one disease.

10.5.5 Rule—As we learn more and more about the pathogenesis of diseases, new therapies will be targeted against the most sensitive precursor lesions, not against the fully developed disease.

Brief Rationale—When we successfully eliminate a precursor lesion, we eliminate all the diseases that may develop from the common precursor.

11.1.1 Rule—When you graph the frequency of occurrence of a rare disease against the age of the individuals that develop the disease, there is usually one clear peak.

Brief Rationale—Rare diseases often result from a single mutation that enters the germline at the time of conception. The process by which the gene mutation leads to a clinical disease will require roughly the same length of time, in most affected individuals, producing a smooth, single peak, when disease occurrences are graphed against age of occurrence.

11.1.2 Rule—Bimodality, when it occurs, is more often observed in the rare diseases than the common diseases.

Brief Rationale—Because there are many occurrences of a common disease, second peaks (i.e., subpopulations with separate peak occurrence with age) are likely to be masked by the large number of occurrences of the larger peak. Because the total number of individuals with a rare disease is small, a relatively small subpopulation, with its own specific age of disease occurrence, is likely to produce a visible second peak when the data are graphed.

11.1.3 Rule—A disease that can be separated into biological subsets, based on a quantifiable trait such as age, can be interpreted as an aggregate of separate diseases, each with a smaller occurrence rate than the original disease.

Brief Rationale—By definition, a disease is a pathologic condition that is biologically distinct from other pathologic conditions.

11.1.4 Rule—Single gene mutations may account for small subsets of common diseases, but they do not account for large subsets of common diseases.

Brief Rationale—All the single gene disease mutations are rare. If this were not so, we would expect to see Mendelian inheritance, typical for monogenic diseases, among the common diseases; but we do not.

11.1.5 Rule—Rare diseases that are subsets of common diseases often occur in a younger population than the cases occurring in the larger set of individuals with so-called sporadic disease.

Brief Rationale—Rare diseases are typically germline, monogenic diseases that occur in young individuals.

11.1.6 Rule—In a bimodal disease wherein the disease occurs in two age groups, young and old, the strongest likelihood of finding an effective treatment resides in the younger age group.

Brief Rationale—The younger age group is more likely to have a monogenic or oligogenic cause of the disease, and this often translates into a targeted cure (see Glossary item, Oligogenic inheritance). The older age group is likely to develop disease after the accumulation of multiple epigenetic, genetic, and environmental alterations, making it difficult to find an effective treatment.

11.1.7 Rule—Polygenic diseases always have a non-Mendelian pattern of inheritance.

Brief Rationale—Each gene variant in a polygenic disease was inherited independently from one another. Hence, the set of genes that together constitute the polygenic cause of disease were not present as a complete set in either parent. Hence, inheritance cannot be assigned to either parent. Hence, the inheritance pattern in non-Mendelian.

11.1.8 Rule—Most dominantly inherited mutations that cause early death are caused by *de novo* mutations.

Brief Rationale—Non-lethal mutations occurring in germline cells have the opportunity of entering the population germ pool if the offspring have reproductive success [8]. Not so for lethal mutations. If a mutation causes death during development (*in utero*), or during early life, prior to sexual maturation, then there is no way that the mutation could have come through inheritance from the prior generation. The mutation must have arisen *de novo* in the affected individual.

11.1.9 Rule—Genetic diseases caused by *de novo* mutations tend to produce severe clinical consequences.

Brief Rationale—Two reasons apply. First, *de novo* mutations have never gone through the process of natural selection by which lethal inherited mutations are eliminated from the general population. Second, *de novo* mutations that produce disease are likely to be dominant genes, because the likelihood of inheriting bi-allelic *de novo* mutations is exceedingly unlikely. Dominant genes that produce disease tend to involve structural proteins, because non-structural (e.g., enzymatic) proteins are likely to be compensated for by the normal allele. Mutated structural genes can play havoc among the building blocks of tissue. Hence, *de novo* disease mutations tend to produce serious pathology.

11.1.10 Rule—Life-threatening rare monogenic diseases never become common diseases.

Brief Rationale—If a lethal monogenic disease were to become common, then evolutionary pressure would tend to exclude the gene from the population over time. In the process, we would expect the disease to become less and less common, and more and more rare.

11.1.11 Rule—Common diseases have contributing causes other than gene alterations.

Brief Rationale—There is discordance in the disease occurrences among monozygotic twins. As both twins have the same genes, differences in disease occurrences must have a non-genetic cause.

11.1.12 Rule—Genetic diseases with multiple genetic variants (e.g., retinitis pigmentosa) are among the most common of the rare genetic diseases.

Brief Rationale—As previously discussed, common diseases are common because many different causes and pathways lead to the same clinical phenotype (e.g., many paths lead to a heart attack). Many rare diseases are rare because only one particular mutation can cause the rare disease. If we have a disease that is monogenic, and thus many different alleles of a single gene may lead to the disease (i.e., allelic heterogeneity), or for which mutations in any of several different genes may cause the disease (i.e., locus heterogeneity), then we might expect to see disease incidences that are higher than most rare diseases but lower than any common disease.

11.1.13 Rule—When a mutation is deleterious and widespread, it is likely that the mutation also has some useful purpose.

Brief Rationale—Otherwise, the mutation would have been selected against, thus lowering the number of affected individuals.

11.1.14 Rule—When you have a disease with many different possible causes, it is probably not a rare disease.

Brief Rationale—Rare diseases usually result from a single genetic alteration (e.g., inherited condition), or from a single exposure to a specific agent (e.g., *in utero* exposure to thalidomide). As the number of possible causes of a disease rises, so does the incidence of the disease. Eventually, the disease ceases to be rare.

11.4.1 Rule—With the exception of cancer, new mutation plays no direct role in the development of common diseases.

Brief Rationale—Mutations occur infrequently, whereas genetic variation occurs universally. A common disease that occurs in billions of individuals cannot be explained on the basis of mutation.

11.4.2 Rule—Phenotype is influenced by pre-existing gene variations in the human population much more often than it is influenced by new mutations.

Brief Rationale—The rate of new mutations is low; the number of pre-existing gene variants is high.

11.4.3 Rule—Natural selection does not strongly apply to common diseases, except for the infectious diseases.

Brief Rationale—The most common diseases tend to be ailments that have persisted in the human population for millions of years, affecting millions or billions of individuals in the interim. If natural selection was doing its job, these common diseases would have become rare.

11.4.4 Rule—Most common diseases occur more often in men than in women; not so for the rare diseases.

Brief Rationale—The same rule seems to apply to all mammals. Females have lower incidences of the common diseases and live considerably longer than males. The survival of mammalian species requires that children be protected and nurtured for a prolonged period. Women, particularly mothers and grandmothers, are caregivers to children; hence, their prolonged survival is a desirable trait. Nature has no special imperative to keep men living beyond their reproductive years.

12.1.1 Rule—The same genes that cause monogenic rare diseases are found in the sporadically occurring diseases for which there is phenotypic overlap.

Brief Rationale—Because there are many instances of a common disease, it would seem likely that a gene that is known to cause a particular clinical phenotype, in the case of a monogenic disease, is likely to contribute to at least some cases of a polygenic disease that has a similar phenotype.

12.1.2 Rule—Uncommon presentations of common diseases are sometimes rare diseases, camouflaged by a common clinical phenotype.

Brief Rationale—Common diseases tend to occur with a characteristic clinical phenotype and a characteristic history (e.g., risk factors, underlying causes). Deviations from the normal phenotype and history are occasionally significant. Rare diseases may produce a disease that approximates the common disease; the differences being subtle findings revealed to the most astute observers.

12.1.3 Rule—Rare gene variants account for the bulk of the genetic component of common diseases.

Brief Rationale—Current GWAS studies indicate that commonly occurring gene variants do not account for the bulk of the genetic component of diseases. If common genes account for a small fraction of the genetic component of disease, it seems reasonable to suspect that rare gene variants play a large role in the common diseases [9,10].

12.2.1 Rule—We know more about the pathogenesis of rare diseases than we know about the pathogenesis of common diseases.

Brief Rationale—Each common disease has many causes and many pathways that contribute to the fully developed clinical phenotype. Because many cellular events are happening at once, there really is no way to design a controlled experiment that can determine the consequences of altering a single component of the system. Hence, the common diseases are all somewhat inscrutable.

12.2.2 Rule—Common diseases are aggregates of the individual pathogenic pathways that account for the rare diseases.

Brief Rationale—Because every pathway is a product of gene expression, and because virtually every gene of functional importance is a candidate for a rare disease, it is reasonable to assume that each of the many pathways that participate in the phenotypic expression of a common disease will be expressed in one or more of the 7000+ rare diseases.

12.2.3 Rule—Any polygenic disease can be replicated by a monogenic disease.

Brief Rationale—The phenotype associated with a polygenic disease converges toward a physiologically permissible outcome. Because there is a monogenic disease affecting virtually every pathway available to cells, it is likely that each common disease will be replicated by at least one monogenic disease that converges to the same clinical phenotype.

13.1.1 Rule—Rare diseases are the sentinels that protect us from common diseases.

Brief Rationale—A few new cases of a rare disease will raise the suspicions of astute public health workers and can warn us that the general population has been exposed to a new or growing environmental hazard.

13.1.2 Rule—When a new toxin is introduced to a population, individuals with a rare disease will be among the first to succumb to its ill-effects.

Brief Rationale—Many of the rare diseases have mild variants that cause minimal pathology under normal circumstances. When physiological systems are overwhelmed or otherwise stressed by a toxin, a clinical phenotype may emerge.

13.1.3 Rule—When a common disease occurs in a young individual, in the absence of a diagnosed inherited cancer syndrome, then an environmental cause should be suspected.

Brief Rationale—Common diseases typically occur in middle-aged and elderly patient populations. When a common disease occurs in a young person, and it is not caused by an inherited syndrome, then a high exposure to an environmental agent is likely.

13.2.1 Rule—Any specialized diagnostic techniques applied to a common disease will likely draw on knowledge obtained from one or more rare diseases.

Brief Rationale—Genes and pathways that lead to rare diseases are the pathogenetic building blocks of common diseases. We can expect that these same genes and pathways will serve as new diagnostic markers.

13.2.2 Rule—Every diagnostic gene or pathway is a potential drug target.

Brief Rationale—Pathways and the genes that code for pathway proteins that are crucial to the pathogenesis of disease are the logical targets of therapeutic agents.

13.3.1 Rule—Rare diseases are easier to treat than common diseases.

Brief Rationale—Rare diseases have simple genetic defects, have little heterogeneity, and have few metabolic options with which they can evade targeted treatments.

13.3.2 Rule—Every chemical with a known biological activity (i.e., an effect on some biological function) is useful in the treatment of one or more diseases.

Brief Rationale—If a drug has a biological action on a biological pathway, and if the biological pathway is involved in one disease or another, the drug will likely have some effect on diseases that depend on those pathways. Because there are many thousands of diseases, the odds are that an agent that modifies the activity of any pathway will have some value in one or more diseases. As you might expect, chemicals that modify cellular pathways are seldom "non-toxic." The most we can hope for is that at the doses prescribed, their beneficial actions will surpass their toxic actions.

13.3.3 Rule—All drugs that are safe and effective against rare diseases will be used in the treatment of one or more common diseases.

Brief Rationale—The rare diseases, as an aggregate group, comprise every possible pathogenic pathway available to cells. Hence, pathogenic pathways that are active in the common diseases will be active in one or more rare diseases. Agents that target pathways in the rare diseases are candidate treatments for the common diseases with which they share active pathways.

13.3.4 Rule—It is much more useful to treat a disease pathway than it is to treat the individual gene mutation or its expressed protein.

Brief Rationale—Many different diseases may respond to a drug that targets a pathogenic pathway, while only one genetic variant of one rare disease is likely to respond to a drug that targets the disease-causing gene or its expressed protein.

13.3.5 Rule—Common diseases and rare diseases that share a pathway are likely to respond to the same pathway-targeted drug.

Brief Rationale—Pathogenesis (i.e., the biological steps that lead to disease) and clinical phenotype (i.e., the biological features that characterize a disease) are determined by cellular pathways. If a pathway has a crucial role in the development of disease, then you would have reason to hope that drugs that disrupt the pathway will alter the progression and the expression of the disease, whether the disease is common or rare.

14.2.1 Rule—Clinical trials are the best method ever developed to determine whether a drug is safe and effective for a particular purpose in a particular target population. Nonetheless, **clinical trials cannot provide the clinical guidance we need to develop all of the new medications that will be needed to conquer the common diseases**.

Brief Rationale—We simply do not have the money, time, and talent to perform all the anticipated clinical trials for the common diseases.

14.2.2 Rule—Clinical trials for common diseases have limited value if the test population is heterogeneous; as is often the case.

Brief Rationale—Abundant evidence suggests that most common diseases are heterogeneous, composed of genotypically and phenotypically distinct disease populations, with each population responding differently with the clinical trial.

14.2.3 Rule—Clinical trials for the rare diseases are less expensive, can be performed with less money, and provide more definitive results than clinical trials on common diseases.

Brief Rationale—Common diseases are heterogeneous and produce a mixed set of results on subpopulations. This in turn dilutes the effect of a treatment and enlarges the required number of trial participants. Rare diseases are homogeneous, thus producing a uniform effect in the trial population, and lowering the number of trial participants required to produce a statistically convincing result.

14.2.4 Rule—Clinical trials on common disease can be reduced to one or more trials of a subtype of the disease.

Brief Rationale—The heterogeneity of populations with a common disease allows trialists the freedom to design small trials for subsets of individuals who have a particular genotype (i.e., a gene marker or a gene expression profile), a particular mode of inheritance (i.e., Mendelian), or a distinguishing clinical phenotype (e.g., early onset disease).

14.3.1 Rule—For the common diseases of humans, there are no adequate animal models.

Brief Rationale—The common diseases are complex, the end result of many genetic and environmental factors. There is no reason to expect that a complex set of factors interacting in humans could be replicated in an animal.

14.3.2 Rule—Animals can model some of the rare human diseases.

Brief Rationale—Because many of the rare diseases are caused by mutations in one gene, a defect in an orthologous gene may produce a disease in an animal model that resembles the human phenotype.

14.3.3 Rule—Animals cannot serve as models for human responses to treatment.

Brief Rationale—The responses of an organism to a drug are species specific and complex, and are determined by traits that have evolved over time to maximize the survival of the species.

DO-IT-YOURSELF RULES

"All generalizations are false."

—Self-referential paradox

Here are some additional rules that were omitted from the chapters. You might enjoy providing your own rationales, assuming you agree with the assertions.

Rule—Mutations of structural genes tend to produce Mendelian dominant disorders.

Brief Rationale—Hint: How does a builder compensate for poor structural material?

Rule—X-chromosome disorders occur more frequently than Y-chromosome disorders.

Brief Rationale—Hint: How big is the Y-chromosome?

Rule—A man with an X-linked disorder will not pass the mutation to his sons.

Brief Rationale—Hint: Genetically normal males have a Y-chromosome; where does it come from?

Rule—Inherited mitochondriopathies are always inherited from the mother and every child of the mother (male and female) will inherit the mutation, if not the disorder.

Brief Rationale—Hint: How does the zygote get its mitochondria?

Rule—Men with X-linked dominant disorders generally have a more severe case than women with the same disorder.

Brief Rationale—Hint: Women are X-chromosome mosaics.

Rule—A susceptibility gene that occurs in 100% of the population cannot be detected by any of the standard analytic techniques known to medical science.

Brief Rationale—Hint: Think about the methodologies that we employ to identify disease-causing gene variants.

Rule—All commonly occurring polymorphisms are benign or, at worst, have low pathogenicity.

Brief Rationale—Hint: How would natural selection deal with commonly occurring gene variants of high pathogenicity?

Rule—Inherited disorders of highly complex systems (e.g., vision, hearing, immunity, coagulation, nerve conduction) tend to have a great deal of genetic heterogeneity.

Brief Rationale—Hint: How many different ways can a complex system be broken?

Rule—Diseases caused by gain-of-function mutations tend to have no allelic or locus heterogeneity.

Brief Rationale—Hint: There are many mutations that can reduce the functionality of a protein, but are there multiple mutations that will produce added functionality of a protein?

Rule—Human cells often tolerate huge increases in the number of chromosomes in a cell. Decreases in chromosomal number are not well tolerated.

Brief Rationale—Hint: What chromosomes, if any, are optional for cell survival?

Rule—In the common cancers, genetic instability precedes the mutational activation of oncogenes.

Brief Rationale—Hint: What sequence of events is more likely: that genetic instability will produce a mutation in an oncogene, or that a mutation in an oncogene will cause genetic instability?

REFERENCES

1. Adams J. The complexity of gene expression, protein interaction, and cell differentiation. Nat Educ 1:1, 2008.
2. Heintzman ND, Hon GC, Hawkins RD, Kheradpour P, Stark A, Harp LF, et al. Histone modifications at human enhancers reflect global cell-type-specific gene expression. Nature 459: 108–112, 2009.
3. Hill AVS. Evolution, revolution and heresy in the genetics of infectious disease susceptibility. Philos Trans R Soc Lond B Biol Sci 367:840–849, 2012.
4. Cheng MH, Anderson MS. Monogenic autoimmunity. Annu Rev Immunol 30:393–427, 2012.
5. Miller DM, Rossini AA, Greiner DL. Role of innate immunity in transplantation tolerance. Crit Rev Immunol 28:403–439, 2008.
6. Alberts DS. Reducing the risk of colorectal cancer by intervening in the process of carcinogenesis: a status report. Cancer J 8:208–221, 2002.
7. Berman JJ. Precancer: The Beginning and the End of Cancer. Jones & Bartlett, Sudbury, 2010.
8. Eyre-Walker A, Keightley PD. The distribution of fitness effects of new mutations. Nat Rev Gen 8:610–618, 2007.
9. Pritchard JK. Are rare variants responsible for susceptibility to complex diseases? Am J Hum Genet 69:124–137, 2001.
10. McClellan J, King M. Genomic analysis of mental illness: a changing landscape. JAMA 303:2523–2524, 2010.

Glossary

1-gene-to-many-diseases—Various alterations in a single gene can result in several different diseases. For example, the ALAS2 gene codes for delta-aminolevulinate synthase-2. A gain-of-function mutation in the ALAS2 gene causes X-linked erythropoietic protoporphyria. A deficiency of the enzyme results in insufficient hemoglobin production in red cells and causes X-linked sideroblastic anemia. Several different diseases may result from mutations that cause graded losses in gene activity. For example, Lesch–Nyhan syndrome and Kelley–Seegmiller syndrome both result from mutations in the HGPRT gene. In the Kelley–Seegmiller syndrome, the deficiency of hypoxanthine guanine phosphoribosyltransferase is less than that observed in Lesch–Nyhan syndrome, and the symptoms are milder. See Allelic to.

Ab initio—Latin term meaning from the beginning. In disease biology, it refers to a process that begins much the same way that it ends, without going through a series of consecutive steps, over time, leading to the final condition.

Adaptive immunity—Immunity in which the response adapts to the specific chemical properties of foreign antigens. Adaptive immunity is a system wherein somatic T cells and B cells are produced, each with a unique and characteristic immunoglobulin (in the case of B cells) or T-cell receptor (in the case of T cells). Through a complex presentation and selection system, a foreign antigen elicits the replication of a B cell that produces an antibody whose unique immunoglobulin attachment site matches the antigen. Antigen–antibody complexes may deactivate and clear circulating antigens or may lead to the destruction of the organism that carries the antigen (e.g., virus or bacteria). The process of producing unique proteins requires that recombination and hypermutation take place within a specific gene region. Recombinations yield on the order of about a billion unique somatic genes, starting with one germinal genome. This process requires the participation of recombination activating genes (RAGs). The acquisition of an immunologically active recombination activating gene is presumed to be the key evolutionary event that led to the development of the adaptive immune system, present in all jawed vertebrates (gnathostomes). In addition, a specialized method of processing immunoglobulin heavy chain mRNA transcript accounts for the high levels of secretion of immunoglobulin proteins by plasma cells [1]. As one might expect, inherited mutations in RAG genes cause immune deficiency syndromes [2,3]. See Intrinsic immunity and Innate immunity.

Rare Diseases and Orphan Drugs. http://dx.doi.org/10.1016/B978-0-12-419988-0.00025-0

Age-adjusted incidence—An age-adjusted incidence is the crude incidence of disease occurrence within an age category (e.g., age 0–10 years, age 70–80 years), weighted against the proportion of persons in the age groups of a standard population. When we age-adjust incidence, we cancel out the changes in the incidence of disease occurrence, in different populations, that result from differences in the proportion of people in different age groups. For example, suppose you were comparing the incidence of childhood leukemia occurrences in two populations. If the first population has a large proportion of children, then it will likely have a higher number of childhood leukemia in its population, compared with another population with a low proportion of children. To determine whether the first population has a true, increased rate of leukemia, we need to adjust for the differences in the proportion of young people in the two populations. See Incidence.

Aging (alternate spelling, ageing)—A chronic degenerative process that occurs in cells that have lost the ability to divide, while retaining their functional obligations. Such cells include neurons, chondrocytes (i.e., cartilage cells), muscle cells, and cells of the eye lenses. Cells that maintain the ability to divide, indefinitely, such as epithelial lining cells of ducts, mucosal surfaces, glands, and epidermis do not suffer from the degenerative changes associated with aging cells (i.e., nobody dies from an old colon or an old liver). For the purposes of this book, aging is considered a disease differing from other diseases only in its inevitability.

Allele—One of a pair of matched genes on paired chromosomes, wherein each of the matched genes is a variant of the other (i.e., each is a different allele of the gene). In most cases one allele comes from the father, the other from the mother.

Allelic heterogeneity—Occurs when different mutations within the different alleles of a gene can yield the same clinical phenotype. For example, hundreds of different alleles of the cystic fibrosis gene can yield the same phenotype [4]. Additionally, a study of 424 families with members affected by hemophilia B found 167 different allelic mutations of the disease gene [5]. Allelic heterogeneity should not be confused with two diseases being allelic to one another. When two biologically distinct diseases are caused by different mutations in the same gene, the two diseases are said to be allelic to one another. See Locus heterogeneity and Phenotypic heterogeneity.

Allelic to—One genetic disease is allelic to another genetic disease if both are caused by mutations of different alleles of the same gene (i.e., in different inherited forms of the gene). For example, distal myopathy with rimmed vacuoles is allelic to hereditary inclusion body myopathy. Each results from a different loss-of-function mutation in different alleles of the gene encoding UDP-N-acetylglucosamine 2-epimerase/N-acetylmannosamine kinase [6]. Whenever a gene associated with two or more distinct diseases is mapped to the same physical location in the genome, then the cause of the diseases may be due to allelic variation, or to contiguous gene defects (i.e., defects in several genes located in close proximity to one another). See Phenotypic heterogeneity.

Alternative RNA splicing—A normal mechanism whereby one gene may code for many different proteins [7]. In humans, about 95% of genes that have multiple exons are alternately spliced. It has been estimated that 15% of disease-causing mutations involve splicing [8,9]. Cancer cells are known to contain numerous splicing variants that are not found in normal cells [10,11]. Normal cells eliminate most abnormal splicing variants through a post-transcriptional editing process. Alternative RNA splicing may result from mutations in splice sites or from spliceosome disorders. In hereditary thrombocythemia, characterized by an overproduction of platelets, there is a mutation in the gene coding for thrombopoietin protein. Wiestner and coworkers have shown that the gene mutation leads to mRNAs with shortened untranslated regions that are more efficiently translated than the transcripts that lack the mutation. This causes the overproduction of the thrombopoietin, which in turn induces an increase in platelet production [12]. See Spliceosome.

Aneuploidy—The presence of an abnormal number of chromosomes (for the species) in a cell. Most cancers contain aneuploid cells; an observation that holds true for virtually every poorly differentiated cancer. Aneuploidy is seen less often in benign tumors and well-differentiated tumors. Aneuploidy is also found in epithelial precancers and other growing lesions that can sometimes regress spontaneously (e.g., keratoacanthoma). These observations have prompted speculation that chromosomal instability and the acquisition of aneuploidy is an underlying cause of the cancer phenotype (i.e., tumor growth, invasion into surrounding tissues, and metastases). Such causal associations invite skepticism, particularly in the realm of cancer biology, as virtually every cellular process and constituent of cancer cells has been shown to deviate from the norm. Nonetheless, there is good reason to suspect that aneuploidy is at least a factor in tumor development, as mutations that cause aneuploidy are associated with a heightened risk of cancer (e.g., Brca1 gene mutations [13] and mutations of mitotic checkpoint genes [14]). Others have warned that aneuploidy, by itself, may not cause cancer [15]. Aneuploidy may need to be accompanied by other factors associated with genetic instability, such as the accumulation of DNA damage, cytogenetic abnormalities, and reduced cell death [15]. As usual, a rare disease helps to clarify the role of aneuploidy in carcinogenesis. Mosaic variegated aneuploidy syndrome-1 (MVA1) is caused by a homozygous or compound heterozygous mutation in the BUB1B gene, which encodes a key protein in the mitotic spindle check point. This disease is characterized by widespread aneuploidy in more than 25% of the cells of the body, and a heightened risk of developing childhood cancers (e.g., rhabdomyosarcoma, Wilms tumor, and leukemia). **Because the underlying cause of mosaic variegated aneuploidy syndrome-1 is a gene that produces aneuploidy, and because such aneuploidy is an early event (i.e., congenital) that precedes the development of cancer and that is found in the developed cancer cells, then it is reasonable**

to infer that aneuploidy is closely associated with events that lead to cancer. See Mutator phenotype, Carcinogenesis, Cytogenetics, and Karyotype.

Angiogenesis—The formation of new vessels. Angiogenesis in the adult organism always refers to the growth of small vessels, not arteries and veins. The large vessels in the human body develop *in utero*. Tumor cells must receive oxygen from blood; hence, every invasive and growing solid tumor is capable of inducing angiogenesis. As the tumor grows, so do the vessels feeding the tumor. The vessels arise from non-neoplastic connective tissue and are induced to grow by angiogenesis factors secreted by the tumor cells.

Anonymous variation—A genetic variation for which there is no change in gene function. Today, the bulk of the 3 billion base-pair sequence comprising the human genome cannot be assigned to any particular function; a randomly occurring mutation is likely to be anonymous; hence, it is assumed that most SNPs are anonymous. Other commonly encountered anonymous markers include the microsatellites, for which there occur variations in the length of repeated sequences within the microsatellites, but these variations cannot be assigned to a gene or to a particular function. Mutations that occur in somatic, post-mitotic cells (i.e., cells that will never divide) are, for all practical purposes, anonymous and undetectable. A mutation must be passed to a population of progeny cells before it can do much damage and before it can be detected by current molecular biological techniques. Some types of mutations are difficult to find, even when they occur in large numbers of cells. For example, when a mutation is a duplicated exon, the alteration cannot be detected by methods that find base sequence alterations.

Anticipation—The phenomenon by which an offspring develops an inherited disease at a younger age than the age at which the parent developed the disease. In most cases, anticipation is associated with an expansion of the trinucleotide repeat in the inherited gene causing the disease. The expansion of trinucleotide repeats is a common occurrence within the genome, and may have any of several consequences: (1) producing disease via a gain-of-function mutation within a gene coding for a protein (e.g., Huntington disease); (2) producing disease via a loss-of-function mutation (e.g., myotonic dystrophy); (3) producing anticipation in a pre-existing disease-causing gene, possibly by altering the level of expression; and (4) producing no discernible biological effect. Examples of diseases that may display anticipation include: Behçet's disease, Crohn disease, dyskeratosis congenita, fragile X syndrome, Friedreich ataxia (rare cases), Huntington disease, myotonic dystrophy, and spinal cerebellar ataxias (several forms). Why such expansions occur is not well understood.

Aplastic anemia—A profound reduction in circulating blood cells, resulting from the loss of bone marrow stem cells. Severe or prolonged cases of aplastic anemia have a high mortality rate. A reduction of all blood cell lineages (whether profound or mild) is called pancytopenia. When an isolated lineage is reduced, the anemia is named after the cell type involved: reticulocytopenia, immature red cell decline; neutropenia, neutrophil decline; thrombocytopenia, platelet decline; lymphopenia, decline in lymphocytes. See Stem cell.

Apoptosis—Apoptosis is a coordinated cellular activity leading to cell death. Alternate terms are cell suicide and programmed senescence. During apoptosis, chromosomal DNA is broken into small fragments, the nucleus shrinks (i.e., karyopyknosis), and the cytoplasmic membrane blebs out.

Association (statistical)—In the context of diseases, an association is anything that happens to occur more frequently in the presence of a disease than occurs in the absence of the disease. Even when we know that one thing is associated with another thing, it can be very difficult to express the association in a manner that is mechanistically useful. For example, in 2000, Concorde, a supersonic transport jet, crashed on take-off from Charles de Gaulle Airport, Paris. Debris left on the runway, possibly a wrench, flipped up and tore the underside of the hull. All passengers were killed in the subsequent few seconds as the plane exploded and crashed. What is the association here? Is it, "debris associated with jet crash," or do we need to be more specific, "wrench associated with jet crash"? Do jets need to be afraid of wrenches in general, or only with wrenches that are left out on the runway: "wrench on runway associated with jet crash"? If the association contains an implied mechanism that ties an object with a result, wouldn't we need to confine the association to wrenches that are actually run over by the jet, because if the jet tires miss the wrench, the wrench would not flip up and tear the underside of the plane. This would make the assertion: "wrenches that are run over by a tire and flip upwards are associated with jet crashes." It can be very difficult to develop a sensible way to describe associations. The problem is magnified many times when we are dealing with gene polymorphisms (i.e., gene variants found in a population) associated with diseases that have various causes, poorly understood pathogeneses, and complex phenotypes. Like so many scientific observations, associations serve as clues, not answers.

Ataxia telangiectasia—Also known as Louis–Bar syndrome and as Border–Sedgwick syndrome, and caused by a mutation of the ATM gene, resulting in a defect in DNA repair. Cells of individuals with ataxia telangiectasia are highly vulnerable to radiation toxicity. The clinical phenotype consists of cerebellar ataxia (i.e., a body movement disorder secondary to cerebellar impairment), telangiectases (i.e., small focal vascular malformations), immune deficits predisposing to ear, sinus, and lung infections, and a predisposition to malignancy (e.g., lung, gastric, lymphoid, and breast cancers).

Autonomous growth—The growth of normal cells is highly controlled in most adult animals, so that every tissue contains about the same number of cells from day to day. Such controlled growth is referred to a non-autonomous, because each dividing cell is restricted from growing continuously or in a manner that is not somehow matched against a nearly constant tissue-specific number. Cancer cells, which increase in number every day, are said to grow autonomously, and free of the restraining influences of humoral or other external factors. Of course, no cell growth is truly autonomous. Cells in a tumor require a vascular blood supply. Some tumors are hormone responsive, and when the hormone is withdrawn or blocked, the tumor may stop growing, or may shrink. Such tumors exhibit non-autonomous growth. Some gastric maltomas (i.e., a type of lymphoma arising from mucosa-associated lymphoid tissue) will regress completely after the patient is treated for *Helicobacter pylori* infection, the presumed cause of the maltoma. These regressed maltomas would be considered non-autonomous [16].

Basal cell carcinoma—Basal cell carcinoma is the most common skin cancer, with about 600,000 new cases occurring each year in the U.S. They occur as small, smooth patches on sun-exposed skin (e.g., face, arms, neck). A basal cell carcinoma seldom, if ever, metastasizes from its primary site of growth. Along with squamous carcinoma of skin, occurrences of these two conditions equal the occurrences of all other cancers, combined. Neither basal cell carcinoma of skin nor squamous cell carcinoma of skin accounts for many human deaths. Neither of these extremely common tumors is recorded in hospital-based cancer registries, and when statistics are compiled on the incidences of cancers, these tumors are usually excluded. The omission of these two tumors from cancer registries produces an under-representation, by about 50%, of the true biological burden of cancer in the human population.

Blast (blast cell)—A term usually reserved for the dividing cells of the bone marrow (i.e., a hematopoietic stem cell). The cytologic lineages of the bone marrow (i.e., neutrophils, monocytes, lymphocytes, red blood cells, plasma cells, and megakaryocytes) undergo a graded series of morphologic changes as they mature from blast cells to fully bone marrow cell. Blast cells are confined under normal circumstances to the bone marrow. Blast cells are found in the circulation in acute leukemias and in so-called blast transformation of chronic myelogenous leukemia (i.e., a shift from indolent disease to an aggressive leukemia).

Blood pressure—Refers to the pressure exerted by the blood on the walls of arteries that are accessible to external palpation. The blood pressure oscillates due to the repetitive pumping action of the heart. The peak in blood pressure is the systolic value, normally about 120 mmHg, and the trough is the diastolic value, normally about 80 mmHg. It is important to note that the trough value is not zero, or anything close to zero, indicating that there is an intrinsic tension imposed by blood on arterial walls. Sphygmomanometers (blood pressure cuffs) inflate to constrict arteries to a pressure higher than systolic, at which point blood flow through the artery is severely reduced. As the cuff pressure is released, there comes a point when the systolic pressure exceeds the inflated pressure. At that point pressure waves of blood flow through the artery, producing characteristic sounds heard with the assistance of a stethoscope (i.e., Korotkoff sounds). The sounds begin at the systolic blood pressure value and they continue as the cuff continues to deflate, until the diastolic blood pressure value is reached, at which point the cuff's resistance to the flow of pumped blood is zero, and the Korotkoff sounds cease.

Brain attack—At present, the politically correct term for "stroke" and "cerebrovascular accident" is "brain attack"; a term that emphasizes an analogy with "heart attack." Both terms are ill-conceived as neither the brain nor the heart does much attacking. "Attacked brain" or "attacked heart" would be preferable linguistically, but neither sounds right when spoken.

Bronchogenic carcinoma—Cancers arising from the pulmonary bronchus and its branches, rather than from the alveoli (oxygen-exchanging sacs). Most of the common cancers of the lung are bronchogenic. The non-bronchogenic tumors account for fewer than 10% of lung cancers and can be considered rare cancers. The bronchogenic cancers are adenocarcinoma, squamous carcinoma, small cell carcinoma, and their various undifferentiated and mixed variants. About 90% of cases of bronchogenic carcinoma arise in smokers. It is safe to presume that some additional percentage of individuals with bronchogenic carcinoma who are non-smokers may have been exposed to second-hand smoke in the workplace or at home (i.e., more than 90% of bronchogenic carcinoma is linked to cigarette smoking or to secondary inhalation of cigarette smoke). Broncho-alveolar carcinoma, alternately known as bronchioloalveolar carcinoma or as alveolar carcinoma, is a non-bronchogenic lung cancer arising from cells in or near the alveolar sacs. Pure broncho-alveolar carcinoma (i.e., broncho-alveolar carcinoma that is not admixed with adenocarcinoma of bronchogenic origin) and atypical adenomatous hyperplasia, the putative precancer for bronchioloalveolar carcinoma, do not seem to be linked to cigarette smoking [17]. See Undifferentiated tumor.

But-for—From the field of law, the "but-for" test attempts to determine whether a sequence of actions leading to an event could have happened without the occurrence of a particular underlying action or condition. In the realm of death certification, the underlying cause of death satisfies the "but-for" test (i.e., but for the condition, the sequence of events leading to the individual's death would not have occurred). See Proximate cause and Underlying cause of death.

Cancer progression—The acquisition of additional properties of the malignant phenotype over time. Progression is achieved through a variety of mechanisms (e.g., genetic instability [18], epigenetic instability, and aberrant cell death regulation) and results in the eventual emergence of subclones that have growth advantages over other cells in the same tumor. The presence of subclones of distinctive phenotype and genotype within a single tumor accounts for tumor heterogeneity [19]. Tumors that grow without accumulating changes in genotype or phenotype tend to be benign (i.e., benign tumors do not progress or their rate of progression is much less than that observed in malignant tumors). See Tumor heterogeneity.

Cancer-causing syndrome—There are many inherited conditions that are associated with susceptibility to multiple types of cancers. Cancers that arise in these syndromes often occur in children or at an age earlier than the average age of occurrence of their sporadic equivalents. A few examples of eponymic (named for a person) cancer syndromes are: Bloom syndrome, Carney syndrome, Cowden syndrome, Fanconi anemia, Li–Fraumeni syndrome, Lynch cancer family syndrome, Muir–Torre syndrome, Von Hippel–Lindau syndrome.

Candidate gene approach—One of several methods whereby the gene that causes a disease may be discovered. In the candidate gene approach, the researcher begins with some insight into the disease, and the various pathways and metabolic activities that are affected. The researcher chooses a candidate gene to study, based on knowledge of the function of the gene, and a suspicion that alterations in the gene might play an important role in the pathogenesis of the disease. She studies the sequence of the candidate gene in DNA samples from a set of people with the disease, and compares her findings with the sequence of the gene in DNA samples from a set of people who do not have the disease. Consistent differences between the gene in the disease-carrying individuals and the control subjects would suggest that the gene contributes to the development of the disease. Finding a disease association for a candidate gene does not tell us whether other, unexamined, genes may play an important role in disease development. Conversely, failing to find an association does not rule out the presence of an association that was not detected in the gene sequence (e.g., a defect in any of the processes that regulate the transcription, assembly, or deployment of the final gene product).

Carcinogen—The term "carcinogen" is used differently by different people. Confusion arises because carcinogenesis is a multi-step process that can be modified at many different biological stages. Some people use the term "carcinogen" to mean a chemical, biological, or physical agent that, when exposed to normal cells, will result in the eventual development of cancers (i.e., the carcinogen acting as the underlying cause of the cancer). Sometimes, the term "complete carcinogen" is used to emphasize the self-sufficiency of the agent as the primary underlying cause of a cancer. Others in the field use the term "carcinogen" to mean anything that will increase the likelihood of tumor development. An agent that causes an increase in the number of tumors that are produced by a complete carcinogen, or an agent that must be followed or preceded with another agent for tumors to occur, or a process that increases the number of cancers occurring in a population known to be at high risk of cancer due to an inherited condition, would all be considered carcinogens under this alternate definition.

Carcinogenesis—The cellular events in a multi-event process that leads to cancer, equivalent to the pathogenesis of cancer. Carcinogenesis in adults is a long process that involves the accumulation of genetic and epigenetic alterations that confer the malignant phenotype to a clone of cells. The envisioned sequence of events that comprise carcinogenesis begins with initiation, wherein a carcinogen damages the DNA of a cell, producing a mutant clonal founder cell that yields a group of cells that have one or more subtle (i.e., morphologically invisible) differences from the surrounding cells (e.g., less likely to senesce and die, more likely divide, less genetically stable, better able to survive in an hypoxic environment). After a time, which could easily extend into years, subclones of the original clone emerge that have additional properties that are conducive to the emergence of the malignant phenotype (e.g., new mutations that confer growth or survival advantage, greater ability to grow in hypoxic conditions). The process of continual subclonal selection continues, usually for a period of years, until a morphologically distinguishable group of cells appear: the precancer. Subclonal cells from the precancer eventually emerge, having the full malignant phenotype (i.e., the ability to invade surrounding tissues and metastasize to distant sites). The entire process can take decades.

Carrier—In the field of genetics, a carrier is an individual who has a disease-causing gene that does not happen to cause disease in the individual. For example, individuals with one sickle cell gene are typically not affected by sickle cell disease, which usually requires homozygosity (i.e., both alleles having the sickle cell gene mutation) for disease expression. When two carriers mate, they pass the homozygous state to offspring with a likelihood of 25%. As another example, carriers may also have a low-penetrance disease gene. In such cases, offspring with the same gene defect as the carrier may develop disease. In the field of infectious diseases, a carrier is an individual who harbors an infectious organism, but who suffers no observable clinical consequences. If the carrier state is prolonged, and if infectious organisms cross to other individuals, a single carrier can produce an epidemic.

Cause of death—In the case of a natural death (i.e., not homicidal and not accidental) a cause of death is one item from a standard list of medical conditions known to produce death in humans [20]. The term "causes of death" implies that multiple different conditions may contribute to a person's death, and the term does not provide a clue as to which condition meets "but-for" criteria. See But-for and Underlying cause of death.

Cause of death error—Cause of death data comes from death certificates [21]. Death certificate data have many deficiencies [22,23]. The most common error occurs when a mode of death (i.e., the way that an individual dies) is listed as the cause of death. For example, cardiac arrest is not a cause of death, though it appears incorrectly as the cause of death on many death certificates. An international survey has shown very little consistency in the way that death data are collected [24]. Most death certificates are completed without benefit of an autopsy (i.e., without using the most thorough and reliable medical procedure designed to establish the causes of death). In the absence of an autopsy, a death certificate expresses a clinician's reasonable judgment at the time of a patient's death. See Cause of death.

Cell type—The number of different kinds of cells in an organism varies based on how you choose to categorize and count them, but most would agree that there are at least 200 different cell types in the adult body. The number of cell types that appear for a short period during *in utero* development, then disappear before birth, is not included in the count. It can be difficult to assign a cell type to a fetal cell whose precise function cannot be specified. Every cell type in the body has the same genome as every other cell; the differences between one cell type and another are determined by the epigenome. Because differentiated cells under normal conditions do not change their cell type (e.g., a hepatocyte does not transform into a neuron), and because cell types of a given lineage produce other cells of the same lineage (e.g., a dividing hepatocyte produces two hepatocytes), then we can infer that the epigenome is heritable among somatic cell types.

Channelopathy—Disorders of the electrical systems in humans, all of which depend on the depolarization and repolarization of electrical current (i.e., the flux of charged molecules) across ion channels (e.g., sodium channel, potassium channel, chloride channel, calcium channel). Ion channels are found on the membranes of specialized cells. Disorders of these channels are termed channelopathies, and encompass a wide range of neural, cardiac, and muscular disorders and always play at least a contributing role in common seizures and arrhythmias. Specific rare conditions in which channel disorders play a principal role, in at least some forms of the disease, include: Alternating hemiplegia of childhood, Bartter syndrome, Brugada syndrome, congenital hyperinsulinism, cystic fibrosis, Dravet syndrome (severe myoclonic epilepsy of infancy), episodic ataxia, erythromelalgia (Mitchell disease), generalized epilepsy with febrile seizures plus, familial hemiplegic migraine, hyperkalemic periodic paralysis, hypokalemic periodic paralysis, long QT syndrome, malignant hyperthermia, mucolipidosis type IV, myasthenia gravis, myotonia congenita, neuromyotonia, non-syndromic deafness, paramyotonia congenita, retinitis pigmentosa, short QT syndrome, and Timothy syndrome.

Chemokine—A cytokine that stimulates white blood cells to move to a tissue target. An allele of the beta-chemokine receptor 5 (CCR5) gene seems to confer a high level of protection against HIV infection. In a study of over 1200 individuals at risk for HIV infection, the homozygous allele was always absent from infected individuals. Among the individuals at high risk of HIV infection who remained infection free, the homozygous allele was found in 3.6% of the population [25].

Chromosomal disorder—Disorders associated with physical abnormalities in the physical structure of the chromosome. An example is found in fragile X syndrome. In this disease, a not uncommon cause of mental retardation, fragile sites are inherited as poorly condensed regions of the chromosome. Under experimental conditions, these regions break easily. Fragile sites have been associated with CCG repeats. Other examples of chromosomal disorders include Pelger–Huet anomaly and Roberts syndrome.

Cis-acting—A gene regulation function that is exerted by some segment of genetic material on another segment of genetic material. In most instances, a short sequence of DNA regulates the transcriptional activity of a nearby gene that codes for a protein. The cis-acting sequence is typically activated or inactivated by some diffusible molecule that attaches to the cis-acting sequence. Cis-acting processes apply to RNA as well as to DNA. The regulation of alternative splicing of mRNAs employs proteins that bind to cis-acting sites on pre-mRNA. See Trans-acting.

Clinical trial—Before a drug can be approved for use, it must undergo and pass three phases of a clinical trial. Phase 1 is the safety phase; the drug must be safe for humans. Phase 2 is the effectiveness phase; the drug must have some desired biological effect. Phase 3 is the large, expensive trial wherein individuals are tested against a control group treated with a placebo or with the standard-of-care medication. Phase 3 trials are very expensive to conduct, and many trials are negative (i.e., fail to indicate that the drug is effective in a phase 3 trial) or demonstrate only incremental success. Of the successful phase 3 trials, a significant number of drugs will eventually be withdrawn, because their effectiveness in clinical practice could not meet the earlier expectations observed in the phase 3 trial results [26]. Clinical trials are experiments, and like any other experiment they must be repeated over and over in various settings before they can be trusted. Large clinical trials, of the kind designed for common diseases, are impractical for the rare diseases, for which it is never possible to accrue a large number of individuals who have a rare disease. The topic of clinical trials for rare diseases is discussed in some detail in Section 14.2. See Preclinical trial.

Collagenopathy—A variety of clinical conditions involving genetic alterations of the various collagen genes or other genes involved in the complex processes of collagen synthesis. Lists of clinical collagenopathies usually include Ehlers–Danlos syndrome, osteogenesis imperfecta, familial aneurysmal disorders or aortic dissection disorders, Caffey disease (infantile cortical hyperostosis), and Bruck syndrome. Some of the non-collagen genes involved in collagenopathies include the ACTA2 gene (thoracic aortic aneurysms and aortic dissection) and the PLOD2 gene (procollagen-lysine dioxygenase 2 involved in Bruck syndrome), and familial aneurysm disorders (e.g., smad3, tgfbr1, tgfbr2, and tgfb2). There are over 17 genes coding for the different species of collagen molecules. A list of collagen disorders, arranged by collagen gene, is included here:

- COL1. Osteogenesis imperfecta, Ehlers–Danlos syndrome, types 1, 2, 7
- COL2. Hypochondrogenesis, achondrogenesis type 2, Stickler syndrome, Stickler syndrome membranous vitreous type, Marshall syndrome, spondyloepiphyseal dysplasia congenita, spondyloepimetaphyseal dysplasia, Strudwick type, Kniest dysplasia, osteoarthritis with mild chondrodysplasia, Czech dysplasia
- COL3. Ehlers–Danlos syndrome types 3, 4 (Sack–Barabas syndrome)
- COL4. Alport syndrome, porencephaly-1

- COL5. Ehlers–Danlos syndrome types 1, 2
- COL6. Bethlem myopathy, Ullrich congenital muscular dystrophy
- COL7. Epidermolysis bullosa dystrophica, recessive dystrophic epidermolysis bullosa, Bart syndrome, transient bullous ermolysis of the newborn, classic dystrophic epidermolysis bullosa pruriginosa, non-syndromic congenital nail disorder-8
- COL8. Fuchs' dystrophy 1
- COL9. Multiple epiphyseal dysplasia 2,3,6, Stickler syndrome 5
- COL10. Schmid metaphyseal chondrodysplasia
- COL11. Weissenbacher–Zweymuller syndrome, otospondylomegaepiphyseal dysplasia, Stickler syndrome type 3, fibrochondrogenesis-2, and a form of non-syndromic hearing loss
- COL17. Bullous pemphigoid (includes an autoimmune disease in which antibodies react with the COL17 transmembrane protein in epidermal keratinocytes [27])

Commensal—A symbiotic relationship between two organisms in which one of the organisms benefits and the other is unaffected under normal conditions. A commensal may become an opportunistic pathogen when the host provides a physiologic opportunity for disease, such as malnutrition, advanced age, immunodeficiency, overgrowth of the organism (e.g., after antibiotic usage), or some mechanical portal that introduces the organism to a part of the body that is particularly susceptible to the pathologic expression of the organism, such as an indwelling catheter, or an intravenous line. In addition, a commensal relationship between bacteria and an animal parasite may produce a pathogenic relationship in the parasite's host. For example, the bacterium *Wolbachia pipientis* happens to be an endosymbiont that infects most members of the filarial Class Onchocercidae [28]. *Onchocerca volvulus* is a parasitic filarial nematode in humans. The filaria migrate to the eyes, causing river blindness, the second most common infectious cause of blindness worldwide [29]. *Wolbachia pipientis* lives within *Onchocerca volvulus*, and it is the *Wolbachia* organism that is responsible for the inflammatory reaction that leads to blindness. Hence, *Wolbachia pipientis* is a commensal in *Onchocerca volvulus* and a pathogen in humans simultaneously. Treatment for river blindness may involve a vermicide to kill *Onchocerca voluvulus* larvae, plus an antibiotic to kill *Wolbachia pipientis*.

Complex disease—A somewhat vague term often indicating that the pathogenesis of a disease cannot be understood. The presumption is that our lack of understanding is not based on our failure to discover the underlying cause of the disease. Our lack of understanding is based on our discovery that there are so many different factors to consider that it is impossible to understand the pathogenesis in a way that we can fully grasp. When the development of a disease involves numerous environmental factors, some known and others assumed, as well as multiple genetic and epigenetic influences, we have no way of fully understanding how all of these factors interact with one another, and we have no way of fully describing the biological steps that lead to the clinical expression of disease. Likewise, we have no way of predicting how different individuals with a complex disease will respond to treatment. In this book, we use the term "common disease" interchangeably with "complex disease." Without exception, all of the clinically significant common diseases of humans are complex. The rare diseases tend to be simple, though there are exceptions, particularly for the "more common" of the rare diseases. **As a rule of thumb, complexity rises in direct proportion to the frequency of occurrences of a disease.**

Congenital disorder of glycosylation (CDG)—A group of congenital, multi-organ syndromes caused by post-translational defects in protein glycosylation [30]. The steps in post-translational glycosylation are complex, and may involve systems that move nascent proteins from the endoplasmic reticulum to other sites (e.g., Golgi apparatus). In such cases, there may be overlap between the congenital disorders of glycosylation and the vesicular trafficking disorders. Many different disorders of glycosylation have been identified, involving N-glycosylation, O-glycosylation, or both. Because there are so many different inherited glycosylation disorders, and because these disorders tend to produce neurologic symptoms and multi-organ impairments, physicians should always include congenital disorders of glycosylation in the differential diagnosis when evaluating infants with otherwise unexplained multi-organ involvement or neurologic abnormalities [30]. See Vesicular trafficking disorder.

Contig disease—See Contiguous gene deletion syndrome.

Contiguous gene deletion syndrome—A syndrome caused by abnormalities of two or more genes that are located next to each other on a chromosome. When the abnormality is a deletion, a contiguous gene syndrome is equivalent to a microdeletion syndrome. See Microdeletion.

Convergence—As applied to diseases, convergence occurs when different genes, cellular events, exposures, and pathogenetic mechanisms all lead to the same clinical phenotype. Convergence is a phenomenon that is observed in virtually every common disease. In the case of systemic responses to injury, convergence seems to have evolutionary origins. The organism evolves to respond in an orchestrated way to a variety of pathologic stimuli (e.g., systemic inflammatory response syndrome [31]). Convergence is also observed in rare diseases that have genetic heterogeneity (e.g., multiple causes for epidermolysis bullosa, retinitis pigmentosa, long QT syndrome). It would seem that for any given species, the variety of pathologic responses is limited.

COPD—An abbreviation for chronic obstructive pulmonary disease. COPD covers a range of lung disorders characterized by airway damage. COPD is a common sequela of chronic cigarette abuse.

Copy-number—It is possible to produce a rare genetic disease without actually producing a mutation in a gene; simply changing the number of genes can be sufficient [32]. Charcot–Marie–Tooth disease is an inherited neuropathy caused by duplication of a 1.5 megabase segment on chromosome 17. No altered protein is produced. The clinical phenotype is caused by an increased gene dosage. See Gene dosage.

CpG island—DNA methylation is a form of epigenetic modification that does not alter the sequence of nucleotides in DNA. The most common form of methylation in DNA occurs on cytosine nucleotides, most often at locations wherein cytosine is followed by guanine. These methylations are called CpG sites. CpG islands are concentrations of CpG dinucleotides that have a GC content over 50% and that range from 200 base pairs (bp) to several thousand bp in length. There are about 29,000 to 50,000 CpG islands and most of these are associated with a promoter [33]. Various proteins bind specifically to CpG sites. For example, MECP2 is a chromatin-associated protein that modulates transcription. MECP2 binds to CpGs; hence, alterations in CpG methylation patterns can alter the functionality of MECP2. Mutations in MECP2 cause RETT syndrome, a progressive neurologic developmental disorder and a common cause of mental retardation in females. It has been suggested that the MECP2 mutation disables normal protein–epigenome interactions [34].

Cyanobacteria—The most influential organisms on earth, cyanobacteria were the first and only organisms to master the biochemical intricacies of photosynthesis (more than 3 billion years ago). Photosynthesis involves a photochemical reaction that uses carbon dioxide and water, and releases oxygen. All photosynthesizing life-forms are either cyanobacteria, or they are eukaryotic cells (e.g., algae, plants) that have acquired chloroplasts (an organelle created in the distant past by endosymbiosis between a eukaryote and a cyanobacterium). Before the emergence of oxygen-producing cyanobacteria, Earth's atmosphere had very little oxygen.

Cytopenia—A reduction in the normal number of cells of a particular type. The term is usually applied to hematopoietic cells (i.e., marrow-derived blood cells). Anemia is a cytopenia of red blood cells. Thrombocytopenia is a cytopenia of platelets (i.e., thrombocytes). Neutropenia is a cytopenia of lymphocytes. A pancytopenia refers to a reduction of all the different types of cells of hematopoietic lineage.

***De novo* germline mutation**—In the context of this book, *de novo* mutations are new (i.e., Latin, *de novo,* anew) disease-causing mutations found in the germline of organisms (i.e., in every somatic cell of the organism) that were not present in the germline of either parent. A *de novo* mutation may result as a new mutation in a differentiated germ cell of either parent (i.e., it was not present in all of the cells of the parent, but appeared as a mutation in the specific parental germ cell that contributed to the offspring), or it may be a new mutation in the zygote (i.e., ovum fertilized by sperm) or in an early embryonic cell. Examples of some rare diseases that are caused by *de novo* gene mutations would include:

Baraitser–Winter syndrome, characterized by central nervous system and facial malformation; Borhing–Opitz syndrome, characterized by intellectual disability plus congenital malformations; CHARGE syndrome, an acronym for coloboma of the eye, heart defects, atresia of the nasal choanae, retardation of growth and/or development, genital or urinary abnormalities, and ear abnormalities and deafness, and a leading cause of combined deafness and blindness in newborns; Kabuki syndrome, characterized by intellectual disability and congenital anomalies; KBG syndrome, characterized by a disease-typical facial dysmorphism, macrodontia of the upper central incisors, costovertebral anomalies and developmental delay; and Schinzel–Giedion syndrome, characterized by distinctive facial features, neurological problems, and organ and bone abnormalities. See Somatic mosaicism.

Differentiation—The term "differentiation," as it is used by pathologists, refers to the cellular process that makes one cell different from other cells, and capable of being identified as a particular named cell type (e.g., red blood cell, neutrophil, hepatocyte, spermatocyte, neuron, etc.). Every cell in the body has the same genetic sequence in their DNA. The reason why one cell develops into a neutrophil and another cell develops into a neuron is due to the epigenome; the set of modifications to the chemical and physical structure, not the sequence, of the genome. Such modifications are cell-type specific, so that every neutrophil looks and acts like every other neutrophil, and does not look or act like neurons or gut lining cells, or muscle cells. An individual organism can be identified by his or her genomic sequence, which is unique, with some few exceptions: identical twins and organisms that reproduce asexually by division of a somatic cell. The cell types within an organism can be distinguished by their epigenome. See Epigenome and Erasure.

Digenic disease—Digenic diseases require mutations in two genes to produce the complete clinical phenotype. There are several rare diseases that are known or suspected to be digenic. Several different forms of Usher disease, combined retinitis pigmentosa and hearing loss, are digenic. A digenic cause of several forms of long QT syndrome, a type of heart arrhythmia, has also been reported [35]. Kallman syndrome, a form of hypogonadotropic hypogonadism, is suspected to be digenic [35]. Digenic diseases often have a variable clinical phenotype, even among family members with the disease. Mice with digenic diabetes have a non-Mendelian pattern of inheritance, typical of a polygenic familial disease [36]. As a group of disorders, the inherited digenic disease occupies an intermediate niche, between monogenic diseases and polygenic diseases. See Monogenic disease and Polygenic disease.

DNA methylation—DNA methylation is a chemical modification of DNA that does not alter the sequence of nucleotide bases. It is currently believed that DNA methylation plays a major role in cellular differentiation, controlling which genes are turned on and which genes are turned off in a cell, hence determining a cell's "type" (e.g., hepatocyte, thyroid follicular cell, neuron).

Because cells of a particular cell lineage divide to produce more cells of the same lineage, DNA methylation patterns must be preserved with each somatic cell generation. The cellular process by which DNA is modified and controlled without altering the sequence of nucleotide bases is called epigenomics, and the collection of such modifications in DNA constitutes the epigenome. About 1% of DNA is methylated in human somatic DNA, and DNA methylation occurs primarily on cytosine, usually at locations for which cytosine is followed by guanine, and designated as "CpG." See CpG island.

DNA mutation rate—In most normal tissues, the DNA mutation rate is quite low. In humans, point mutations (i.e., mutations that occur in a single nucleotide base within the genome) occur with a frequency of about 1 to 3×10^{-8} per base [37–39]. This estimate is in line with estimates from other labs, all of which are somewhat speculative. Cancer cells have genetic instability. Cancer cells from the same individuals with low rates of mutation in normal cells had rates that were about a hundred-fold higher, with an average of 210×10^{-8} mutations per base pair [40]. See Mutator phenotype.

DNA repair—When damage occurs in DNA, the cell has three options: (1) do nothing and risk that the damaged DNA will be replicated by cell division and passed to somatic daughter cells. If the DNA damage occurs in a germ cell, the genetic alteration may be passed to the progeny as a new, stable mutation in the human gene pool; (2) eliminate the defect by killing the cell that harbors the damaged DNA, employing a cellular suicide process known as apoptosis; (3) repair the damaged DNA. Several tumor suppressor genes regulate normal DNA repair mechanisms. The inactivation of tumor suppressor genes may lead to genetic instability and the likelihood that an increase in the cellular mutation rate will ultimately initiate carcinogenesis.

Dormancy—The period from the time that a metastatic cell has seeded to a site that is non-adjacent to the primary tumor and the time that the seeded focus grows to a clinically detectable mass. Dormancy has a variable length, varying from days to decades. We know very little about the pathways that control dormancy. Most people who die from cancer succumb to their metastatic lesions; the primary cancer seldom kills. If we had a method that prolonged the dormancy of metastatic foci, it would have enormous medical benefit to individuals with cancer.

Dyserythropoiesis—A dysfunctional form of blood cell formation in which there is excessive cell death of precursor and differentiated blood cells, often leading to pancytopenia. The death of precursor blood cells forces hematopoietically active tissues (i.e., blood-forming tissues such as bone marrow) to produce more and more precursor cells, compensating for high cell death rates. This leads to the expansion of hematopoietic tissue, sometimes resulting in blood cell formation in sites other than the bone marrow, such as spleen, lymph nodes, and liver. Ineffective hematopoiesis is a near-synonym for dyserythropoiesis. HEMPAS (hereditary erythroblastic multinuclearity with positive acidified-serum test), also known as congenital dyserythropoietic anemia type II (CDAN2), is an inherited dyserythropoiesis caused by a mutation in the SEC23B gene.

Dysplasia—The term means abnormal growth, and it is used in different ways in different biomedical specialties. Developmental biologists and pediatricians use the term "dysplasia" to refer to organs or parts of organs that have not grown properly. Stunted growth of an organ, or morphologically abnormal tissues within an organism, would be types of developmental dysplasia. Oncologists (i.e., cancer specialists) use the term "dysplasia" to mean cellular atypia characteristic of neoplastic cells. Cellular dysplasia is found in precancers, cancers, and benign tumors.

Ectoderm—There are three embryonic layers that eventually develop into the fully developed animal: endoderm, mesoderm, and ectoderm. The ectoderm gives rise to the skin epidermis and the skin appendages (hairs, sebaceous glands, breast glandular tissue).

Endoderm—There are three embryonic layers that eventually develop into the fully developed animal: endoderm, mesoderm, and ectoderm. The endoderm forms a tube extending from the embryonic mouth to the embryonic anus. The mucosa of the gastrointestinal tract, the glandular cells of the liver and pancreas, and the lining cells of the respiratory system all derive from the endoderm.

Enhancer—A site on DNA that binds to trans-acting protein factors to enhance the transcription of genes. Unlike promoters, enhancers do not need to be close to their target genes. Enhancers play a major role in the control of gene expression [41]. See Promoter.

Epigenetic instability—The condition in which the normal epigenetic modifications are progressively changing, within one cell or from one cell generation to another. Epigenomic instability, like genomic instability, is a near-constant feature of tumor progression. Because cellular differentiation is under epigenetic control, the loss of tumor cell differentiation observed with tumor progression is presumably due to epigenetic instability. Likewise, cancer cells that have an unstable epigenome may inactivate or activate a variety of disease genes in surprising ways. For example, epigenetic instability may produce cancer cells with inactivated Werner syndrome gene, the same gene that causes a premature aging syndrome when it occurs in the germline cells of an organism [42]. In similar fashion, cancer cells may have epigenetic inactivation of the lamin A/C gene, the same gene that, when inactivated in germline cells, causes a form of cardiomyopathy [43,44].

Epigenome—At a minimum, the epigenome consists of the non-sequence modifications to DNA that control the expression of genes. These modifications include DNA methylations, histones, and non-histone nuclear proteins. Beyond this minimalist definition, there are expanded versions of the definition that would include any conformational changes in DNA that influence gene expression, as well as protein interactions that influence gene expression. As used in this book, the terms "epigenome" and "epigenetics" apply exclusively to non-sequence alterations in chromosomes that are heritable among somatic cell lineages. In general, the epigenome controls differentiation and the biological characteristics of the different somatic cell types of the body.

Epistasis—The condition under which the effect of a gene is influenced by another gene. For example, a gene may be active only when a particular allele of one or more additional genes is also active. Because dependencies among genes are built into cellular systems, the role of epistasis in the penetrance of disease genes and the pathogenesis of disease phenotypes is presumed to be profound. For example, there are at least 27 epistatic interactions among genes associated with Alzheimer disease [45]. Epistatic interactions can be synergistic or antagonistic [46]. See Penetrance and Genome wide association studies.

Epithelial cell—Epithelial cells are polyhedral cells that typically line a surface or a lumen (i.e., an empty gland or duct that leads to a surface). Examples of epithelial lineages are the skin, the mucosa lining the alimentary tract, and the cells that line ducts. Many glandular organs are composed predominantly of epithelial cells (e.g., liver, lungs, kidneys, thyroid). Epithelial cells fit tightly together as polyhedral units fastened together by specialized cell junctions (e.g., desmosomes). Tumors arising from epithelia, and composed of neoplastic epithelial cells, account for over 95% of the cancers that occur in humans.

Erasure—Every cell in an organism has the same genome. The distinctions between the different types of cells in an organism are determined by the epigenome. When cells differentiate into particular cell types (e.g., hepatocyte, muscle cells, ductal cells, etc.), they obtain a cell-type-specific epigenome. Germ cells differ from other somatic cells because they contribute to a totipotent and undifferentiated zygote (i.e., the fusion cell produced by sperm and ovum). Somehow, the highly specific epigenome, passed into the zygote by the differentiated germ cell, must be erased in germline cells, so that the development of a new organism can occur. In theory, erasure removes all of the epigenetic patterns of the differentiated germ cell [47,48]. See Imprinting.

Etiology—The cause of a disease. See Pathogenesis.

Exome sequencing—Also known as targeted exome capture, exome sequencing is a relatively new laboratory technique wherein only the exons (the sections of DNA that code for proteins) are sequenced, sparing analysts from dealing with the non-coding regions of DNA [49]. In the human genome, there are only about 180,000 exons, accounting for about 1% of the genome, and about 85% of known disease-causing mutations [49].

Forme fruste—From the French, a crude or unfinished form; plural formes frustes. A term used by diagnosticians and applied to difficult cases wherein a patient presents with some of the features of a recognized disease or syndrome, but who does not quite fit the accepted diagnostic criteria. The clinical presentation is said to be the forme fruste (i.e., wrong, incomplete, or unfinished form) of the disease. In the context of a rare disease, the forme fruste may present as a near-syndrome, lacking one or more of the definitive features of a set of inherited abnormalities. In many, if not all, cases, studying the forme fruste will help us to understand the classic form of a disease. For example, geneticists reported a child who presented with renal angiomyolipoma, a rare tumor sometimes found in patients with tuberous sclerosis. Several years later, the same patient developed cystic disease in the contralateral kidney, a condition often associated with polycystic kidney disease. Genetic analysis demonstrated a contiguous gene deletion involving both the TSC2 gene for tuberous sclerosis and the PKD1 gene for polycystic kidney disease. The patient's phenotype was the forme fruste of two rare diseases, but genetic analysis proved that the presentation fitted a contiguous gene syndrome [50].

Founder effect—Occurs when a specific mutation enters the population through the successful procreational activities of a founder and his or her offspring, who carry the founder's mutation. When all of the patients with a specific disease have an identical mutation, the disease may have been propagated through the population by a founder effect. This is particularly true when the disease is confined to a separable subpopulation, as appears to be the case for Navaho neurohepatopathy, in which the studied patients, all members of the Navaho community, have the same missense mutation. Not all diseases characterized by a single gene mutation arise as the result of a founder effect. In the case of cystic fibrosis, a dominant founder effect can be observed within a genetically heterogeneous disease population. One allele of the cystic fibrosis gene accounts for 67% of cystic fibrosis cases in Europe. Hundreds of other alleles of the same gene account for the remaining 33% of cystic fibrosis cases [4]. See Gain-of-function mutations.

Gain-of-function—Occurs when a mutation produces a new type of functionality for a gene. It should be noted that the new functionality gained by such mutations is seldom beneficial. It represents a "gain" only in the restricted sense that the mutated gene does something that is different from normal. Most mutations in a gene produce no effect or they reduce the functionality or the expression (e.g., the quantity of expressed protein) of the gene. It is unusual for a mutation to produce a gain in function, and it turns out that most gain-of-function mutations are unique to the disease they cause. For example, everyone with sickle cell disease, caused by a gain-of-function mutation, has precisely the same point mutation causing glutamic acid to be replaced by valine in the sixth position of the beta-globin chain in hemoglobin. Other diseases wherein a particular gain-of-function mutation accounts for most or all affected individuals are hemochromatosis and achondroplasia. Nephrogenic syndrome of inappropriate antidiuresis is an exception to the general rule, being caused by one of two gain-of-function mutations in the same gene.

Gene mutation rate—See DNA mutation rate.

Gene regulation—Gene expression is influenced by many different regulatory systems, including the epigenome (e.g., chromatin packing, histone modification, base methylation), transcription (e.g., transcription factors, DNA promoter sites, DNA enhancer sites, trans-acting factors), post-transcription (splicing, RNA silencing, RNA polyadenylation, mRNA stabilizers), translation (e.g., translation initiation factors, ribosomal processing), and post-translational protein modifications. Mutations in any of the genes that control or participate in any of these regulatory mechanisms may contribute to a disease phenotype. Moreover, anything that modifies any regulatory process (e.g., environmental toxins, substrate availability, epistatic genes) can influence gene regulation; hence, can produce a disease phenotype. See Regulatory DNA element and Regulatory RNA element.

Genetic heterogeneity—In the context of genetic diseases, the term refers to diseases that can be expressed by any one of multiple allelic variants in a gene or by any one of multiple different genes that carry disease-producing alleles (locus heterogeneity). Tuberous sclerosis is an example of the latter. This inherited disease can be caused by a mutation in the TSC1 gene on chromosome 9q34, which codes for hamartin; or the TSC2 gene on chromosome 16p13, which codes for tuberin. Retinitis pigmentosa is a disease with enormous genetic heterogeneity, and can result from allele heterogeneity or from locus heterogeneity. **When a rare disease demonstrates genetic heterogeneity, we are provided with an opportunity to learn how a common pathogenesis develops from different genes.** Genetic heterogeneity should be contrasted with the concept of genetic pleiotropism, in which one gene may be responsible for several different functions or disorders. See Pleiotropism, Locus heterogeneity, Oligogenic inheritance, and Allelic variants.

Genetic instability—The process whereby the genome accumulates genetic alterations (e.g., SNPs, GSVs) over time. Low levels of unrepaired DNA damage are an inescapable feature of living cells. The older the cell, the more mutations might be found [51]. Many cancers have a high rate of genetic instability. Mutations that arise in germ cells are sometimes passed onto progeny [52]. See Mutator phenotype, SNP, and Genomic structural variation.

Genetic surplus disorder—Mutations that expand the genome or that produce an increased dosage of one or more genes can produce rare diseases [53]. Examples are: Charcot–Marie–Tooth disease, an inherited neuropathy caused by a duplication of a segment of chromosome 17 [32]; and Down syndrome, caused by an extra chromosome 21. In addition, there is a group of rare diseases characterized by trinucleotide repeats. About half of the studied trinucleotide repeat disorders demonstrate repeated CAG sequences. CAG codes for glutamine; thus, CAG repeats produce a polyglutamine protein tract. Examples of polynucleotide repeat disorders are: dentatorubropallidoluysian atrophy, fragile

X syndrome, Friedreich ataxia, Huntington disease, myotonic dystrophy, spino-bulbar muscular atrophy, several forms of spinocerebellar ataxia.

Genome—The collected assortment of an organism's hereditary information, encoded as DNA. For humans, this would mean the set of chromosomes found in a somatic cell, plus the DNA from one of the cell's mitochondria. In practice, when an organism's genome is sequenced, a haploid set of chromosomes is examined, and the mitochondrial DNA is omitted. See Haploid.

Genome wide association study (GWAS)—A method to find common SNPs (single nucleotide polymorphisms) that are statistically associated with a polygenic disease. The methodology involves hybridizing DNA from individuals with disease, as well as individuals from a control group, against a DNA array of immobilized fragments of DNA known to contain commonly occurring SNPs (i.e., allele-specific oligonucleotides). The SNPs that hybridize against the DNA extracted from individuals with disease (i.e., the SNPs matching the case samples) are compared with the SNPs that hybridize against the controls. SNPs that show a statistical difference between case samples and control samples are said to be associated with the disease. Of course, there are many weaknesses to this approach; one being that differences in SNPs do not necessarily imply any functional variance in the gene product [54]. In addition, differences in SNPs may lead to statistically valid results that nonetheless have no relevance to the pathogenesis of disease [55]. Aside from false-positive GWAS associations, the methodology is virtually guaranteed to miss valid SNP associations, simply because SNP arrays are not exhaustive (i.e., do not contain all 50 million SNPs), and are limited to a selected set of commonly occurring polymorphisms. For example, a rare variant of the APOE gene has been shown to be strongly correlated with longevity [56]. This variant, because it is not included among the common APOE variants included in SNP arrays, would have been missed by a GWAS study. True associations are those that can be found repeatedly from laboratory to laboratory, and that can be shown to have pathogenetic relevance. To date, very few disease-associated SNPs found in GWAS studies have met these criteria. It has been suggested that the GWAS studies, *in toto*, have had little scientific merit and have been misleading [57]. A sympathetic evaluation of GWAS studies is that they help us to see recurrent sets of pathway genes involved in diseases. Knowing that a related set of genes seems to implicate a pathway in the development or expression of a common disease has great value [58]. By focusing attention on a pathway, scientists can start to dissect the important events in the pathogenesis of a disease. If the pathway is known to be disrupted in a monogenic disease, particularly when the monogenic disease replicates the phenotype of a common disease, then an effective new treatment, aimed at the pathway, may be feasible.

Genomic structural variation (GSV)—A variation in the structure of chromosomes, usually involving stretches of DNA. GSVs include alterations in karyotype or cytogenetic alterations observable with special techniques, as well as changes too small to see with a microscope, such as small deletions, insertions, single nucleotide polymorphisms (SNPs), larger insertions, inversions, and translocations. GSVs would also include duplications and other copy-number alterations as well as gene conversions [59]. GSVs among different individuals in the human population occur frequently, and may account for more phenotypic variations in the human population than do SNPs [60]. Several databases assist scientists in search of GSVs: Ensembl genome database, NCBI dbSNP database, The Genomic Association Database and SNPedia, Varietas [61]. For examples of GSV disorders, see Microdeletion and Copy-number.

Germline—The germline consists of the cells that derive from the fertilized egg of an organism. All of the somatic cells (i.e., the cells composing the body), as well as the germ cells of the body (oocyte and spermatozoa), arise from the same germline. The extra-embryonic cells (e.g., placental cells) have the same germline as the somatic cells. An inherited condition can be described as being in the germline; in every cell that derives from the fertilized egg. The word "germline" has confused many students, who use the term "germline cell" interchangeably with "germ cell." The confusion is exacerbated by the usual sequence whereby a mutation enters the organism's germline via an inherited mutation present in a parental "germ cell." It is best to think of a germline mutation by its functional definition, a mutation passed to every cell in an organism, and not by its somewhat inaccurate mechanistic definition, a mutation passed from a parental germ cell. See *De novo* germline mutation.

GSV—See Genomic structural variation.

GWAS—See Genome wide association studies.

Hamartoma—Hamartomas are benign tumors that occupy a peculiar zone lying between neoplasia (i.e., a clonal expansion of an abnormal cell) and hyperplasia (i.e., the localized overgrowth of a tissue). Some hamartomas are composed of tissues derived from several embryonic lineages (e.g., ectodermal tissues mixed with mesenchymal tissue). This is almost never the case in cancers, which are clonally derived neoplasms wherein every cell is derived from a single embryonic lineage. Tuberous sclerosis is an inherited hamartoma syndrome. The pathognomonic lesion in tuberous sclerosis is the brain tuber, from which the syndrome takes its name. Tubers of the brain consist of localized but poorly demarcated malformations of neuronal and glial cells. Like other hamartoma syndromes, the germline mutation in tuberous sclerosis produces benign hamartomas as well as carcinomas; indicating that hamartomas and cancers are biologically related. Hamartomas and cancers associated with tuberous sclerosis include cortical tubers of brain, retinal astrocytoma, cardiac rhabdomyoma, lymphangiomyomatosis (very rarely), facial angiofibroma, white ash leaf-shaped macules, subcutaneous nodules, cafe-au-lait spots, subungual fibromata, myocardial rhabdomyoma, multiple bilateral renal angiomyolipoma, ependymoma, renal carcinoma, subependymal giant cell astrocytoma [62].

Another genetic condition associated with hamartomas is Cowden syndrome, also known as multiple hamartoma syndrome. Cowden syndrome is associated with a loss of function mutation in PTEN, a tumor suppressor gene. Features that may be encountered are macrocephaly, intestinal hamartomatous polyps, benign hamartomatous skin tumors (multiple trichilemmomas, papillomatous papules, and acral keratoses), dysplastic gangliocytoma of the cerebellum, and a predisposition to cancers of the breast, thyroid and endometrium.

Haploid—From Greek haplous, "onefold, single, simple." The chromosome set of a gamete. In humans, this would be 23 chromosomes; one set of unpaired autosomes (chromosomes 1 to 22) plus one sex chromosome (X- or Y-chromosome).

Haploinsufficiency—Occurs when one of two alleles of a required gene is inactivated and the other allele does not express sufficient quantities of the gene product to maintain normal cellular functionality. In the field of carcinogenesis, haploinsufficiency may result in a heightened susceptibility to cancer if one copy of a tumor suppressor gene is inactivated and the other copy cannot provide sufficient functionality to suppress tumorigenesis [63,64].

Haplotype—A set of DNA polymorphisms that tend to be inherited together, often as a result of their close proximity on a chromosome. It is often used in a restricted sense to refer to a set of SNPs that are statistically associated with one another on a chromosome. A related term, "haplogroup," refers to a subpopulation of individuals that share a common ancestor and a haplotype.

Histopathology—Pathologists render diagnoses by examining biopsied specimens. Sampled tissues are fixed in formalin and embedded in paraffin (wax). Thin slices of the paraffin-embedded tissues are mounted on glass slides and stained so that the cellular detail can be visualized under a microscope. A histopathologic diagnosis is based on finding specific cellular alterations that characterize diseases (see Figure G.1).

Homeobox—Genes that code for transcription factors involved in anatomic development in animals, fungi, and plants. Hox genes are homeobox genes found in animals that determine the axial relationship of organs. Mutations of homeobox genes are associated with remarkably specific, often isolated, anatomic alterations. Examples are: MSX2 homeobox gene mutation, which produces enlarged parietal foramina; PITX1 homeobox gene mutation, which produces Rieger syndrome (hypodontia and malformation of the anterior chamber of the eye including microcornea and muscular dystrophy); PITX3 homeobox gene mutation, which produces anterior segment dysgenesis of the eye, moderate cataracts, and anterior segment mesenchymal dysgenesis; NKX2.5 homeobox gene, which produces atrial septal defect and atrioventricular conduction defects; SHOX homeobox (short stature homeobox) gene mutation causes Leri–Weill dyschondrosteosis (deformity of distal radius, ulna, and proximal

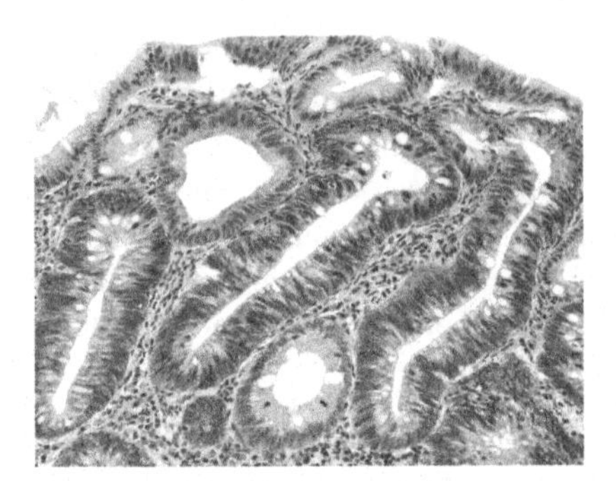

Figure G.1 Histopathologic section of biopsy of a colon adenoma. The specimen has been stained with hematoxylin and eosin, which colors the nuclei blue and the cytoplasm pink. The pathologist can inspect individual cells, as well as the architecture of the tissue. In many cases, the pathologist can render a specific diagnosis based largely on observations of histologic sections. See color plate at the back of the book. (*Source: Dr. G. William Moore, for the U.S. Department of Veterans Administration, and released into the public domain.*)

carpal bones as well as mesomelic dwarfism). The reason why homeobox mutations tend to produce diseases in isolated anatomic locations or involve some specific function probably results from the coordinated regulatory activity of the individual homeobox genes. For example, one gene might regulate the synthesis of a group of proteins exclusively involved in growth of particular skull bones; another homeobox gene might regulate proteins involved in insulin production. Disorders caused by alterations in homeobox genes include: aniridia, Axenfeld–Rieger syndrome, branchiootorenal syndrome, coloboma, combined pituitary hormone deficiency, congenital central hypoventilation syndrome, congenital fibrosis of the extraocular muscles, congenital hypothyroidism, craniofacial-deafness-hand syndrome, enlarged parietal foramina, hand-foot-genital syndrome, Langer mesomelic dysplasia, Leri–Weill dyschondrosteosis, microphthalmia, Mowat–Wilson syndrome, nail–patella syndrome, forms of non-syndromic deafness, non-syndromic holoprosencephaly, Partington syndrome, Potocki–Shaffer syndrome, renal coloboma syndrome, septo-optic dysplasia, Turner syndrome, Waardenburg syndrome, Wilms tumor aniridia genitourinary anomalies and mental retardation syndrome, Wolf–Hirschhorn syndrome, X-linked infantile spasm syndrome, X-linked lissencephaly.

Homozygosity—Occurs when only one allele of a gene is expressed in cells. This may occur when both of the inherited alleles of a gene (the maternally derived allele and the paternally derived allele) are identical to each other. It may also result when the expression of one of the inherited alleles is unattained or lost, in which case homozygosity is said to result from loss of heterozygosity.

Host—The organism in which the infectious agent resides.

Immune system—In humans, there are three known host defense systems that recognize and destroy foreign organisms: intrinsic, innate, and adaptive. See Intrinsic immunity, Innate immunity, and Adaptive immunity.

Imprinting—Early in mammalian embryogenesis, the pattern of epigenetic modifications (e.g., methylations) inherited from the paternal and maternal gametes is erased, forcing the embryo to develop its own unique pattern of methylations. This process of epigenome erasure is necessary; otherwise, the embryonic germline would have a differentiated epigenome, and the normal process of gradual epigenetic modifications, applied throughout embryogenesis, could not occur. Erasure is not a totally thorough process. There are about 100 known genes that retain their parental epigenetic patterns. Retention of parental epigenetic patterns is known as imprinting. When imprinted genes contain disease-causing mutations, the disease that develops will express a phenotype that is influenced by paternal lineage. For example, Prader–Willi syndrome is a genetic disease characterized by growth disorders (e.g., low muscle tone, short stature, extreme obesity, and cognitive disabilities). Angelman syndrome is a genetic disease characterized by neurologic disturbances (e.g., seizures, sleep disturbances, hand-flapping), and a typifying happy demeanor. Both diseases can occur in either gender and both diseases are caused by the same microdeletion at 15q11-13. When the microdeletion occurs on the paternally derived chromosome, the disease that results is Prader–Willi syndrome. When the microdeletion occurs on the maternally derived chromosome, the disease that results is Angelman syndrome. Another example is the NOEY2 tumor suppressor gene, which is imprinted in females and which contributes to some cases of breast and ovarian cancers [65]. See Loss of imprinting and Epigenomic syndrome.

In situ—Latin for "in its place." When referring to a cancer, *in situ* implies that the cancer has not invaded surrounding tissues and has not metastasized to lymph nodes or to distant organs. The term "*in situ* epithelial neoplasm" (i.e., neoplasms arising from mucosal surfaces, epidermis, or glandular tissues) is virtually synonymous with the alternate terms "intraepithelial neoplasm" or "epithelial precancer." See Precancer.

Incidence—The number of new cases of a disease occurring in a chosen time interval (e.g., 1 year), expressed as a fraction of a predetermined population size (e.g., 100,000 people). For example, if there were 10 new cases of a rare disease occurring in a period of 1 year, in a population of 50,000 people, then the incidence would be 20 cases per 100,000 persons per year. See Age-adjusted incidence and Prevalence.

Infectious disease—A disease caused by an organism on or in the human body. The term "infectious disease" is sometimes used in a way that excludes diseases caused by parasitic animals. In this book, the term "infectious disease" is all-inclusive.

Inflammasome—A protein complex expressed by white blood cells that activates an inflammatory process. Some inflammasome proteins are caspase 1 and 5, and NALP. The inflammasome promotes other inflammatory cytokines and is part of the innate immune system. See Innate immunity.

Informed consent—Subjects who are put at risk in an experimental study must first confirm consent. To this end, researchers must provide prospective human subjects with a consent form that informs the subject of the purpose and risks of the study, and discloses any information that might reasonably affect the participant's decision to participate, such as financial conflicts of interest among the researchers. The informed consent must be understandable to laymen, must be revocable (i.e., subjects can change their mind and withdraw from the study), must not contain exculpatory language (i.e., no waivers of responsibility for the researchers), must not promise any benefit for participation, and must not be coercive.

Initiation—In the field of cancer, the term "initiation" refers to the inferred changes in cells following exposure to a carcinogen that may eventually lead to the emergence of a cancer in the cell's descendants. Though we know much about the many possible changes that can occur in cells exposed to carcinogens, the essential and defining changes that begin the process of carcinogenesis are still unknown. The process that begins with initiation and extends to the emergence of a cancer is called carcinogenesis. In molecular biology, the term "initiation" has a distinctly different meaning, referring instead to the necessary molecular events that allow a process (e.g., replication, transcription, or translation) to begin.

Initiation factor—Synonymous with translation factor; not to be confused with cancer initiation. See Translation factor.

Innate immunity—An ancient and somewhat non-specific immune and inflammatory response system found in plants, fungi, insects [66], and most multicellular organisms. This system recruits immune cells to sites of infection, using cytokines (chemical mediators). Innate immunity includes the complement system, which acts to clear dead cells. It also includes the macrophage system, also called the reticuloendothelial system, which engulfs and removes foreign materials. Examples of rare, monogenic disorders of the innate immune system include: familial Mediterranean fever; TNF receptor-associated periodic syndrome; hyperimmunoglobulin D syndrome; familial cold autoinflammatory syndrome; Muckle–Wells syndrome; neonatal-onset multi-system inflammatory disease, also known as chronic infantile neurologic, cutaneous, and arthritis syndrome; pyogenic arthritis, pyoderma gangrenosum, and acne; Blau syndrome; early-onset sarcoidosis, and Majeed syndrome [67]. See Adaptive immunity and Inflammasome.

Intellectual property—Intangible products (e.g., methods, preparative processes, certain types of information) owned by their creator (i.e., a human or corporate entity). The owner has the right to determine how the intellectual property can be used and distributed. Protections for intellectual property come in three forms: (1) copyrights; (2) patents; and (3) secrecy (e.g., hiding the intellectual property from the public). Intellectual property can be sold outright, essentially transferring ownership to another entity. Alternately, intellectual property can be retained by the creator who permits its limited use to others via a legal contrivance (e.g., license, contract, transfer agreement, royalty, usage fee, and so on). The legal rules that apply to intellectual property may influence the cost and availability of new diagnostic tests and therapeutic interventions for the rare diseases.

Intermediate host—Same as secondary host. An organism that contains a parasite for a period of time during which the parasite matures in its life cycle, but in which maturation does not continue to the adult or sexual phase. Maturation to the adult or sexual phase only occurs in the primary or definitive host. A parasitic eukaryotic organism may have more than one intermediate host. The survival advantages offered to the parasite by the intermediate host stage may include the following: to provide conditions in which the particular stages of the parasite can develop, which are not available within the primary host; to disseminate the parasite (e.g., via water or air) to distant sites; to protect the immature forms from being eaten by the adult forms; to protect the parasite from harsh conditions that prevail in the primary host; to protect the parasite from external environmental conditions that prevail when the parasite leaves the primary host. See Primary host.

Interstitial deletion—See Contiguous gene deletion and Microdeletion.

Intra-tissue genetic heterogeneity—Refers to the expression of different gene variants in different cells within the same organism or lesion (i.e., the cells directly involved in the disease process). The term is most often applied to cancers, wherein subclones of cell emerge, each with a unique genotype and phenotype. The expression of different forms of the same gene in different cells is a type of somatic mosaicism. The full extent of intra-tissue genetic heterogeneity in normal tissues of the body is not known. Hypothetically, somatic mosaicism may play a significant role in the development of polygenic or multifactorial diseases [68].

Intraepithelial neoplasm—A term applied to an early stage of growth of epithelial tumors, primarily tumors that arise from the epithelial cells, such as those found on epithelial surfaces lining various tissues (such as the mucosal lining of the gastrointestinal tract or the epidermal surface of the skin). The word intraepithelial conveys the idea that the malignant cells reside within the epithelium, and have not invaded down into the underlying connective tissues. Intraepithelial neoplasms are subtypes of precancers. See *In situ* and Precancer.

Intrinsic immunity—A cell-based (i.e., not humoral) anti-viral system that is always "on" (i.e., not activated by the presence of its target, as seen in adaptive immunity and innate immunity) [69]. Intrinsic immunity is a newly discovered immune response system, and there is much we need to learn about this type of immunity. Intrinsic immunity has been studied for its role in controlling retrovirus infections (e.g., HIV infection). It is known that intrinsic immunity is

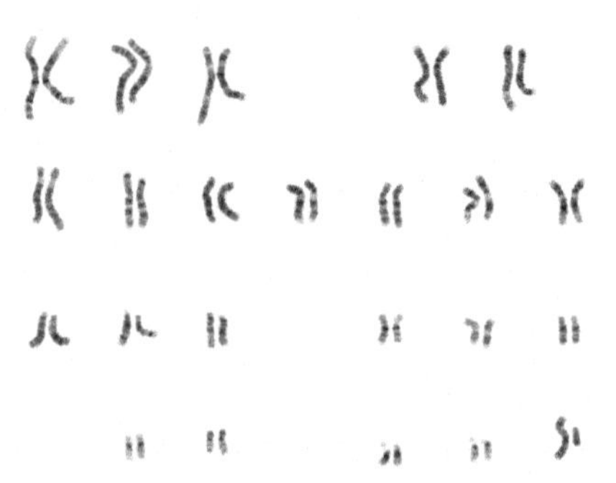

Figure G.2 Karyogram of normal human male paired chromosomes. The chromosomes are banded by light and dark staining areas. The dark staining corresponds to areas of dense heterochromatin. The locations of the bands are characteristic for each chromosome. Variations in banding can indicate cytogenetic aberrations, such as translocations of segments from one chromosome to another. (*Source: U.S. Human Genome Research Institute.*)

not restricted to retroviruses, but its role in blocking infection by other classes of virus is something of a mystery. See Innate immunity and Adaptive immunity.

Invasion—In the field of cancer, invasion occurs when tumor cells move into and through normal tissues. All tumors that can metastasize can also invade, and, for this reason, it is inferred that invasion is involved in the process of metastasis. For metastasis to occur tumor cells invade through the walls of lymphatic and blood vessels, thus gaining access to the general circulation; likewise, tumor cells invade through vessels at the site of distant seeding. The opposite assertion is not true; tumors that invade do not necessarily metastasize. Examples of non-metastasizing invasive tumors include basal cell carcinoma of skin and most tumors arising within the brain.

Karyotype—From the Greek root karyon, meaning nucleus, the karyotype is a standard shorthand describing the chromosomal complement of a cell. The normal karyotype of a human male diploid somatic cell is 46 XY, a somewhat confusing way to express that there are two sets of 23 chromosomes, producing a total complement of 46 chromosomes, which includes one X and one Y sex chromosome (see Figure G.2). The normal female karyotype is 46 XY. Abnormalities in karyotype are described using the International System for Human Cytogenetic Nomenclature (ISCN).

Koch's postulates—A set of observations and experimental requirements proposed by Heinrich Hermann Robert Koch in the late 1800s, intended to prove that a particular organism causes a particular infectious disease. For the experimentalist, the most important of the Koch's postulates require the extraction of the organism from a lesion (i.e., from diseased, infected tissue), the isolation and culture of the organism in the laboratory, and the consistent reproduction of the lesion in an animal inoculated with the organism. Over the ensuing century, some modifications to Koch's original postulates were necessary to accommodate our expanding experience with infectious agents and our increasing awareness of the limits of biological causality [70]. As an example, *Helicobacter pylori* is known to cause gastric lymphoma, but *H. pylori* fails Koch's postulates. It is currently presumed that *H. pylori* lymphoma arises as a consequence of chronic *H. pylori* infection with gastritis. *H. pylori*-associated lymphoma cells do not contain *H. pylori* bacteria (i.e., the organism cannot be consistently isolated from lymphoma cells), and the gastric injection of cultured *H. pylori* is not likely to induce stomach cancers. In his thoughtful paper, Inglis discusses that biological causation is an elusive concept. Inglis proposes the concept of the "priobe," referring to a biological agent that is a necessary and sufficient cause of a series of events that eventually leads to a disease (i.e., an underlying or antecedent cause) [70].

Locus heterogeneity—Also known as non-allelic heterogeneity, occurs when mutations in different genes can produce the same disease. For example, mutations in c-KIT or PDGFR alpha can lead to gastrointestinal stromal tumor. Mutations in the gene encoding the protein hamartin or the gene encoding the protein tuberin can produce the disease tuberous sclerosis. Carney complex can be caused by mutations in the PRKAR1A gene on chromosome 17q23-q24, or it may be caused by a mutation in chromosome 2p16. Both types of mutation produce the same clinical phenotype, which carries an increased risk of developing several types of tumors, including cardiac myxoma. Locus heterogeneity is a special case of the broader concept of genetic heterogeneity. See Genetic Heterogeneity, Allelic heterogeneity, and Oligogenic inheritance.

Loss of imprinting (LOI)—During normal embryogenesis, genes are imprinted with epigenetic modifications that suppress the expression of various genes. When there is an acquired loss of this normal imprinting, the affected gene may be overexpressed. The first discovered example of loss of imprinting was the overexpression of insulin-like growth factor-2 (Igf2) in Wilms tumor. In this case, a mutation disrupts the H19 imprinting control region that would normally silence the maternally inherited Igf2 allele.

Malignant—A disease that can kill its host. Cancers are often malignant if left untreated. Severe hypertension is referred to as malignant hypertension if the untreated condition has a high likelihood of producing stroke or renal failure.

Mendelian inheritance—A pattern of inheritance observed for traits that are determined by genes contributed by the mother or the father.

Metabolic syndrome—The combination of obesity plus hypertriglyceridemia plus low levels of HDL cholesterol plus hypertension plus hyperglycemia (i.e., prediabetes or diabetes). Metabolic syndrome occurs in nearly one in four adults in the U.S. and carries an increased risk of death from a variety of common causes, including heart attacks [71].

Microdeletion—Microdeletions are cytogenetic abnormalities that typically span several megabases of DNA. Microdeletions are too small to be visible with standard cytogenetics, but they can often be detected with FISH (fluorescent *in situ* hybridization). All of the microdeletion syndromes are rare diseases, and they typically arise as *de novo* germline aberrations (i.e., not inherited from mother or father, in most instances). Conditions that occur rarely and sporadically to produce a uniform set of phenotypic features in unrelated subjects may be new cases of microdeletion syndromes [72]. DiGeorge syndrome is a typical microdeletion disease, with a germline 22q11.2 deletion encompassing about 3 million base pairs on one copy of chromosome 22, containing about 45 genes. Neurofibromatosis I sometimes occurs as a microdeletion syndrome involving a region of chromosome 17q11.2 that includes the NF1 gene. Microdeletion disorders are a subtype of contig disorders (i.e., contiguous gene disorder). Examples of microdeletion syndromes include: cri du chat, Kallman syndrome, Miller–Dieker syndrome, Prader–Willi/Angelman syndrome, retinoblastoma, Rubinstein–Taybi syndrome, Smith–Magenis syndrome, steroid sulfatase deficiency (ichthyosis), velocardiofacial syndrome (also known as DiGeorge syndrome), Williams–Beuren syndrome, and Wolf–Hirschhorn syndrome. See Genomic structural variation and *De novo* germline mutation.

MicroRNA—Small but abundant species of RNA that regulate gene expression by pairing with complementary sequences of mRNA. Such complementation usually causes silencing of the mRNA. It is estimated that humans have more than 1000 different microRNA, also called miRNA species [73]. A form of autosomal dominant hearing loss is caused by mutations in MIRN96 microRNA. See Regulatory RNA element.

Microsatellite—Also known as simple sequence repeats (SSRs), microsatellites are DNA sequences consisting of repeating units of 1–4 base pairs. Microsatellites are inherited and polymorphic. Within a population there may be wide variation in the number of repeats at a chosen microsatellite locus. Friedreich ataxia, a neurodegenerative disease characterized by ataxia and an assortment of neurologic and muscular deficits is an example of a microsatellite disease. A common molecular abnormality of Friedreich ataxia is a GAA trinucleotide repeat expansion within an intron belonging to the gene encoding frataxin. Normal levels of frataxin are apparently necessary for the health of nerve cells and muscle cells [74]. See Microsatellite instability and Anonymous variation.

Microsatellite instability—When there is a deficiency of proper mismatch repair (a type of DNA repair), DNA replication is faulty, and novel microsatellites appear in chromosomes. This phenomenon is called microsatellite instability and it occurs in some types of cancers, particularly colon cancers arising in hereditary non-polyposis colorectal cancer syndrome.

Mitochondria—Self-replicating organelles wherein respiration, the production of cellular energy from oxygen, occurs. As far as anyone knows, the very first eukaryote came fully equipped with a nucleus, one or more undulipodia, and one or more mitochondria. Similarities between mitochondria and eubacteria of Class Rickettsia suggest that the eukaryotic mitochondrium was derived from an ancestor of a modern rickettsia. All existing eukaryotic organisms, even the so-called amitochondriate classes (i.e., organisms without mitochondria), contain vestigial forms of mitochondria (i.e., hydrogenosomes and mitosomes) [75–78]. A single eukaryotic cell may contain thousands of mitochondria, as is the case for human liver cells, or no mitochondria, as is the case for human red blood cells. The control of mitochondrial number is determined within the nucleus, not within the mitochondrion itself. The mitochondria in a human body are descended from mitochondria contained in the maternal oocyte; hence, mitochondria have a purely maternal lineage. See Mitochondriopathy.

Mitochondriopathy—A disease whose underlying cause is mitochondrial pathology (i.e., dysfunctional mitochondria, or an abnormal number of mitochondria). Mitochondriopathies can be genetic or acquired. Most of the genetic mitochondriopathies are caused by nuclear gene mutations. Though mitochondria have their own genes, the mitochondrial genome codes for only 13 proteins of the respiratory chain. All the other proteins and structural components of the mitochondria are coded in the nucleus. Mitochondriopathies can involve many different organs and physiologic processes. Mitochondrial defects affecting muscles include myopathy (i.e., weakness), fatigue, and lactic acidosis. The peripheral and central nervous systems disorders include: polyneuropathy, leucencephalopathy, brain atrophy, epilepsy, upper motor neuron disease, ataxia, and extrapyramidal side effects. Endocrine manifestations may include hyperhidrosis, diabetes, hyperlipidemia, hypogonadism, amenorrhea, delayed puberty, and short stature. Heart damage may include conduction abnormalities, heart failure, and cardiomyopathy. Ocular changes may include cataract, glaucoma, pigmentary retinopathy, and optic atrophy. Hearing changes may include deafness, tinnitus, and vertigo. Gastrointestinal disorders may include dysphagia, diarrhea, liver disease, motility disorder, pancreatitis, and pancreatic insufficiency. Renal disease may include renal failure and cyst formation. Blood cells may develop sideroblastic anemia. A mitochondriopathy should be in the differential work-up for any unexplained multi-system disorder, especially those arising in childhood [79].

Mitosis—The phase in the cell cycle of somatic cells (i.e., not germ cells) wherein the replicated chromosomes condense and separate to form two daughter cells (see Figure G.3).

Mitotic—Relating to mitosis. See Post-mitotic.

Monoclonal gammopathy of undetermined significance (MGUS)—MGUS is the precancer for multiple myeloma. It consists of a clonal proliferation of plasma cells that all produce an identical immunoglobulin molecule. The cumulative effect of a clone of plasma cells, all synthesizing the same immunoglobulin

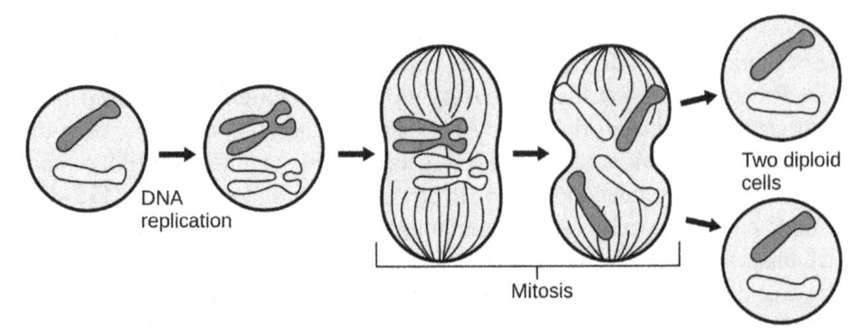

Figure G.3 Schematic of mitosis, indicating how chromosomes replicate and migrate into daughter cells. See color plate at the back of the book. (*Source: Wikipedia, and created as a U.S. government public domain work by the National Center for Biotechnology Information, part of the National Institutes of Health.*)

species, is a distinctive protein spike on blood examined by a technique known as electrophoresis, which separates out different molecular species of proteins in serum samples. MGUS is a common condition found in the elderly and may occur in about 1% of the population over 70 years of age. Progression of MGUS to multiple myeloma is infrequent, with a conversion of about 1–2% per year. Because MGUS occurs in an elderly population, the chance of MGUS progressing to myeloma within the lifespan of the patient is quite low. Still, it seems as though every case of multiple myeloma follows a preceding MGUS [80].

Monogenic disease—Disease caused by an alteration in a single gene. It should be noted that a single gene may have pleiotropic effects on multiple cellular pathways, on different cell types, and at different stages of organismal development. Hence, monogenic diseases may have complex phenotypes. This is sometimes true for altered regulatory elements (e.g., genes that code for transcription factors). It should also be noted that every monogenic disease will have polygenic modifiers. The expression of a normal or mutant gene depends on complex interactions with cellular machinery, and those interactions will depend on epigenetic and genetic conditions that vary among individuals. In general, most inherited rare diseases are monogenic disorders; common diseases seem to be polygenic. Some of the specific types of genetic alterations that account for many monogenic rare diseases are: deletions (e.g., Duchenne muscular dystrophy), frame-shift mutations (e.g., factor VIII and IX deficiencies), fusions (e.g., chronic myelogenous leukemia, hemoglobin variants), initiation and termination codon mutations (e.g. alpha thalassemia), inversions (a type of beta thalassemia), nonsense mutations (familial hypercholesterolemia), point mutations (e.g., sickle cell disease, glucose-6-phosphate dehydrogenase deficiency), promoter mutations (a type of thalassemia), RNA processing mutations, including splice mutations (e.g. phenylketonuria) [81]. Not all rare diseases are monogenic and not all common diseases are polygenic. There are examples of rare diseases that are digenic (i.e., caused by two genes), and there are examples of common diseases for which a small subset of individuals has a non-syndromic monogenic form of a common disease (e.g., MODY 4, also known as monogenic diabetes [82]). There are also examples of common diseases for which a syndromic monogenic disease accounts for a small subset of individuals who have the common disease (e.g., Werner syndrome, which produces a complex disease phenotype, including diabetes). See Polygenic disease, Digenic disease, Non-syndromic disease, Syndromic disease, and Mody 4.

Monte Carlo simulation—A mathematical technique developed in 1946 by John von Neumann, Stan Ulam, and Nick Metropolis [83]. The technique employs computers to calculate the outcomes of probabilistic events by generating random numbers and using the resultant values, constrained within a model system, to simulate repeated biological trials [84,85]. As discussed in Section 11.2, Monte Carlo simulations can be used to model polygenic diseases or pathways composed of proteins coded by polymorphic genes.

Mutagen—A chemical that produces alterations in the genetic sequence of DNA molecules. With few exceptions, carcinogens (i.e., chemicals that cause cancer) are mutagens. There are some mutagens that have not been shown to be carcinogens, and these chemicals tend to be so highly reactive with cellular molecules (e.g., lipids, proteins, RNA, etc.) that they cannot effectively reach nuclear DNA, the target molecule, or they kill cells rather than inflicting heritable damage.

Mutator phenotype—One of the hallmarks of cancer is genetic instability. The common cancers typically have thousands of genetic mutations, and these mutations perturb virtually every aspect of cellular physiology. It is hypothesized that during carcinogenesis, cells acquire a mutator phenotype that increases the rate at which genetic aberrations occur in cells, thus raising the likelihood that a cell will emerge with an oncogenic mutation that confers a malignant or premalignant phenotype [40].

Myelodysplasia—Synonym for myelodysplastic syndrome. See Myelodysplastic syndrome.

Myelodysplastic syndrome—The myelodysplastic syndromes, formerly known as preleukemias, are several closely related diseases characterized by anemia, pancytopenia, disordered myeloid growth and maturation, the appearance of circulating blast cells, chromosomal abnormalities, and the frequent progression to leukemia. The myelodysplasias have a bimodal age distribution. Most cases occur in the elderly. Rare instances of myelodysplasia occur in children. A secondary type of myelodysplasia occurs following a bout of aplastic anemia or following chemotherapy for some other neoplasm.

Myeloproliferative disorder—A blood disorder characterized by a clonal expansion of hematopoietic cells. These would include all of the non-lymphoid leukemias, plus the non-neoplastic diseases characterized by an increased number of normal or abnormal circulating blood cells (e.g., essential thrombocythemia, polycythemia vera). When cells of lymphoid lineage proliferate in the blood, the analogous term, lymphoproliferative disorder, is applied.

Natural selection—A tendency for favorable heritable traits to become more common over successive generations. The traits are selected from pre-existing genetic variations among individuals in the population. The genetic variations may take the form of genetic sequence variations (e.g., SNPs) or genetic structural variations. See SNP and Genomic structural variation.

Neglected disease—A near-synonym for rare disease. The assumption is that the rare diseases are often neglected: underfunded, underdiagnosed, undertreated, and generally lacking the kind of support systems available to the more common diseases. Though support for the rare diseases has been increasing, it is reasonable to conclude that with about 7000 rare diseases to consider, most of these conditions will be, to some extent, neglected. Terms that are commonly used interchangeably are: "rare disease," "orphan disease," and "neglected disease."

Neoplasm—Neoplasm means "new growth" and is a near-synonym for "tumor." Neoplasms can be benign or malignant. Leukemias, which grow as a population of circulating blood cells, and which do not generally produce a visible mass (i.e., do not produce a tumor), are included under the general term "neoplasm." Hamartomas, benign overgrowths of tissue, are generally included among the neoplasms, as are the precancers, which are often small and scarcely visible.

Nomenclature—A nomenclature is a specialized vocabulary containing all, or nearly all, terms that comprehensively cover a well-defined field of knowledge. For example, there may be a nomenclature of genes or genetic mechanisms, cellular pathways, or rare diseases. Some nomenclatures are ordered alphabetically. Others are ordered by synonymy, wherein all synonyms and plesionyms (near-synonyms) are collected under a canonical (best or preferred) term. Indexes prepared from medical records or scholarly documents can be harmonized under a nomenclature that collects synonymous terms under a canonical term.

Non-inherited genetic disease—A significant but unquantified portion of genetic diseases is non-inherited; occurring from *de novo* (new) mutations in the germ cells of the affected individuals. In humans, point mutations (i.e., mutations that occur in a single nucleotide base within the genome) occur with a frequency of about 1 to 3×10^{-8} per base [37,38]. There are many types of mutational alterations other than point mutations (e.g., mutations in microsatellites) [86]. Our knowledge of the likelihood of most of these alternate types of mutation is limited. In many cases *de novo* mutations cause lethal genetic diseases that occur in children, through the action of a dominant gene (i.e., one gene copy that produces the disease). Such diseases are seldom inherited because they cannot be conserved in the population; those with the gene die early in life, without passing the gene to progeny. Sometimes, non-inheritance accounts for some proportion of cases that would otherwise occur as dominant gene disorders. Neurofibromatosis is an example of a disease that occurs through autosomal dominant inheritance in about half of the cases. The other half of occurrences are *de novo* mutations incurring in the affected individual. In general, *de novo* mutations are often suspected as the cause of diseases that occur in early childhood for which no other cause (e.g., no evidence of familial or parental inheritance, infectious etiology, or environmental influences) can be determined (e.g., autism) [87]. See *De novo* mutation.

Non-Mendelian inheritance—Inheritance by virtue of a combination of polymorphisms (i.e., gene variants) that are prevalent within a family. In non-Mendelian inheritance, it is possible that neither the mother nor the father will carry the complete set of gene variants that cause inherited disease, but that the combination of disease-causing gene variants will occur in an offspring, through meiotic recombination. Generally, Mendelian inheritance is monogenic; non-Mendelian inheritance is polygenic. Diseases that have Mendelian patterns of inheritance tend to be less common than diseases with polygenic inheritance.

Non-syndromic disease—A disease that affects a single organ or function, unaccompanied by abnormalities of other organs or physiologic systems. A syndromic disease is a constellation of pathologic features associated with a single disease or condition, involving multiple organs or physiologic systems. Non-syndromic deafness affects hearing and no other structures or functions. When inherited deafness is syndromic, it is accompanied by other abnormalities, possibly involving facial structure or nerve function.

Nuclear atypia—Refers to variations from normal nuclear morphology. The term is typically applied to cancer cells and precancer cells, whose nuclei look different from normal nuclei. Cancerous and precancerous nuclei are larger than normal nuclei, with irregular shape (i.e., not oval or round or smooth), with indentations in the nuclear membrane, coarse chromatin, areas of light and dark within the nucleus, and enlarged, irregularly shaped nucleoli. Traditionally, pathologists were taught that genetic changes accounted for the atypia present in cancer cells. This opinion was based on several observations and assumptions: (1) cancer cells were known to contain genetic abnormalities, such as increased numbers of chromosomes, missing pieces of chromosomes, duplicated pieces of chromosomes, and translocated pieces of chromosomes; (2) the genetic material of the cell resides in the nucleus, and it was natural to assume that morphologic changes to the nucleus would result in damage to DNA; and (3) little was known about the epigenome when the genetic code was broken in the late 1950s. In recent years, cancer biologists have been rethinking the cause of cellular atypia in cancer, focusing their attention on the role of the epigenome. Observations that support an epigenetic cause of nuclear atypia include the following: (1) some cancers have marked atypia with little or no genetic instability (e.g., rhabdoid tumor [88]); (2) profound changes in nuclear morphology can be produced by alteration in a single protein, as is seen in rare Pelger–Huet anomaly [89], suggesting that marked changes in gene sequence are not necessary to produce misshapen nuclei; (3) the common histologic stains with which we assess nuclear atypia bind to the histone and non-histone proteins of the epigenome; hence, the morphologic abnormalities of cancer nuclei reflect changes in the nuclear distribution of epigenetic constituents.

Obligate intracellular organism—An obligate intracellular organism can only reproduce within a host cell. Obligate intracellular organisms can include species from various classes of organisms, but the term applies to all viruses and all members of Class Chlamydia. Examples of other genera that contain obligate intracellular species include: *Coxiella, Leishmania, Plasmodia, Rickettsia, Toxoplasma, Trypanosoma.* Such organisms have adapted to life within human cells, shucking most of their genetic material, and opting to survive on the cellular machinery provided by their hosts.

Off-label—Refers to the use of a drug to treat a condition other than the condition for which the drug was awarded FDA approval. The term "off-label" comes from the section of the pill-bottle label that describes the intended use or uses of the drug. Treatments other than those described on the drug's label are "off-label." The FDA does not regulate the practice of medicine. Hence, physicians who use good professional judgment to prescribe an off-label treatment for an FDA approved drug may do so. Prudent physicians will not prescribe off-label until such uses are supported by credible, repeated studies published in highly regarded medical journals. It is impossible to know with any precision or confidence the prevalence of off-label treatments in the rare diseases. Still, it is a commonly held view that about 90% of all treatments for rare diseases are conducted without specific FDA approval. Medicare and private insurers, at present, tend to pay for off-label uses of drugs, particularly when these drugs are used to treat rare diseases.

Oligogenic inheritance—In the context of genetic diseases, occurs when the expression of several genes (i.e., not one gene and not many genes) produces a disease phenotype. If two genes are required, the term "digenic disease" applies. Macular degeneration may qualify as a common disease with oligogenic inheritance. A few gene variants present in the general population may account for 70% of the risk of developing age-related macular degeneration [90,91], the third leading cause of blindness worldwide [92]. Examples of oligogenic rare diseases are Bardet–Biedl syndrome [93] and Williams–Beuren syndrome [94]. See Allelic heterogeneity and Locus heterogeneity.

Oncogene—Genes that are key drivers of the malignant phenotype in cancer cells. Oncogenes need to be activated before they can contribute to the phenotype of cancer cells. Prior to activation, oncogenes are called proto-oncogenes. Activation usually involves mutation or amplification (i.e., an increase in gene copy-number), translocation, or fusion with an actively transcribed gene, or some sequence of events that increases the expression of the gene product. Some retroviruses contain oncogenes and can cause tumors by integrating their oncogene into the host genome. See Tumor suppressor gene.

Opportunistic infection—Opportunistic infections are diseases that do not typically occur in healthy individuals, but which can occur in individuals who have a physiologic status favoring the growth of the organisms (e.g., diabetes, malnutrition). Sometimes, opportunistic infections occur in patients who are very old or very young. Most often, opportunistic infections occur in immune-compromised patients. A disease may increase susceptibility to specific types of organisms. For example, diabetics are more likely to contract systemic fungal diseases than are non-diabetic individuals. Some opportunistic infections arise from the population of organisms that live within most humans, without causing disease under normal circumstances (i.e., commensals). The concept of an opportunistic organism is, at best, a gray area of medicine, as virtually all of the organisms that arise in immune-compromised patients will, on rare occasions, cause disease in immune-competent patients (e.g., *Cryptococcus neoformans*). Moreover, the so-called primary infectious organisms that produce disease in normal individuals will tend to produce a more virulent version of the disease in immunosuppressed individuals (e.g., *Coccidioides immitis*). Examples of organisms that cause opportunistic infections are: *Acinetobacter baumanni*, *Aspergillus* sp., *Candida* sp., *Clostridium difficile*, *Cryptococcus neoformans*, *Cryptosporidium parvum*, *Cytomegalovirus*, herpes zoster, *Histoplasma capsulatum*, human herpesvirus-8, *Pneumocystis jirovecii*, polyomavirus JC, *Proteus* sp., *Pseudomonas aeruginosa*, *Streptococcus pyogenes*, *Toxoplasma gondii*. See Commensals.

Orphan drug—A drug that is helpful to a small number of people, usually indicating a drug developed for individuals who have a rare disease. In the past, the term "orphan drug" was applied to existing drugs that were not marketed due to their perceived unprofitability. Today, the term is generally applied to any drug that happens to have a limited market. An in-depth discussion of this topic is found in Chapter 14.

Orthodisease—A genetic disease in humans caused by an alteration of a gene that is orthologous to a known gene in a different species. Though we can expect phenotypes to diverge among species with orthologous genes, the pathways perturbed by orthologous genes will tend to be conserved [95]. By studying the role of orthologous genes in different species, we may learn something about the pathogenesis of the human disease. For example, the 2013 Nobel Prize in Physiology or Chemistry was awarded for work on vesicular transport disorders (see Section 10.4). Much of the progress in this area came from studies of human inherited transport disorders [96]. Because vesicular transport is an ancient cellular system, researchers could dissect the transport pathway in mutant orthologous genes from yeast [97].

Parasite—A parasite is an organism that lives and feeds in or on its host. In common usage, the term "parasite" was often reserved for multicellular animals that are parasitic in humans and other animals, and at other times the term was extended to include the once-called one-celled animals (i.e., members of the class formerly known as Protoctista). We now know that members of Class Protoctista are not one-celled animals. Furthermore, some of the once-called one-celled animal parasites are now known to be members of Class Fungi (e.g., *Pneumocystis jiroveci*). As we know more and more about classes of organisms, the term "parasite" seems to have diminishing biologic specificity and utility. In this book, the term "parasite" refers to any infectious organism. See Protozoa.

Pareto's principle—Also known as the 80/20 rule, Pareto's principle holds that a small number of items account for the vast majority of observations. For example, a small number of rich people account for the majority of wealth. Just two countries, India plus China, account for 37% of the world population. Within most countries, a small number of provinces or geographic areas contain the majority of the population of a country (e.g., east and west coastlines of the U.S.). A small number of books, compared with the total number of published books, account for the majority of book sales. Likewise, a small number of diseases account for the bulk of human morbidity and mortality. For example, two common types of cancer, basal cell carcinoma of skin and squamous cell carcinoma of skin, account for about 1 million new cases of cancer each year in the U.S. This is approximately the sum total for all other types of cancer combined. We see a similar phenomenon when we count causes of death. About 2.6 million people die each year in the U.S. [98]. The top two causes of death account for 1,171,652 deaths (596,339 deaths from heart disease and 575,313 deaths from cancer [99]), or about 45% of all U.S. deaths. All of the remaining deaths are accounted for by more than 7000 conditions. Sets of data that follow Pareto's principle are often said to follow a Zipf distribution, or a power law distribution. These types of distributions are not tractable by standard statistical descriptors because they do not produce a symmetric bell-shaped curve. Simple measurements such as average and standard deviation have virtually no practical meaning when applied to Zipf distributions. Furthermore, the Gaussian distribution does not apply, and none of the statistical inferences built upon an assumption of a Gaussian distribution will hold on data sets that observe Pareto's principle. See Zipf distribution.

Pathogenesis—The successive changes in tissues and their cells that eventually lead to the expression of a disease. The term "pathogenesis" is occasionally confused with the term "etiology." The term "etiology" refers to the cause of the disease. The term "pathogenesis" is the process leading to the disease, set in motion by some etiologic agent or event. It must be noted that knowing the defective gene underlying a disease (i.e., the genetic alteration without which the disease would not have occurred) is a far cry from understanding the pathogenesis of the disease. For example, a form of congenital neutropenia, a disease characterized by low white blood cells in newborns, is caused by mutations in the gene that codes for neutrophil elastase [100]. You would think that after identifying the gene, its gene product, and knowing the role of the gene product in a target cell (i.e., metabolizing elastase in neutrophils) that we would be well on our way to claiming that we understand the pathogenesis of this disease. Not so. Neutrophil elastase may have any number of cellular substrates, and may be involved in many different cellular pathways. The steps through which an altered neutrophil elastase may eventually lead to neutropenia cannot be directly inferred. It is easy to find rare diseases whose causes are known, but whose pathogenesis is obscure. For example, a mutation in the gene encoding fumarate hydratase is the underlying cause of HLRCC, hereditary leiomyomatosis and renal cell cancer. A mutation in the gene encoding magnesium transporter-1 causes XMEN, the acronym for X-linked immunodeficiency with magnesium defect, Epstein–Barr virus infection, and neoplasia. What is the pathogenesis that connects these gene defects with their clinical phenotype? For thousands of rare diseases, their underlying genetic causes have been identified, but we have much to learn about the events that lead from an altered gene to the expression of a characteristic clinical phenotype.

Pathologist—Pathology is the study of disease. Broadly speaking, pathologists are individuals whose careers are devoted to the study of disease. In a sense, pathology is all of medicine minus the hands-on patient care. In the past 150 years, the field of pathology has become specialized. Medical pathologists are physicians who perform diagnostic tests on tissue samples (e.g., biopsies and excised tissues), or cells (e.g., exfoliated cells in urine, scraped cells from cervix, aspirated cells from fine needle samplings of tissues), or on blood and other body fluids. Within the specialty of medical pathology, there are numerous subspecialties (e.g., molecular diagnostics, clinical pathology, surgical pathology, cytopathology, dermatopathology, forensic pathology). Research pathologists are scientists who study diseases in laboratories.

Pathway—According to traditional thinking, a pathway is a sequence of biochemical reactions, involving a specific set of enzymes and substrates that produce a chemical product. The classic pathway was the Krebs cycle. It was common for students to be required to calculate the output of the cycle (in moles of ATP) based on stoichiometric equations employing known amounts of substrate. As we have learned more and more about cellular biology, the term "pathway" has acquired a broader and more complex meaning. It can apply to activities (not just chemical products). One pathway may intersect or subsume other pathways. Furthermore, a pathway may not be constrained to an anatomically sequestered area of the cell, and the activity of a pathway may change from cell type to cell type or may change within one cell depending on the cell's physiologic status. Still, the term "pathway" is a convenient conceptual device to organize classes of molecules that interact with a generally defined set of partner molecules to produce a somewhat consistent range of biological actions.

Throughout this book, the term pathway will apply to cellular actions produced by groups of interacting molecules. In most cases, the pathways are not named; the existence of the pathway is inferred whenever a complex cellular activity occurs (e.g., replication of DNA, post-translational modifications of a protein, synthesis of a molecule, DNA repair, apoptosis). See Apoptosis and Pathway-driven disease.

Pathway-driven disease—Diseases with similar clinical phenotypes can often be grouped together according to shared pathways. Examples would include the channelopathies, ciliopathies, and lipid receptor mutations. At this point, our ability to sensibly assign diseases to pathways is limited because the effects of a mutation in a single gene may indirectly affect many different pathways, and those pathways may vary from cell type to cell type. Syndromes involving multiple pathways and multiple tissues occur frequently when the mutation involves a regulatory element, such as a transcription factor [101]. One transcription factor may regulate pathways in a variety of cell types with differing functions and embryologic origins. Nonetheless, whenever one pathway has a dominant role in the pathogenesis of a group of diseases, we can begin to ask how we might develop diagnostic tests and treatments that apply to the rare members and the common members of the group.

Penetrance—An individual may have a gene that is necessary for the expression of a particular disease; yet the individual may not express that disease. In medical genetics, the penetrance is the proportion of individuals with the mutation who exhibit clinical symptoms. There are several reasons why the penetrance of a disease-causing gene may be significantly lower than 100%. Some diseases, particularly the common diseases, are polygenic. It may take many genes to produce the disease phenotype. Epistasis may also modify penetrance; one gene may be influenced by a particular allele of another gene. Environmental and epigenetic factors can also influence gene function. An inherited mutation may require environmental triggers (e.g., excessive sunlight exposure in porphyria cutanea tarda) or conditional physiological conditions (e.g., fatigue or starvation preceding the hyperbilirubinemia associated with Gilbert syndrome) to fully express the clinical trait. Such factors may influence whether a disease-causing mutation is expressed as disease (i.e., its penetrance), and may also influence the age at which the disease emerges, or the severity of the disease, or phenotype of the disease (i.e., which clinical problems will develop). Because we cannot know all the factors that may influence penetrance, it is safe to say that "penetrance" is a word invented to describe observations that we do not understand. The concept of disease penetrance serves as a reminder that an inherited mutation can be the underlying cause of a disease, but additional factors will have important roles in the series of events that lead to a clinical phenotype. See Epistasis.

Peroxisome disorder—Eukaryotic organisms contain small organelles that are involved in the catabolism of very long chain fatty acids. In humans, peroxisomes are also involved in the synthesis of various phospholipids essential to the brain. Hence, inherited peroxisomal disorders are neurologic disorders that are accompanied by developmental and organ (e.g., liver) dysfunctions. Examples of peroxisome disorders include Zellweger syndrome, rhizomelic chondrodysplasia punctata, neonatal adrenoleukodystrophy, and infantile Refsum disease [102].

Pharmacogenomic—Pharmacogenomics refers to pharmacologic studies wherein the entire genome is examined and correlations are pursued among sets of genes. Drug response predictions based on gene expression profiles could be described with the term "pharmacogenomics." Tests on a single gene, or on several individual genes, intended to predict the response to a drug might be described with the term "pharmacogenetics." The central dictum of pharmacogenomics seems to be that every individual's genome is unique; hence, every disease occurring in an individual is unique; hence, every response to treatment is unique for each individual; hence, every occurrence of disease deserves to be treated with a medication designed for the unique individual. Underlying these hypotheses is the assumption that the key elements of an individual's disease are captured in the unique sequence of the individual's genome. This assumption short-changes the complexity of genetics. Biological systems have multiple dependencies, and the genome is one player among many. Furthermore, the artifactual distinction between "etics" and "omics" creates a dichotomy where none exists. A gene belongs in a genome and a genome contains genes; they are interrelated and co-dependent concepts. As it happens, the terms "pharmacogenetics" and "pharmacogenomics" are commonly used interchangeably. In this book, the term "pharmacogenetics" is used throughout the chapters.

Phenocopy disease—A non-hereditary disease or condition produced by an exogenous factor that replicates a disease produced by a genetic mutation. Examples would include acquired porphyria due to alcohol abuse; acquired Parkinson-type syndrome due to anti-psychotic medications; myopathy produced by AZT (i.e., azidothymidine), a drug that interferes with mitochondrial DNA replication; Antabuse (disulfiram), an acetaldehyde dehydrogenase inhibitor that induces alcohol intolerance, mimicking genetic diseases of alcohol metabolism. Phenocopy diseases provide important clues to the pathogenesis of rare and common diseases. The drug that produces a phenocopy disease may share the same pathway employed in a specific genetic disease or in a common disease whose phenotype overlaps with the genetic disease. Pharmacologic treatments for the phenocopy disease may apply to pathways operative in the genetic form of the disease or in the common diseases. For a full discussion, see Section 9.5.

Phenotype—The set of observable traits and features in an organism. The normal phenotype would be the complete set of morphologic and physiologic patterns that characterize the organism. As used herein, a "disease phenotype" is the set of observable traits and features that characterize a disease. Geneticists typically think in terms of a process in which a genotype (i.e., the genetic composition of an organism) is expressed as a phenotype (i.e., observable features in organisms), and wherein genetic errors are expressed as disease phenotypes.

Phenotypic heterogeneity—Occurs when a given genotype may produce different phenotypes. In the context of disease, it occurs when a gene mutation may produce any of several different clinical disorders. If you look closely enough at any genetic disease, phenotypic heterogeneity is a universal phenomenon. No two individuals will ever express the same exact set of clinical problems; even identical twins.

Pleiotropic (alternate spelling, pleiotrophic)—See Pleiotypia.

Pleiotypia—Refers to an effect wherein one gene influences more than one phenotypic trait. A gene that has a pleiotypic effect is said to be pleiotropic (alternate spelling, pleiotrophic). An example of pleiotypia is found in X-linked heterotaxy-1. Heterotaxy is a developmental disorder in which one or more organs are found in abnormal locations. X-linked heterotaxy-1 is characterized by situs inversus, wherein the positions of the major organs are reversed along the body axis. Because normal development requires the customary positioning of organs, X-linked heterotaxy-1 is accompanied by complex cardiac defects, and splenic defects. All these changes are caused by a single alteration in a single gene: ZIC3.

Pluripotent stem cell—The ability to yield, after cell divisions, differentiated cell types from any of the three embryonic layers (i.e., endoderm, ectoderm, and mesoderm). Pluripotent stem cells differ from totipotent stem cells as they do not yield cells of extra-embryonic type (e.g., trophoblasts). It is now possible to induce the formation of human pluripotent stem cells from cultured fibroblasts treated with a cocktail of transcription factors [103,104]. See Totipotent stem cell.

Polycythemia—Polycythemia is an increase in the number of circulating red blood cells. An increase in red blood cells can occur as a response to a physiologic stimulus (e.g., chronic anoxia, high-altitude living). When polycythemia occurs as an intrinsic defect of red blood cells, it is called primary polycythemia. The term polycythemia vera, or "true" polycythemia, is reserved for a clonal disorder of the red blood cell lineage in cells that have a genetic aberration that drives proliferation. The cause of polycythemia vera is a mutation of the JAK2 gene occurring in a single hematopoietic stem cell from which a clonal expansion eventually raises the number of circulating erythrocytes [105]. Mutations in JAK2 are associated with a variety of myeloproliferative conditions, including myelofibrosis, and at least one form of hereditary thrombocythemia [106,105].

Polygenic disease—A disease whose underlying cause involves alterations in multiple genes. See Monogenic disease, Digenic disease, and Oligogenic disease.

Polymorphism—The term "polymorphism" can have several somewhat different meanings in various fields of biology. In this book, polymorphism refers to genetic polymorphism, indicating that variants of a gene occur in the general population. A polymorphism is usually restricted to variants that occur with an occurrence frequency of 1% or higher. If a variant occurs at a frequency of less than 1%, it is considered to be sufficiently uncommon that it is probably not steadily maintained within the general population. All commonly occurring polymorphisms are assumed to be benign or, at worst, of low pathogenicity; the reasoning being that natural selection would eliminate frequently occurring polymorphisms that reduced the fitness of individuals within the population. Nonetheless, different polymorphisms may code for proteins with at least some differences in functionality.

Post-mitotic—Refers to fully differentiated cells that have lost the ability to divide. For example, the epidermis of the skin has a basal layer of cells that are capable of dividing to produce a post-mitotic cell and another basal cell. The post-mitotic cells sit atop the basal cells to flatten out and lose their nucleus as the cells rise through the epidermal layers. The top layer of the epidermis sloughs off the body and is replaced by post-mitotic epidermal cells in the next lower layer. This cycle of cell renewal from the bottom and cell sloughing from the top is typical of most epithelial surfaces of the body (e.g., epidermis of skin, gastrointestinal tract, and glandular organs). Aside from epithelial surfaces, post-mitotic cells arise from populations of mitotic cells that have exhausted their regenerative potential. One theory of aging holds that certain cell types of the body (e.g., fibroblasts) have a limited number of mitotic cell cycles. When a predetermined number of cell cycles have elapsed, cells cannot divide further, becoming post-mitotic. See Mitosis.

Post-translational protein modification—Much happens between the moment when the amino acid sequence of a protein is translated from an RNA template and the moment when the fully modified protein, in its optimal conformation, arrives at its assigned station. Errors in the post-translational process, including timing errors (i.e., the proper sequence of events that lead to the finished product), can have negative consequences. An example of a rare disease caused by a defect in a post-translational process is congenital disorder of glycosylation type IIe, caused by homozygous mutation in a gene that encodes a component of a Golgi body protein that is involved in post-translational protein glycosylation; the COG7 gene [107]. This rare disease produces a complex disease phenotype in infants, with multiple disturbances in organs and systems plus various anatomic abnormalities. In the few reported cases, death results in a few months. See Vesicular trafficking disorder.

Precancer—Precancers are the lesions from which cancers derive. Most precancers are non-invasive and non-metastatic, so eliminating precancers cures the patient of the cancer that might have eventually developed. Two of the most interesting properties of precancers, lacking in fully developed cancers, are their propensity for spontaneous regression and the ease with which they can be treated and cured. Precancers have fewer of the genetic and epigenetic alterations that accumulate in cells during the long process of carcinogenesis and tumor progression. Like rare diseases, they are genetically and epigenetically simple; at least, they are simpler than the cancers into which they develop. They seem to obey the general rule that diseases with the simplest genetic alterations are the easiest to treat successfully [108].

Precancer regression—One of the properties of precancers is regression. It is not unusual for a precancer to stop growing, or to shrink and disappear [108]. See Spontaneous regression.

Precancerous condition—A condition or event that predisposes a person to the development of a precancer and to the eventual development of a cancer. For example, patients with cirrhosis have a high risk of eventually developing cancer. Over time, the cirrhotic liver becomes nodular, and precancerous lesions develop from the nodules. In some cancers, the precancerous nodules develop into cancer (i.e., hepatocellular carcinoma). Cirrhosis is a precancerous condition, but it is not a precancer. Cirrhosis simply sets the stage for precancers to develop [108].

Preclinical trial—Investigations of drug activity prior to human studies. The term often applies to animal experiments that determine how candidate human drugs are metabolized in animals, and to measure various parameters of animal toxicity. One of the measures that come from animal trials is the "no observable adverse effect level," which is used to calculate a range of dosages that might be used in the earliest clinical trials.

Predictive test—A test that estimates a patient's response to a particular treatment. The terms "predictive test" and "prognostic test" should not be confused with one another. See Prognosis.

Prevalence—A measure of the number of individuals in a population who have a disease at a particular time. Prevalence differs from incidence; the latter indicating the rate of new cases of a disease that occur in a population. Chronic diseases, especially those that persist through the patient's normal lifespan, may have a high prevalence and a low incidence. Diseases that have a short clinical span, such as influenza or the common headache, will have a lower prevalence than incidence. See Age-adjusted incidence.

Primary host—Also called final host or definitive host, the primary host is infected with the mature or reproductive stage of the parasite. In most cases, the mature stage of animal parasite is the stage in which eggs, larvae, or cysts are produced. See Intermediate host [109].

Prognosis—The likelihood that a patient will recover. Prognostic markers are used to produce a quantitative estimate of the likelihood of recovery. The term "prognostic test" is sometimes used interchangeably with the term "predictive test," but the two terms are not equivalent. See Predictive test.

Promoter—The DNA site that binds RNA polymerase plus transcription factors, thus initiating RNA transcription. Examples of promoter mutations causing disease include beta-thalassemia, Bernard–Soulier syndrome, pyruvate kinase deficiency, familial hypercholesterolemia, and hemophilia [110]. Monogenic promoter mutations generally cause disease by reducing the quantity of a normal protein, not by producing altered protein and not by reducing the quantity of multiple proteins. Because the drop in protein production may be small, promoter diseases may be hard to detect. See Transcription element.

Protozoa—Microbiologic nomenclature has many terms that have persisted long after they have outlived their usefulness; "protozoa" is a perfect example. A commonly found definition for protozoa is "one-celled animal," but this is an oxymoron, as all animals are multi-cellular. Over the years, the term protozoa has come to include any one-celled heterotrophic (i.e., lacking chloroplasts) eukaryotic organism. This definition crosses classes (i.e., includes unrelated organisms) and hence has no phylogenetic meaning. With luck, the term "protozoa" will soon disappear from the scientific literature [109].

Proximate cause—The proximate cause of any event is the closest direct action that can be held to be the cause of the event. For example, the rupture of a blood vessel in the lung may be the proximate cause of death, while an invasive lung cancer may have been the underlying cause of death. The erosion of a vessel by tumor cells was one of a sequence of events leading from the underlying cause of death to the proximate cause of death. The underlying cause of death satisfies the "but-for" condition. But for the lung cancer, the vessel would not have eroded, and blood would not have flooded the lung tissue. The proximate cause of death need not be a necessary condition resulting from the underlying cause of death. Had the vessel not ruptured, the individual may have died from an alternate proximate cause (e.g., metastasis, pneumonia). See Underlying cause of death.

Pseudogene—Genes that do not code for proteins. Theories explaining the origin of pseudogenes are many. Some pseudogenes presumably devolved from genes that acquired mutations that rendered the genes non-functional. Other pseudogenes may have been reverse-transcribed into DNA via RNA retrotransposons. Pseudogenes are identified from sequence data by computational algorithms that search for stretches of DNA that have some sequence similarities to functional genes, along with sequences that might render the gene non-functional (e.g., premature stop codons, frameshift mutations, a poly-A tail, the lack of promoters). Though pseudogenes do not code for translated proteins, they may play an important role in disease. The RNA transcribed by a pseudogene, and protein molecules translated from the RNA, may have regulatory or modifier functions acting on a variety of cellular processes. At present, pseudogenes are suspected of playing a role in the dysregulation of cancer cells, and in cell defects found in neurodegenerative disorders [111,112].

Rare disease—As written in Public Law 107-280, the Rare Diseases Act of 2002, "Rare diseases and disorders are those which affect small patient populations, typically populations smaller than 200,000 individuals in the United States" [113]. Since the population of the U.S. is about 314 million, in 2013, this comes to a prevalence of about one case for every 1570 persons. This is not too far from the definition recommended by the European Commission on Public Health; a prevalence less than one in 2000 people.

Regression—Reversal of a disease process. The term is most often applied to the reduction in the size of a cancer. Regression usually results from treatment, or from the removal of a condition that stimulates the growth of tumors. For example, hormone-responsive tumors (e.g., subsets of breast and prostate cancers) may regress under conditions of hormone deficiency. Some cases of helicobacter-induced maltomas regress when the helicobacter infection is treated. Virally-induced tumors that occur in immuno-deficient patients (e.g., herpes virus-8 induced Kaposi sarcoma, human papillomavirus-induced warts) may regress when normal immune status is restored. Finally, some tumors may regress spontaneously. See Spontaneous regression.

Regulatory DNA element—Sites in DNA that bind to specialized proteins (e.g., transcription factors and RNA polymerase) to regulate transcription. Promoters and enhancers are types of regulatory DNA elements.

Regulatory RNA element—Transcribed RNA can influence the subsequent transcription of other RNA species. The various RNA regulatory elements include: antisense RNA (including cis-natural antisense transcript and trans-acting siRNA), long non-coding RNA, microRNA, piwi-interacting RNA, repeat-associated siRNA, RNAi, small interfering RNA, and small temporal RNA. Mutations of regulatory RNA elements may cause disease. For example, miR-96 is expressed exclusively in the inner ear and the eye. Mutations in the miR-96 precursor molecule may cause a rare form of autosomal dominant hearing loss [114].

Retrovirus—An RNA virus that replicates through a DNA intermediate. The DNA intermediate becomes integrated into the host DNA, from which viral RNA is transcribed. When integration of the virus occurs in germ cells, the viral DNA can be inherited. Through this mechanism, the human genome carries a legacy of retroviral DNA, accounting for about 8% of the human genome [115].

RNA splicing—See Alternative RNA splicing.

Russell–Silver syndrome—A growth disorder that produces primordial dwarfism, a rare form of proportionate growth reduction. Because infants are small, but have an otherwise unremarkable physical appearance, they are often undiagnosed until they are about 3 years old. Russell–Silver syndrome is an imprinting disorder, produced by hypomethylation of the H19 to IGF2 regions of chromosome 11p15, the same area involved in Beckwith–Wiedemann syndrome [116]. See Imprinting and Section 9.2.

Secondary host—Synonymous with intermediate host. See Intermediate host.

Silent mutation—A mutation that does not alter phenotype. Silent mutations can occur in non-coding regions or in exons. It has been reported that silent mutations may have subtle effects on the tertiary structure of proteins [117]. See Synonymous SNP.

Single nucleotide polymorphism—See SNP.

SNP—single nucleotide polymorphism. Locations in the genome wherein different individuals are known to have single base differences in DNA sequence. These SNPs are frequent and account for the polymorphisms in DNA. It is currently estimated there are 10–50 million SNPs in the human population, and an SNP occurs about once in every 300 nucleotides [118]. See Genomic structural variations and Polymorphism.

Somatic—From the Greek meaning body, refers to non-germ cells (i.e., not oocytes, not spermatocytes). The somatic cells, then, consist of the differentiated cells of the body and the stem cells in their lineage. Somatic cells cannot undergo meiosis and, under natural conditions, cannot pass acquired mutations to embryonic progeny.

Somatic mosaicism—If a new mutation occurs in an embryonic cell after the zygote has split to produce daughter cells, then the new mutation produces somatic mosaicism; meaning that it will only occur in those somatic cells that are descended from the embryonic cells in which the mutation occurred. If the somatic cells that inherit the new mutation include germ cells (i.e., cells that will differentiate into ova or sperm), then the *de novo* mutation can be passed to the next generation as an inherited germline mutation. Proteus syndrome is an example of a disease that exhibits somatic mosaicism [119]. Presumably, the gene causing Proteus syndrome, if present in the germline, would have been lethal to the embryo. Somatic mosaicism is a particular type of *de novo* mutation. See *De novo* mutation and Intra-tissue genetic heterogeneity.

Spliceosome—In animals, DNA sequences are not transcribed directly into full-length RNA molecules ready for translation into a final protein. There is a pre-translational process wherein transcribed sections of DNA, so-called introns, are spliced together, and a single gene can be assembled into alternative spliced products. Alternative splicing is one method whereby more than one protein form can be produced by a single gene [120]. Cellular proteins that coordinate the splicing process are referred to, in aggregate, as the spliceosome. Errors in normal splicing can produce inherited disease, and it estimated that 15% of disease-causing mutations involve splicing [8,9]. Examples of spliceosome diseases are spinal muscular atrophy and some forms of retinitis pigmentosa [120]. In both diseases, pathology is limited to a specific type of cell; retinal cells and their pigment layer in retinitis pigmentosa, and motor neuron cells in the spinal muscular atrophy. One might expect that mutations in spliceosomes would cause deficiencies in diverse cell types, with multi-organ and multi-system disease (e.g., syndromic disease). That this is not the case is somewhat of a mystery, and the catalyst for much speculation. Faustino and Cooper have categorized splicing diseases into different types, including: those that affect a single gene, those that affect multiple genes, those that cause aberrant splicing that result in unnatural mRNAs, and those that cause the inappropriate expression of natural mRNAs [120]. See Alternative RNA splicing.

Spontaneous regression (in cancer)—For the most part, cancers grow continuously and accumulate additional genetic alterations as they grow. Left untreated, cancer growth leads to death. There are special cases wherein neoplastic growth reverses, and the patient is cured without treatment. This is a common occurrence in precancers. The presumption is that regressing precancerous lesions failed to acquire the properties necessary for sustained, autonomous growth. Spontaneous regression is occasionally encountered in tumors of childhood, particularly neuroblastoma. Spontaneous regression is also encountered in some cases of melanoma. In the case of melanoma, the primary cancer may regress after it has already metastasized widely. In this case, the metastatic lesions continue to grow and may eventually lead to the death of the patient, although the primary site of the melanoma is fully regressed. In all of these cases, the phenomenon of regression would suggest that these particular cancers, always on the verge of regression, might be easier to cure than their non-regressing counterparts. See Regression.

Sporadic—Describes a case occurrence of a disease without any discernible cause, as though by random chance. Rare diseases are typically non-sporadic. Common diseases are typically sporadic, but may contain subsets of disease occurrences that are non-sporadic. An example is schizophrenia. Schizophrenia is a common disease with a prevalence of about 1.1%. This translates to about 51 million individuals worldwide who suffer from this mental disorder. Many cases of schizophrenia occur in families and such cases are considered to be inherited and non-sporadic. Other cases seem to have no familial association and are considered sporadic. Are these sporadic cases caused by environmental factors, or are they caused by *de novo* mutations that arose in the affected individuals? Recent evidence would suggest that many of the so-called sporadic cases arise from new mutations in affected individuals [121]. It can be very difficult to make the distinction between sporadic and non-sporadic disease. For example, the occurrence of a tumor that is thought to be sporadic may turn out to be non-sporadic when a specific cause is found. Likewise, a tumor thought to be non-sporadic may turn out to be sporadic when it is eventually shown that the presumed cause did not apply. For example, an individual who develops breast cancer may have a strong family history of breast cancer, and affected family members may all carry a predisposing BRCA gene mutation. In this case, the patient's tumor is considered to be non-sporadic. If the patient is tested and found to lack a BRCA mutation, then the tumor must be considered to be sporadic, in this case, despite any family predisposition. If additional studies show that the patient has an inherited cancer-causing gene different from the BRCA gene, then the tumor would revert back to being non-sporadic.

Stem cell—A cell that is capable of employing a strategy for cell division that produces two different types of cells: another stem cell plus a cell that is more differentiated than the original stem cell. According to the stem cell theory of development, all of the differentiated cells of the body derive from stem cells, and all of the stem cells derive from more primitive ancestral stem cells. In most cases, fully differentiated cells are incapable of further division, and are sometimes referred to as post-mitotic cells. In the case of skin epidermis, a layer of nearly differentiated stem cells lines the bottom (i.e., basal layer) of the epidermis. Each basal stem cell divides to replace itself and to create a new, non-dividing epidermal cell that slowly keratinizes as it is pushed upward through the layers of the epidermis by new generations of underlying epidermal cells. When the fully keratinized epidermal cell reaches the topmost layer of the skin surface, it sloughs into the environment. Much of what we perceive as "house dust" are fully keratinized epidermal cells wafted into the air. A similar process accounts for top-most gastrointestinal lining cells sloughing into the gut lumen, contributing to the non-bacterial bulk matter in stools.

Syndrome—A syndrome is a constellation of pathologic features associated with a single disease or condition, usually involving multiple organs. Inherited deafness is often syndromic; deafness is accompanied by other abnormalities, possibly involving facial structure or nerve function. Non-syndromic deafness affects hearing and no other structures or functions.

Synonymous SNP—SNPs that have different sequences but which have the same transcriptional meaning because the genetic code contains codon triplet redundancy. For example, guu, guc, gua, and gug all code for the amino acid valine (when translated in the same reading frame). These triplets are synonymous though their sequences are non-identical.

Synteny—An ordering of a group of genes along the chromosome that is the same among related species. Evaluation of cross-species synteny can sometimes establish or discredit animal models of human disease. For example, human oligodendrogliomas are often characterized by losses of chromosomes 1p or 19q. Regions in mouse syntenic to 1p are found on murine chromosomes 3 and 4 and regions on mouse syntenic to 19q are found on murine chromosome 7.

In an experimental mouse oligodendroglioma model, no losses were found in either of these syntenic chromosomal regions, indicating that the cytogenetic markers for human oligodendroglioma are absent from the mouse tumor [122].

Taxonomy—The science of classification, derived from the ancient Greek taxis, "arrangement," and nomia, "method." Naturalists use the word "taxonomy" to include the hierarchy of ancestral organisms and their descendants, and the names assigned to the classes and species of organisms.

Telomere—Chromosomes are built with a long padding sequence of repetitive DNA at the chromosome tips, and this sequence is called the telomere. Animal cells lose a fragment of DNA from the tip of the chromosome with each cell division. This happens because one strand of DNA is replicated as sequential fragments, with each fragment requiring a template sequence beyond its end to initiate replication. The last fragment in the DNA strand has no template for itself and is not replicated. By providing a padding at the tips of chromosomes, the telomere sequence sacrifices fragments of itself for the sake of preserving the coding sequences of the chromosome. As all good things come to an end, the telomere padding exhausts itself after about 50 rounds of mitosis. At this time, the cell ceases further replication and will eventually die. Cells that continually renew throughout life, such as bone marrow stem cells, skin and hair follicle basal layer cells, and intestinal basal crypt cells, retain an enzyme, telomerase, that restores telomere length. When such cells contain a loss of function mutation in genes encoding for components of the telomerase complex, their ability to divide throughout the lifetime of the organism is reduced. A mutation in the telomerase gene causes dyskeratosis congenita, a rare inherited condition in which bone marrow failure frequently occurs [123]. Telomerase gene mutations have also been found in some cases of acquired bone marrow failure [124]. Cancer cells, like bone marrow cells, continuously divide, and have high concentrations of telomerase. The ability of some cancer cells to restore their telomeres contributes to their continuous capacity to divide without limit, a phenomenon sometimes called cancer cell immortality. Telomerase insufficiency has been suggested as a possible cause of spontaneous regression in tumors, a rarely observed phenomenon [125].

Teratogenesis—The biological process that leads to a developmental malformation. Current thinking would suggest that environmental agents that produce developmental malformations do so by killing specific types of embryonic or fetal cells at vulnerable moments in development. Theory holds that normal development requires specific cells fulfilling specific functions at specific times. Interruption of this orchestrated process may result in developmental abnormalities.

Thalassemia—Alpha thalassemia and beta thalassemia are the most common inherited monogenic disorders worldwide. The disorder is characterized by ineffective red blood cell production due to a reduction in the synthesis of the alpha or beta chains of hemoglobin. As is the case with sickle cell disease, the alpha thalassemia trait seems to confer some protection against malaria [126], and this beneficial effect may explain the conservation of the thalassemia gene in those populations wherein malaria is endemic. Also, as with sickle cell disease, one consequence for mated carriers of the thalassemia trait is the increased likelihood of producing offspring with homozygous disease. See Sickle cell disease.

Totipotent stem cell—A stem cell that can produce, after cell divisions, differentiated cells of any type. This would include cells of any of the three embryonic layers (ectoderm, endoderm, and mesoderm), germ cells, and cells of the extra-embryonic tissue (e.g., trophoblasts).

Trans-acting—In molecular biology, a trans-acting agent is usually a regulatory sequence of DNA that acts through an intermediary molecule (i.e., protein or RNA) on some other location of the chromosome or on some other chromosome. A cis-acting agent does not operate through an intermediary molecule.

Transcription factor—A protein that binds to specific DNA sequences to control the transcription of DNA to RNA. Some of the most phenotypically complex rare disease syndromes are caused by single gene mutations that code for transcription factors. Examples include: mutation in the gene encoding transcription factor TBX5 producing Holt–Oram syndrome, consisting of hand malformations, heart defects, and other malformations; and mutation in the gene encoding microphthalmia-associated transcription factor, MITF, producing Waardenburg syndrome 2A, consisting of lateral displacement of the inner canthus of both eyes, pigmentary disturbances of hair and iris, white eyelashes, leukoderma, and cochlear deafness.

Transgene—An experimental procedure wherein a DNA sequence from one organism is added to the genome of another organism. When this involves incorporating the DNA into a fertilized egg (i.e., zygote), the foreign sequence has the opportunity to enter the germline of the developing organism.

Translation factor—Also called initiation factor or translation initiation factor, these factors facilitate the initiation of protein synthesis from mRNA by forming a complex with ribosomal RNA. Mutations in translation factors can cause disease. For example, leukoencephalopathy with vanishing white matter is caused by mutations in any one of five genes encoding subunits of the eukaryotic initiation factor EIF2B gene. Each of these five genes is located on a different chromosome from the others.

Transposable element—See Transposon.

Transposon—Also called transposable elements, transposons are repeated gene sequences that are scattered throughout chromosomes. Some transposons can replicate themselves and move to different locations, and some may contain regulatory elements that can modify gene expression. Transposons are the ancient remnants of retroviruses and other horizontally transferred genes that insinuated their way into the human genome. A transposon is credited with the acquisition of adaptive immunity in animals. The RAG1 gene was acquired as a transposon. This gene enabled the DNA that encodes a segment of the immunoglobulin molecule to rearrange, thus producing a vast array of protein variants [127]. A role for transposons in the altered expression of genes in cancer cells has been suggested [128].

Trinucleotide repeat syndrome—A group of rare diseases are characterized by trinucleotide repeats in DNA. About half of the studied trinucleotide repeat disorders demonstrate repeated CAG sequences. CAG codes for glutamine to produce a polyglutamine tract. Examples of polynucleotide repeat disorders are: dentatorubropallidoluysian atrophy, fragile X syndrome, Friedreich ataxia, Huntington disease, mytotonic dystrophy, spinobulbar muscular atrophy, and several forms of spinocerebellar ataxia.

Tumor heterogeneity—During tumor progression, genetic and epigenetic instability results in subclones of the tumor, each having a genome and epigenome that is slightly different from the genome and epigenome of cells in other subclones of the same tumor [19]. In effect, tumor heterogeneity is a condition wherein lots of different tumors, all deriving from the same original cancer cell, exist within the same tumor. The subclones that have the greatest number of cells will be those that have a growth advantage that can prevail in an enlarging tumor (e.g., enhanced growth under anaerobic conditions, increased invasiveness). Tumors that have progressed to develop a high degree of tumor heterogeneity may be very difficult to treat successfully.

Tumor progression—Synonymous with cancer progression. See Cancer progression.

Tumor suppressor gene—A gene that arrests, delays, or makes less likely one or more of the cellular events involved in the pathogenesis of cancer.

Underlying cause of death—The disease that initiated the clinical events that led to the individual's death. The underlying cause of death is sometimes difficult to choose. If a patient has a severe case of emphysema as well as lung cancer, and dies in respiratory distress, would the underlying cause of death be emphysema or cancer? Assuming that the deceased was a long-time smoker, might the underlying cause of death be "smoking addiction," or maybe "cigarette abuse," leading in turn to emphysema and lung cancer? The World Health Organization, aware of the difficulties in choosing an underlying cause of death, and assigning a sequential list of the ensuing clinical consequences leading to the proximate cause of death, has issued reporting guidelines [129]. Instructions notwithstanding, death certificate data are notoriously inconsistent, giving rise to divergent methods of reporting the diseases that cause death [22,23]. See Proximate cause and But-for.

Undifferentiated tumor—As the result of tumor progression, cancer cells tend to become more aneuploid, with larger nuclei, and have fewer of the morphologic features that characterize normal differentiated cells. For instance, an advanced, unpigmented melanoma may have few of the morphologic features of a normal melanocyte. Loss of differentiation is the morphologic expression of tumor progression. It should be noted that undifferentiated tumors may employ pathways that are highly conserved for their cell lineage (e.g., an undifferentiated melanoma cell may retain metabolic pathways typical of normal, differentiated melanocytes). The so-called undifferentiated cells may be vulnerable to therapeutic agents targeted to lineage-specific pathways. It is also worth noting that some primitive tumors, particularly the primitive tumors of childhood, lack differentiation because they derive from primitive cells, not from differentiated cells. The primitive tumors, such as the primitive neuroectodermal tumors, are *ab initio* undifferentiated lesions that have never progressed from a differentiated cell type to an undifferentiated cell type. There is no reason to expect primitive, undifferentiated tumors of childhood to behave like undifferentiated tumors in adults. See *Ab initio*.

Uniparental disomy—Occurs in the germline when an error in meiosis results in the embryo receiving a chromosome pair or a paired partial chromosome from one parent with no matching contribution from the other parent. Uniparental disomy can also occur as an acquired feature in somatic (i.e., non-germline) cells and is a frequent alteration found in cancers. There are several mechanisms whereby somatic cell mitosis can lead to an acquired uniparental disomy; these involve faulty chromosome migration with or without accompanying translocations when the chromosomes migrate to daughter cells [130]. The acquired or somatic form of uniparental disomy does not actually involve the participation of two parents (as seen in uniparental disomy of germline cells), and is often referred to by the less mechanistic, but more precise, term "copy neutral loss of heterozygosity." Because only one allele is represented in the uniparental disomic gene, there is loss of heterozygosity. Because cells have the normal number of copies of expressed genes, producing a normal gene expression level, the result is "copy neutral."

Vesicular trafficking disorder—Alternately known as protein trafficking disorder, cargo disorder, and vesicular transport disorder. After a protein molecule is translated from mRNA, a complex set of post-translational events must occur for the protein to serve its intended purpose. The protein must be modified (e.g., glycosylated), shaped (e.g., folded), transported from the endoplasmic reticulum into a series of subcellular locations (e.g., Golgi apparatus and cargo vesicle), and delivered to its ultimate location. Such post-translational steps are often divided into disorders of post-translational modification (e.g., congenital disorders of glycosylation) and protein transport disorders. For their work on protein transport mechanisms, the 2013 Nobel Prize in Physiology or Medicine was awarded to James E. Rothman, Randy W. Schekman, and

Thomas C. Sudhof. Much of the progress in this field was based on examining transport gene mutants in yeast cells, as well as inherited transport disorders in humans. Examples of such disorders in humans include: congenital disorder of glycosylation type IIe; combined factor V and factor VIII coagulation factor deficiency; Hermansky–Pudlak syndrome; Chediak–Higashi syndrome; cranio-lenticulo-sutural dysplasia; choroideremia; Warburg micro syndrome and Martsolf syndrome; Bardet–Biedl syndrome-3; Griscelli syndrome types I, II, III; Charcot–Marie–Tooth disease 2a,2b; hereditary spastic paraplegia SPG10, SPG4; Troyer syndrome; spinocerebellar ataxia 5; Lowe oculocerebrorenal syndrome; Usher syndrome type IB; slow progressing amyotrophic lateral sclerosis-8; CEDNIK syndrome; familial hemophagocytic lymphohistiocytosis; limb girdle muscular dystrophy; and Miyoshi myopathy [96].

X-chromosome—The female sex chromosome. Genetically normal human females have paired X-chromosomes, one inherited from the mother and one from the father. In each somatic cell, one X-chromosome is active and the other X-chromosome is inactive. The inactive cell can often be visualized in cytologic preparations as a small dense clump of heterochromatin hugging the nuclear membrane. Which X-chromosome is inactivated will vary from cell to cell; hence, genotypically normal human females are X-chromosome mosaics, composed of cells expressing one of two different X-chromosomes.

Y-chromosome—The male sex chromosome. Genetically normal human males have one X-chromosome and one Y-chromosome. The Y-chromosome contains genes inherited exclusively from the paternal side; hence, Y-chromosome variations and abnormalities serve as clues to paternal lineage.

Zipf distribution—George Kingsley Zipf (1902–1950) was an American linguist who demonstrated that for most languages a small number of words account for the majority of occurrences of all the words found in prose. Pareto's principle is a generalization of the Zipf distribution, as applied to income, populations, disease frequencies, and a wide variety of naturally occurring phenomena. See Pareto's principle.

REFERENCES

1. Borghesi L, Milcarek C. From B cell to plasma cell: regulation of V(D)J recombination and antibody secretion. Immunol Res 36:27–32, 2006.
2. Zhang J, Quintal L, Atkinson A, Williams B, Grunebaum E, Roifman CM. Novel RAG1 mutation in a case of severe combined immunodeficiency. Pediatrics 116:445–449, 2005.
3. de Villartay JP, Lim A, Al-Mousa H, Dupont S, Déchanet-Merville J, Coumau-Gatbois E, et al. A novel immunodeficiency associated with hypomorphic RAG1 mutations and CMV infection. J Clin Invest 115:3291–3299, 2005.
4. Estivill X, Bancells C, Ramos C. Geographic distribution and regional origin of 272 cystic fibrosis mutations in European populations. Hum Mutat 10:135–154, 1997.

5. Green PM, Saad S, Lewis CM, Giannelli F. Mutation rates in humans I: Overall and sex-specific rates obtained from a population study of hemophilia B. Am J Hum Genet 65:1572–1579, 1999.
6. Nishino I, Noguchi S, Murayama K, Driss A, Sugie K, Oya Y, et al. Distal myopathy with rimmed vacuoles is allelic to hereditary inclusion body myopathy. Neurology 59:1689–1693, 2002.
7. Sorek R, Dror G, Shamir R. Assessing the number of ancestral alternatively spliced exons in the human genome. BMC Genomics 7:273, 2006.
8. Pagani F, Baralle FE. Genomic variants in exons and introns: identifying the splicing spoilers. Nat Rev Genet 5:389–396, 2004.
9. Fraser HB, Xie X. Common polymorphic transcript variation in human disease. Genome 19:567–575, 2009.
10. Venables JP. Aberrant and alternative splicing in cancer. Cancer Res 64:7647–7654, 2004.
11. Srebrow A, Kornblihtt AR. The connection between splicing and cancer. J Cell Sci 119:2635–2641, 2006.
12. Wiestner A, Schlemper RJ, van der Maas AP, Skoda RC. An activating splice donor mutation in the thrombopoietin gene causes hereditary thrombocythemia. Nat Genet 18: 49–52, 1998.
13. Xu X, Weaver Z, Linke SP, Li C, Gotay J, Wang XW, et al. Centrosome amplification and a defective G2-M cell cycle checkpoint induce genetic instability in BRCA1 exon 11 isoform-deficient cells. Mol Cell 3:389–395, 1999.
14. Cahill DP, Lengauer C, Yu J, Riggins GJ, Willson JK, Markowitz SD, et al. Mutations of mitotic checkpoint genes in human cancers. Nature 392:300–303, 1998.
15. Weaver BAA, Cleveland DW. The role of aneuploidy in promoting and suppressing tumors. J Cell Biol 185:935–937, 2009.
16. Komoto M, Tominaga K, Nakata B, Takashima T, Inoue T, Hirakawa K. Complete regression of low-grade mucosa-associated lymphoid tissue (MALT) lymphoma in the gastric stump after eradication of Helicobacter pylori. J Exp Clin Cancer Res 25:283–285, 2006.
17. Kitamura H, Okudela K. Bronchioloalveolar neoplasia. Int J Clin Exp Pathol 4:97–99, 2011.
18. Benvenuti S, Arena S, Bardelli A. Identification of cancer genes by mutational profiling of tumor genomes. FEBS Lett 579:1884–1890, 2005.
19. Swanton C. Intratumor heterogeneity: evolution through space and time. Cancer Res 72:4875–4882, 2012.
20. Documentation for the Mortality Public Use Data Set, 1999. Mortality Statistics Branch, Division of Vital Statistics, National Center for Health Statistics, 1999.
21. Frey CM, McMillen MM, Cowan CD, Horm JW, Kessler LG. Representativeness of the surveillance, epidemiology, and end results program data: recent trends in cancer mortality rate. JNCI 84:872, 1992.
22. Ashworth TG. Inadequacy of death certification: proposal for change. J Clin Pathol 44:265, 1991.
23. Kircher T, Anderson RE. Cause of death: proper completion of the death certificate. JAMA 258:349–352, 1987.
24. Walter SD, Birnie SE. Mapping mortality and morbidity patterns: an international comparison. Intl J Epidemiol 20:678–689, 1991.
25. Huang Y, Paxton WA, Wolinsky SM, Neumann AU, Zhang L, He T, et al. The role of a mutant CCR5 allele in HIV-1 transmission and disease progression. Nat Med 2:1240–1243, 1996.
26. Leaf C. Do clinical trials work? The New York Times, July 13, 2013.

27. van den Bergh F, Eliason SL, Burmeister BT, Giudice GJ. Collagen XVII (BP180) modulates keratinocyte expression of the proinflammatory chemokine, IL-8. Exp Dermatol 21:605–611, 2012.
28. Slatko BE, Taylor MJ, Foster JM. The Wolbachia endosymbiont as an anti-filarial nematode target. Symbiosis 51:55–65, 2010.
29. Resnikoff S, Pascolini D, Etyaale D, Kocur I, Pararajasegaram R, Pokharel GP, et al. Global data on visual impairment in the year 2002. Bull WHO 2004(82):844–851, 2004.
30. Lefeber DJ, Morava E, Jaeken J. How to find and diagnose a CDG due to defective N-glycosylation. J Inherit Metab Dis 34:849–852, 2011.
31. Seok J, Warren HS, Cuenca AG, Mindrinos MN, Baker HV, Xu W, et al. Genomic responses in mouse models poorly mimic human inflammatory diseases. Proc Natl Acad Sci USA 110:3507–3512, 2013.
32. Lupski JR, de Oca-Luna RM, Slaugenhaupt S, Pentao L, Guzzetta V, Trask BJ, et al. DNA duplication associated with Charcot-Marie-Tooth disease type 1A. Cell 66:219–232, 1991.
33. Bogler O, Cavenee WK. Methylation and genomic damage in gliomas. In Genomic and Molecular Neuro-Oncology. Zhang W, Fuller GN, eds. Jones & Bartlett, Sudbury, MA, pp. 3–16, 2004.
34. Amir RE, van den Veyver IB, Wan M, Tran CQ, Francke U, Zoghbi HY. Rett syndrome is caused by mutations in X-linked MECP2, encoding methyl-CpG-binding protein 2. Nat Genet 23:185–188, 1999.
35. Pitteloud N, Quinton R, Pearce S, Raivio T, Acierno J, Dwyer A, et al. Digenic mutations account for variable phenotypes in idiopathic hypogonadotropic hypogonadism. J Clin Invest 117:457–463, 2007.
36. Bruning JC, Winnay J, Bonner-Weir S, Taylor SI, Accili D, Kahn CR. Development of a novel polygenic model of NIDDM in mice heterozygous for IR and IRS-1 null alleles. Cell 88:561–572, 1997.
37. Nachman MW, Crowell SL. Estimate of the mutation rate per nucleotide in humans. Genetics 156:297–304, 2000.
38. Roach JC, Glusman G, Smit AF, Huff CD, Hubley R, Shannon PT, et al. Analysis of genetic inheritance in a family quartet by whole-genome sequencing. Science 328:636–639, 2010.
39. Oller AR, Rastogi P, Morgenthaler S, Thilly WG. A statistical model to estimate variance in long term low dose mutation assays: testing of the model in a human lymphoblastoid mutation assay. Mutat Res 216:149–161, 1989.
40. Bierig JR. Actions for damages against medical examiners and the defense of sovereign immunity. Clin Lab Med 18:139–150, 1998.
41. Heintzman ND, Hon GC, Hawkins RD, Kheradpour P, Stark A, Harp LF, et al. Histone modifications at human enhancers reflect global cell-type-specific gene expression. Nature 459:108–112, 2009.
42. Agrelo R, Cheng WH, Setien F, Ropero S, Espada J, Fraga MF, et al. Epigenetic inactivation of the premature aging Werner syndrome gene in human cancer. Proc Natl Acad Sci USA 103:8822–8827, 2006.
43. Agrelo R, Setien F, Espada J, Artiga MJ, Rodriguez M, Pérez-Rosado A, et al. Inactivation of the lamin A/C gene by CpG island promoter hypermethylation in hematologic malignancies, and its association with poor survival in nodal diffuse large B-cell lymphoma. J Clin Oncol 23:3940–3947, 2005.
44. Malhotra R, Mason PK. Lamin A/C deficiency as a cause of familial dilated cardiomyopathy. Curr Opin Cardiol 24:203–208, 2009.

45. Combarros O, Cortina-Borja M, Smith AD, Lehmann DJ. Epistasis in sporadic Alzheimer's disease. Neurobiol Aging 30:1333–1349, 2009.
46. Lobo I. Epistasis: gene interaction and the phenotypic expression of complex diseases like Alzheimer's. Nat Educ 1:1, 2008.
47. Allegrucci C, Thurston A, Lucas E, Young L. Epigenetics and the germline. Reproduction 129:137–149, 2005.
48. Seisenberger S, Peat JR, Hore TA, Santos F, Dean W, Reik W. Reprogramming DNA methylation in the mammalian life cycle: building and breaking epigenetic barriers. Philos Trans R Soc Lond B Biol Sci 368 2013.
49. Choi M, Scholl UI, Ji W, Liu T, Tikhonova IR, Zumbo P, et al. Genetic diagnosis by whole exome capture and massively parallel DNA sequencing. Proc Natl Acad Sci USA 106:19096–19101, 2009.
50. Smulders YM, Eussen BHJ, Verhoef S, Wouters CH. Large deletion causing the TSC2-PKD1 contiguous gene syndrome without infantile polycystic disease. J Med Genet 40:e17, 2003.
51. Dolle MET, Snyder WK, Gossen JA, Lohman PHM, Vijg J. Distinct spectra of somatic mutations accumulated with age in mouse heart and small intestine. PNAS 97:8403–8408, 2000.
52. Crow JF. The high spontaneous mutation rate: is it a health risk? Proc Natl Acad Sci USA 94:8380–8386, 1997.
53. Roberts J. Looking at variation in numbers. The Scientist, March 14, 2005.
54. Ikegawa S. A short history of the genome-wide association study: where we were and where we are going. Genomics Inform 10:220–225, 2012.
55. Platt A, Vilhjalmsson BJ, Nordborg M. Conditions under which genome-wide association studies will be positively misleading. Genetics 186:1045–1052, 2010.
56. Beekman M, Blanch H, Perola M, Hervonen A, Bezrukov V, Sikora E, et al. Genome-wide linkage analysis for human longevity: genetics of healthy aging study. Aging Cell 12:184–193, 2013.
57. Couzin-Frankel J. Major heart disease genes prove elusive. Science 328:1220–1221, 2010.
58. Field MJ, Boat T. Rare diseases and orphan products: accelerating research and development. Institute of Medicine (US) Committee on Accelerating Rare Diseases Research and Orphan Product Development, 2010. The National Academies Press, Washington, DC. Available from: http://www.ncbi.nlm.nih.gov/books/NBK56189/.
59. Kim HL, Iwase M, Igawa T, Nishioka T, Kaneko S, Katsura Y, et al. Genomic structure and evolution of multigene families: "flowers" on the human genome. Int J Evol Biol:917678, 2012.
60. Korbel JO, Urban AE, Affourtit JP, Godwin B, Grubert F, Simons JF, et al. Paired-end mapping reveals extensive structural variation in the human genome. Science 318:420–426, 2007.
61. Paananen J, Ciszek R, Wong G. Varietas: a functional variation database portal. Database, July 29, 2010.
62. OMIM. Online Mendelian Inheritance in Man. Available from: http://omim.org/downloads, viewed June 20, 2013.
63. Izeradjene K, Combs C, Best M, Gopinathan A, Wagner A, Grady WM, et al. Kras(G12D) and Smad4/Dpc4 haploinsufficiency cooperate to induce mucinous cystic neoplasms and invasive adenocarcinoma of the pancreas. Cancer Cell 11:229–243, 2007.
64. Cabelof DC, Ikeno Y, Nyska A, Busuttil RA, Anyangwe N, Vijg J, et al. Haploinsufficiency in DNA polymerase beta increases cancer risk with age and alters mortality rate. Cancer Res 66:7460–7465, 2006.
65. Yu Y, Xu F, Peng H, Fang X, Zhao S, Li Y, et al. NOEY2 (ARHI), an imprinted putative tumor suppressor gene in ovarian and breast carcinomas. Proc Natl Acad Sci USA 96:214–219, 1999.

66. Vilmos P, Kurucz E. Insect immunity: evolutionary roots of the mammalian innate immune system. Immunol Lett 62:59–66, 1998.
67. Glaser RL, Goldbach-Mansky R. The spectrum of monogenic autoinflammatory syndromes: understanding disease mechanisms and use of targeted therapies. Curr Allergy Asthma Rep 8:288–298, 2008.
68. Gottlieb B, Beitel LK, Alvarado C, Trifiro MA. Selection and mutation in the "new" genetics: an emerging hypothesis. Hum Genet 127:491–501, 2010.
69. Yan N, Chen ZJ. Intrinsic antiviral immunity. Nat Immunol 13:214–222, 2012.
70. Inglis TJ. Principia aetiologica: taking causality beyond Koch's postulates. J Med Microbiol 56:1419–1422, 2007.
71. Ford ES, Giles WH, Dietz WH. Prevalence of the metabolic syndrome among US adults: findings from the third national health and nutrition examination survey. JAMA 287:356–359, 2002.
72. Harmon A. The DNA age: searching for similar diagnosis through DNA. The New York Times, December 28, 2007.
73. Bentwich I, Avniel A, Karov Y, Aharonov R, Gilad S, Barad O, et al. Identification of hundreds of conserved and nonconserved human microRNAs. Nat Genet 37:766–770, 2005.
74. Al-Mahdawi S, Pinto RM, Varshney D, Lawrence L, Lowrie MB, Hughes S, et al. GAA repeat expansion mutation mouse models of Friedreich ataxia exhibit oxidative stress leading to progressive neuronal and cardiac pathology. Genomics 88:580–590, 2006.
75. Stechmann A, Hamblin K, Perez-Brocal V, Gaston D, Richmond GS, van der Giezen M, et al. Organelles in blastocystis that blur the distinction between mitochondria and hydrogenosomes. Curr Biol 18:580–585, 2008.
76. Tovar J, Leon-Avila G, Sanchez LB, Sutak R, Tachezy J, van der Giezen M, et al. Mitochondrial remnant organelles of Giardia function in iron-sulphur protein maturation. Nature 426:172–176, 2003.
77. Tovar J, Fischer A, Clark CG. The mitosome, a novel organelle related to mitochondria in the amitochondrial parasite Entamoeba histolytica. Mol Microbiol 32:1013–1021, 1999.
78. Burri L, Williams B, Bursac D, Lithgow T, Keeling P. Microsporidian mitosomes retain elements of the general mitochondrial targeting system. PNAS 103:15916–15920, 2006.
79. Finsterer J. Mitochondriopathies. Eur J Neurol 11:163–186, 2004.
80. Landgren O, Kyle RA, Pfeiffer RM. Monoclonal gammopathy of undetermined significance (MGUS) consistently precedes multiple myeloma: a prospective study. Blood 113:5412–5417, 2009.
81. Weatherall DJ. Molecular pathology of single gene disorders. J Clin Pathol 40:959–970, 1987.
82. Klupa T, Skupien J, Malecki MT. Monogenic models: what have the single gene disorders taught us? Curr Diab Rep 12:659–666, 2012.
83. Cipra BA. The best of the 20th century: editors name top 10 algorithms. SIAM News 33(4) May 2000.
84. Berman JJ. Methods in Medical Informatics: Fundamentals of Healthcare Programming in Perl, Python, and Ruby. Chapman and Hall, Boca Raton, 2010.
85. Berman JJ. Biomedical Informatics. Jones & Bartlett, Sudbury, MA, 2007.
86. Whittaker JC, Harbord RM, Boxall N, Mackay I, Dawson G, Sibly RM. Likelihood-based estimation of microsatellite mutation rates. Genetics 164:781–787, 2003.
87. Veltman JA, Brunner HG. De novo mutations in human genetic disease. Nat Rev Genet 13:565–575, 2012.

88. McKenna ES, Sansam CG, Cho YJ, Greulich H, Evans JA, Thom CS, et al. Loss of the epigenetic tumor suppressor SNF5 leads to cancer without genomic instability. Mol Cell Biol 28:6223–6233, 2008.
89. Wang E, Boswell E, Siddiqi I, Lu CM, Sebastian S, Rehder C, et al. Pseudo-Pelger-Huet anomaly induced by medications: a clinicopathologic study in comparison with myelodysplastic syndrome-related pseudo-Pelger-Huet anomaly. Am J Clin Pathol 135:291–303, 2011.
90. Lotery A, Trump D. Progress in defining the molecular biology of age related macular degeneration. Hum Genet 122:219–236, 2007.
91. Maller J, George S, Purcell S, Fagerness J, Altshuler D, Daly MJ, et al. Common variation in three genes, including a noncoding variant in CFH, strongly influences risk of age-related macular degeneration. Nat Genet 38:1055–1059, 2006.
92. Katta S, Kaur I, Chakrabarti S. The molecular genetic basis of age-related macular degeneration: an overview. J Genet 88:425–449, 2009.
93. Eichers ER, Lewis RA, Katsanis N, Lupski JR. Triallelic inheritance: a bridge between Mendelian and multifactorial traits. Ann Med 36:262–272, 2004.
94. Pober BR. Williams-Beuren syndrome. N Engl J Med 362:239–252, 2010.
95. McGary KL, Parka TJ, Woodsa JO, Chaa HJ, Wallingford JB, Marcottea EM. Systematic discovery of nonobvious human disease models through orthologous phenotypes. Proc Nat Acad Sci USA 107:6544–6549, 2010.
96. Gissen P, Maher ER. Cargos and genes: insights into vesicular transport from inherited human disease. J Med Genet 44:545–555, 2007.
97. Novick P, Field C, Schekman R. Identification of 23 complementation groups required for post-translational events in the yeast secretory pathway. Cell 21:205–215, 1980.
98. The World Factbook. Central Intelligence Agency, Washington, DC, 2009.
99. Hoyert DL, Heron MP, Murphy SL, Kung H-C. Final Data for 2003. National Vital Statistics Report 54, April 19, 2006.
100. Dale DC, Person RE, Bolyard AA, Aprikyan AG, Bos C, Bonilla MA, et al. Mutations in the gene encoding neutrophil elastase in congenital and cyclic neutropenia. Blood 96:2317–2322, 2000.
101. Seidman JG, Seidman C. Transcription factor haploinsufficiency: when half a loaf is not enough. J Clin Invest 109:451–455, 2002.
102. Weller S, Cajigas I, Morrell J, Obie C, Steel G, Gould SJ, et al. Alternative splicing suggests extended function of PEX26 in peroxisome biogenesis. Am J Hum Genet 76:987–1007, 2005.
103. Takahashi K, Tanabe K, Ohnuki M, Narita M, Ichisaka T, Tomoda K, et al. Induction of pluripotent stem cells from adult human fibroblasts by defined factors. Cell 131:861–872, 2007.
104. Okita K, Ichisaka T, Yamanaka S. Generation of germline-competent induced pluripotent stem cells. Nature 448:313–317, 2007.
105. Zhang L, Lin X. Some considerations of classification for high dimension low-sample size data. Stat Methods Med Res 2011. Available from: http://smm.sagepub.com/content/early/2011/11/22/0962280211428387.long, viewed January 26, 2013.
106. Barosi G, Bergamaschi G, Marchetti M, Vannucchi AM, Guglielmelli P, Antonioli E, et al. JAK2 V617F mutational status predicts progression to large splenomegaly and leukemic transformation in primary myelofibrosis. Blood 110:4030–4036, 2007.
107. Ng BG, Kranz C, Hagebeuk EE, Duran M, Abeling NG, Wuyts B, et al. Molecular and clinical characterization of a Moroccan Cog7 deficient patient. Mol Genet Metab 91:201–204, 2007.
108. Berman JJ. Precancer: The Beginning and the End of Cancer. Jones & Bartlett, Sudbury, 2010.

109. Berman JJ. Taxonomic Guide to Infections Diseases: Understanding the Biologic Classes of Pathogenic Organisms. Academic Press, Waltham, 2012.
110. de Vooght KMK, van Wijk R, van Solingel WE. Management of gene promoter mutations in molecular diagnostics. Clin Chem 55:698–708, 2009.
111. Poliseno L. Pseudogenes: newly discovered players in human cancer. Sci Signal 5:5, 2012.
112. Costa V, Esposito R, Aprile M, Ciccodicola A. Non-coding RNA and pseudogenes in neurodegenerative diseases: "The (un)Usual Suspects". Front Genet 3:231, 2012.
113. Rare Diseases Act of 2002, Public Law 107-280, 107th U.S. Congress, November 6, 2002.
114. Mencia A, Modamio-Hoybjor S, Redshaw N, Morin M, Mayo-Merino F, Olavarrieta L, et al. Mutations in the seed region of human miR-96 are responsible for nonsyndromic progressive hearing loss. Nat Genet 41:609–613, 2009.
115. Emerman M, Malik HS. Paleovirology: modern consequences of ancient viruses. PLoS Biol 8:e1000301, 2010.
116. Bartholdi D, Krajewska-Walasek M, Ounap K, Gaspar H, Chrzanowska KH, Ilyana H, et al. Epigenetic mutations of the imprinted IGF2-H19 domain in Silver-Russell syndrome (SRS): results from a large cohort of patients with SRS and SRS-like phenotypes. J Med Genet 46:192–197, 2009.
117. Kimchi-Sarfaty C, Oh JM, Kim IW, Sauna ZE, Calcagno AM, Ambudkar SV, et al. A "silent" polymorphism in the MDR1 gene changes substrate specificity. Science 315:525–528, 2007.
118. Genetics Home Reference. National Library of Medicine. July 1, 2013. Available from: http://ghr.nlm.nih.gov/handbook/genomicresearch/snp, viewed July 6, 2013.
119. Lindhurst MJ, Sapp JC, Teer JK, Johnston JJ, Finn EM, Peters K, et al. A mosaic activating mutation in AKT1 associated with the Proteus syndrome. N Engl J Med 365:611–619, 2011.
120. Faustino NA, Cooper TA. Pre-mRNA splicing and human disease. Genes Dev 17:419–437, 2003.
121. Xu B, Roos JL, Dexheimer P, Boone B, Plummer B, Levy S, et al. Exome sequencing supports a de novo mutational paradigm for schizophrenia. Nat Genet 43:864–868, 2011.
122. Dai C, Celestino JC, Okada Y, Louis DN, Fuller GN, Holland EC. PDGF autocrine stimulation dedifferentiates cultured astrocytes and induces oligodendrogliomas and oligoastrocytomas from neural progenitors and astrocytes in vivo. Genes Dev 15:1913–1925, 2001.
123. Vulliamy T, Beswick R, Kirwan M, Marrone A, Digweed M, Walne A, et al. Mutations in the telomerase component NHP2 cause the premature ageing syndrome dyskeratosis congenita. Proc Natl Acad Sci USA 105:8073–8078, 2008.
124. Yamaguchi H. Mutations of telomerase complex genes linked to bone marrow failures. J Nippon Med Sch 74:202–209, 2007.
125. Bodey B. Spontaneous regression of neoplasms: new possibilities for immunotherapy. Exp Opin Biol Ther 2:459–476, 2002.
126. Wambua S, Mwangi TW, Kortok M, Uyoga SM, Macharia AW, Mwacharo JK, et al. The effect of alpha+-thalassaemia on the incidence of malaria and other diseases in children living on the coast of Kenya. PLoS Med 3:e158, 2006.
127. Kapitonov VV, Jurka J. RAG1 core and V(D)J recombination signal sequences were derived from Transib transposons. PLoS Biol 3:e181, 2005.
128. Lerat E, Semon M. Influence of the transposable element neighborhood on human gene expression in normal and tumor tissues. Gene 396:303–311, 2007.
129. U.S. Vital Statistics System: Major Activities and Developments, 1950–95. Centers for Disease Control and Prevention, National Center for Health Statistics, 1997.
130. Tuna M, Knuutila S, Mills GB. Uniparental disomy in cancer. Trends Mol Med 15:120–128, 2009.

Index

A

B

C

D

E

H

I

J

K

L

M

N

O

P

S

T

U

V

W

X

Y

Z

Color Plates

FIGURE 3.1 A cross-section of a gross specimen of lung, exhibiting the pathology of tuberculous pneumonia. Cavities and blebs are noted in the upper lobe, indicating past destruction of the lung tissue. White areas of consolidated inflammation are present throughout the specimen. Notice the small white, round grains that encircle the areas of inflammation. These are miliary granulomas, areas of chronic inflammation that arise at sites of mycobacterial infection. (*Source: MacCallum WG. A Textbook of Pathology* [25].)

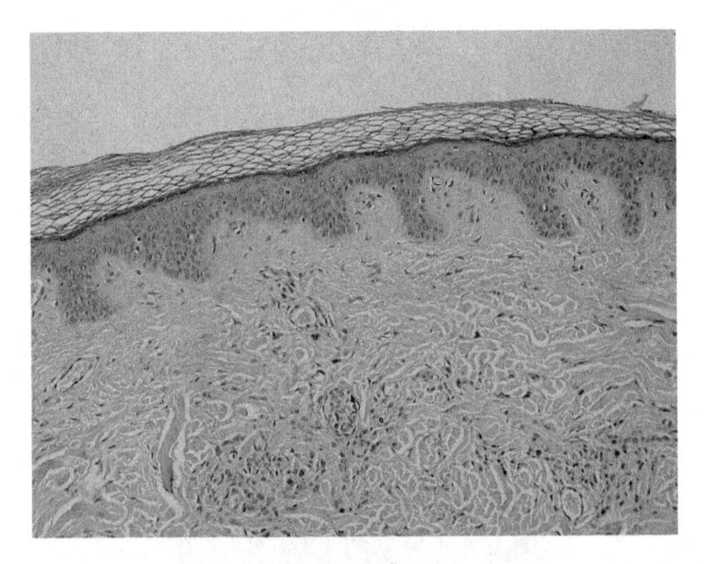

FIGURE 4.1 Skin biopsy showing epidermis overlying the fibrous tissue of the dermis. The epidermis is thinner than the dermis, and contains layers of cells, each layer having characteristic morphologic features. The epidermis has an undulating lower border, with papillae known as rete pegs, jutting down into the dermis. *(Source: Wikimedia Commons, acquired as a public domain image.)*

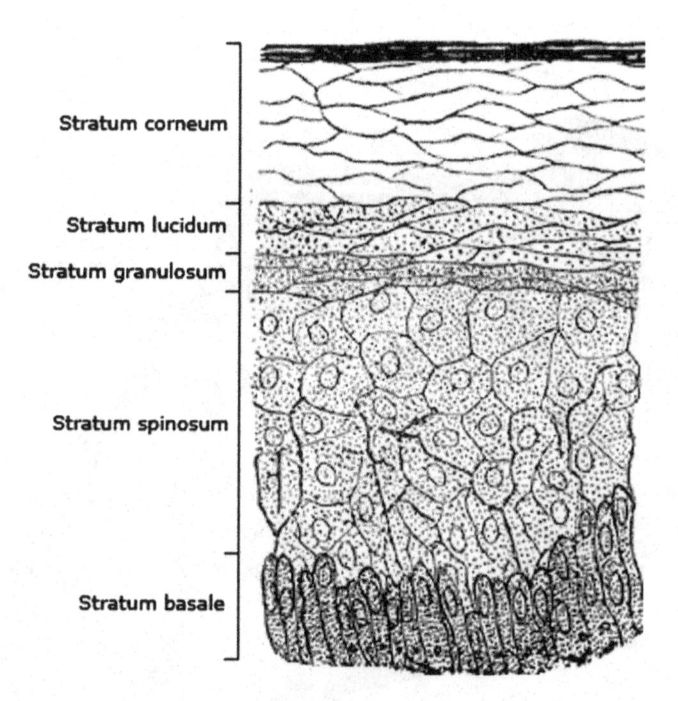

FIGURE 4.2 A graphic representing the various layers of the epidermis. The lowest or basal layer of cells contains the regenerating cells of the epidermis, the only cells of the epidermis capable of cell division. Basal cells, when they divide, produce one basal cell and one post-mitotic cell. The post-mitotic cell is pushed up into the next higher cell layer, and it continues to rise through the epidermis as it is pushed up by successive post-mitotic progeny of the basal layer. As it rises, it flattens out, fills with keratin, and eventually loses its nucleus. At this point, it is little more than a squamous flake, sitting atop the epidermis. The cells at the very top of the epidermis eventually slide off into the air to become floating specks of dust. Common dandruff consists of clumps of squamous cells sloughed from the stratum corneum. *(Source: Wikimedia Commons, acquired as a public domain image.)*

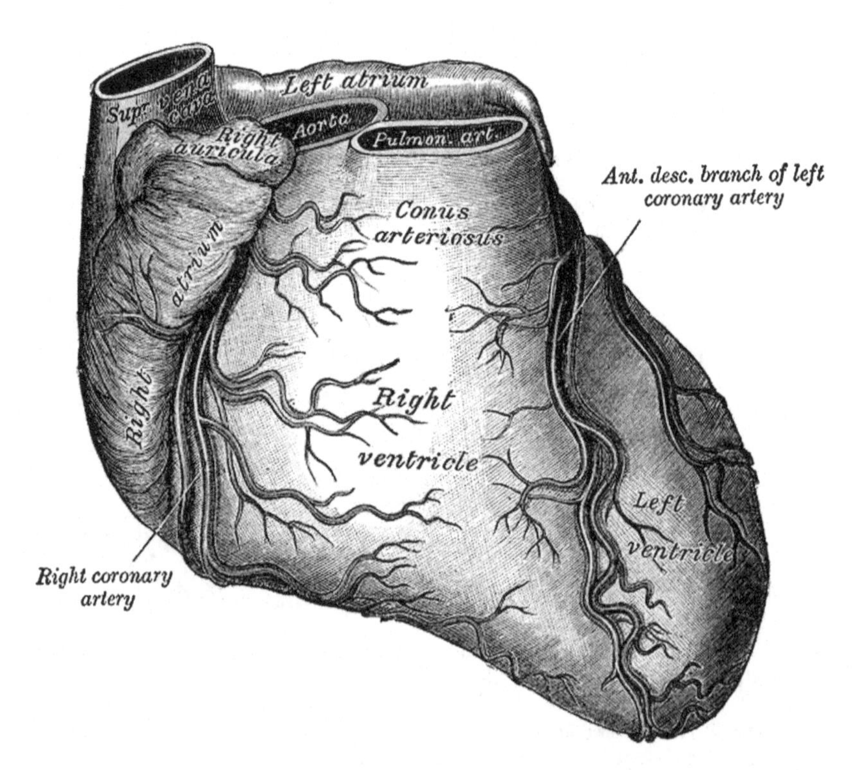

FIGURE 5.2 The coronary arteries travel along the surface of the heart, delivering branches that penetrate into the heart muscle. (*Source: Wikimedia Commons, acquired as a public domain image.*)

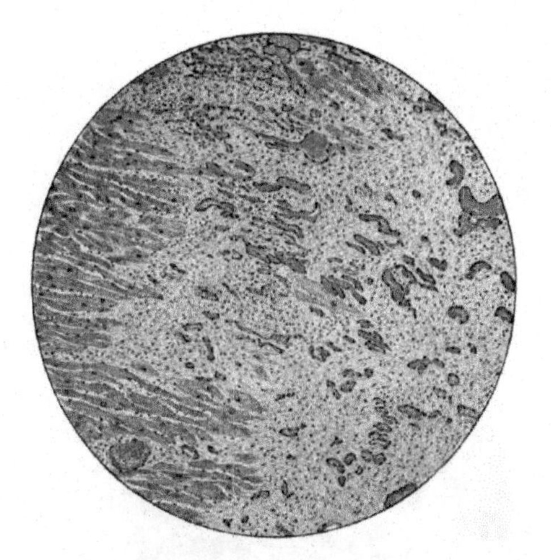

FIGURE 5.4 Histopathology of heart involved by early myocardial infarction. To the left of the image are surviving muscle fibers. To the right, the heart muscle has been replaced by inflammatory cells, vessels, and granulation tissue (i.e., early scar tissue). (*Source: MacCallum WG. A Textbook of Pathology* [1].)

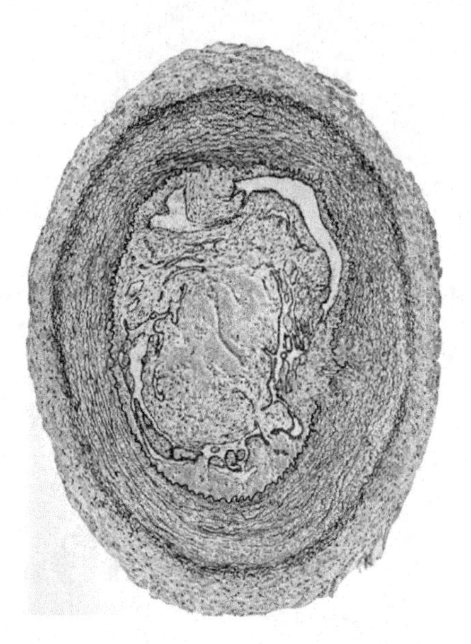

FIGURE 5.6 Thrombus in artery. A fibrinous thrombus fills the lumen of a muscular artery. The artery features a thick media (i.e., muscular wall with circumferentially aligned fibers underlying the wavy intimal lining). Filling the lumen (i.e., the inside of the artery, which would be carrying blood under normal circumstances) is a thrombus. Notice that the thrombus has numerous lined vessels running through it (i.e., the long-thin empty spaces within the thrombus). These vessels within the thrombus arise through a biologic process called re-canalization, in which new vessels grow, and blood flows, at a reduced rate and volume through old thrombi. (*Source: MacCallum WG. A Textbook of Pathology* [1].)

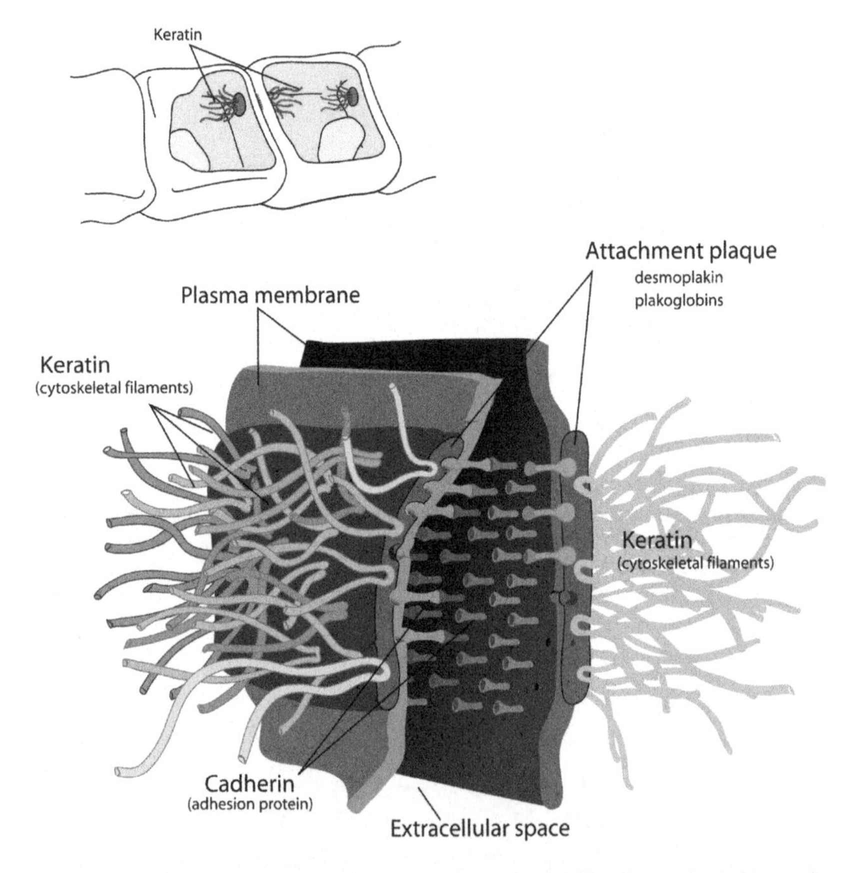

FIGURE 5.9 Graphic of a desmosome. The desmosome is much like a button that holds together the flat-surfaced areas where cells touch one another. The net effect is to produce a permanent epithelial (i.e., polyhedral) network of cells. Some cells are particularly rich in desmosomes, such as keratinocytes, cardiac myocytes, and the cells that ensheath nerve fibers. (*Source: Wikipedia, created and released into the public domain by Mariana Ruiz, Ladyof Hats.*)

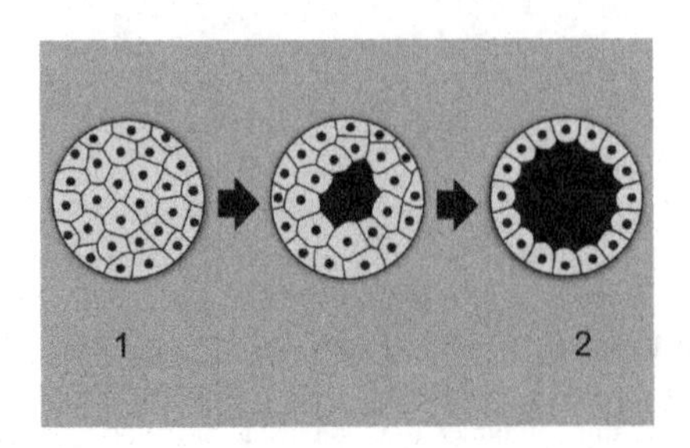

FIGURE 5.10 Graphic of blastulation. The early, solid, embryo secretes fluid into a central viscous, the developing blastula. Blastulation is accomplished with specialized membrane channels that transport ions and water, and with desmosomes, that provide a water-tight boundary between adjacent cells. Among multicellular organisms, the blastulated embryo is found exclusively in members of Class Animalia. (*Source: Wikipedia, created and released into the public domain by Pidalka44.*)

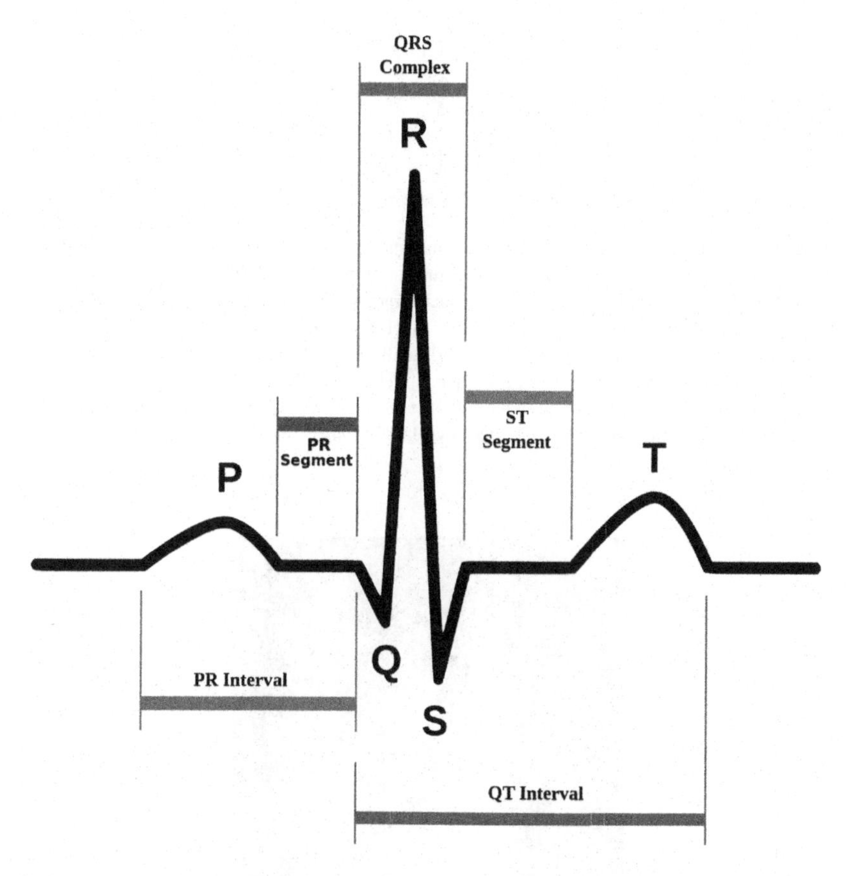

FIGURE 5.11 An electrocardiogram tracing of the electrical conduction through the heart in a single, normal heartbeat. The QRS complex indicates the depolarization of the right and left ventricles. (*Source: Wikipedia, created by Agateller (Anthony Atkielski), and released into the public domain.*)

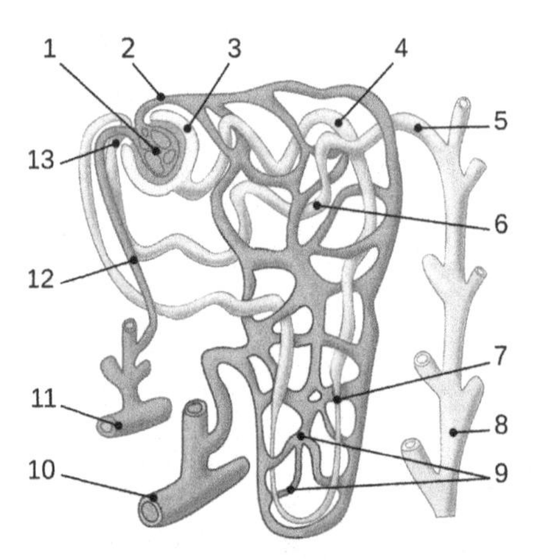

FIGURE 5.12 Illustration of a single nephron, demonstrating specific anatomic components targeted by inherited forms of hypertension: (1) glomerulus; (2) efferent arteriole; (3) Bowman capsule; (4) proximal convoluted tubule; (5) collecting duct, target of Liddle syndrome; (6) distal convoluted tubule, target of Gitelman syndrome; (7) loop of Henle, wherein the thick ascending limb is targeted in Bartter syndrome; (8) Bellini duct; (9) capillaries; (10) arcuate vein; (11) arcuate artery; (12) afferent arteriole; (13) juxtaglomerular apparatus, effector of the renin–angiotensin–aldosterone system an active pathway employed in the pathogenesis of various causes of hypertension, including hypertension associated with renal artery dysplasia. (*Source: Wikipedia, and released into the public domain.*)

FIGURE 6.1 Larval form of the screw-worm, a poorly chosen and misleading common name for an arthropod whose scientific name is *Cochliomyia hominivorax*. (*Source: Wikipedia, produced for the public domain by the Agricultural Research Service, the research agency of the United States Department of Agriculture.*)

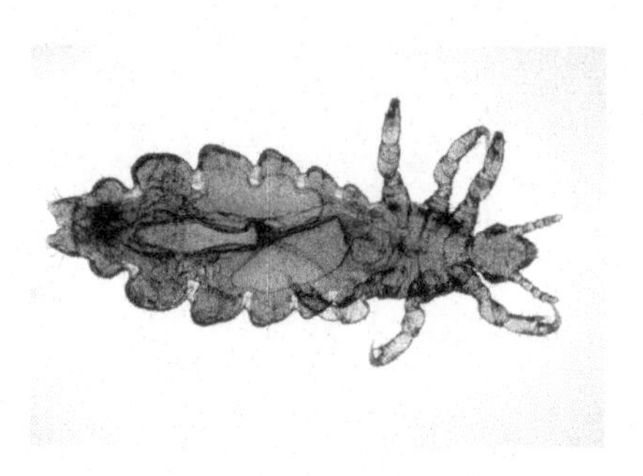

FIGURE 7.1 *Pediculus capitus*, or head louse, an insect specialized to live on the hair of human heads. (*Source: Wikipedia, created and released to the public domain by Dr. Dennis D. Juranek of the U.S. Centers for Disease Control and Prevention.)*

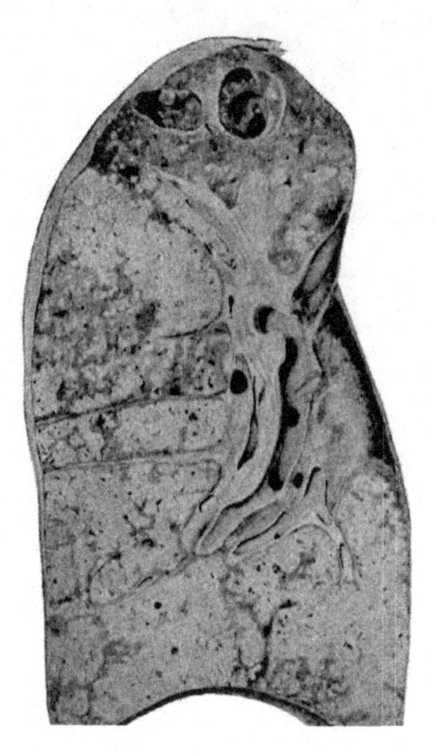

FIGURE 9.2 A cross-section of a gross specimen of lung, cut apex to base. The lung is involved entirely by a consolidating tuberculous pneumonia. Notice a few round blebs in the upper lobe. (*Source: MacCallum WG. A textbook of pathology* [3].)

FIGURE 10.2 Cross-section of gross specimen of liver involved by cirrhosis. In this specimen, the nodules are all small, but of varying size (i.e., each nodule has a slightly different size from its neighbor nodules). Fibrous tissue fills the spaces between nodules, and the capsule covering the liver is thickened. (*Source: MacCallum WG. A Textbook of Pathology* [32].)

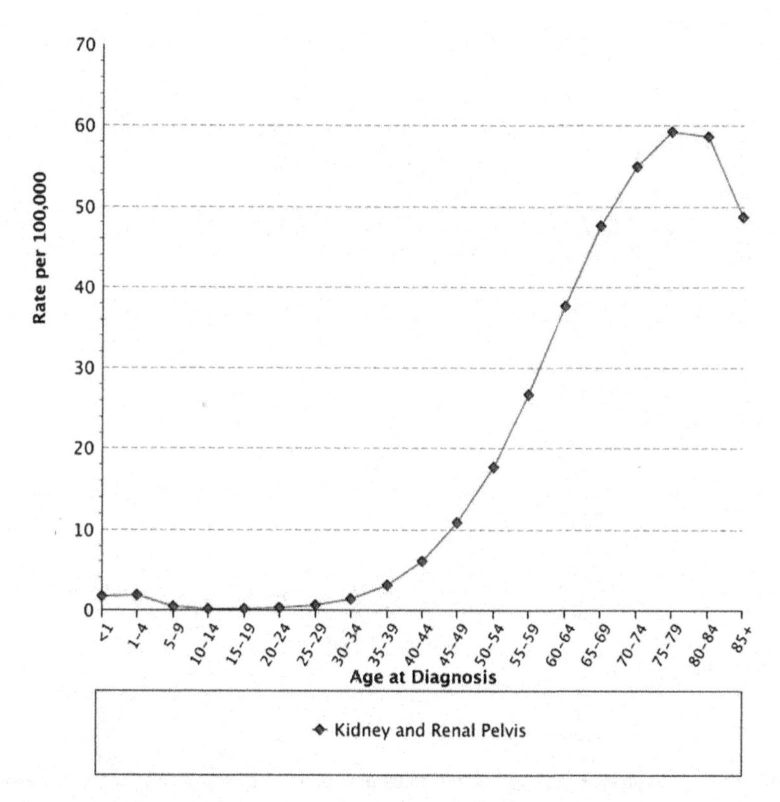

FIGURE 11.1 Graph showing incidence of kidney and renal pelvis cancer by age of occurrence of disease. The incidence of kidney cancer and renal pelvis cancer rises steadily, as age increases, toward a single peak. The graph was generated at the National Cancer Institute's Surveillance, Epidemiology and End Results "Fast Stats" query site. The query input settings were: SEER incidence; Age-specific rates, 1992–2010, All races, Both sexes, All ages, Kidney and renal pelvis. (*Available from: http://seer.cancer.gov/faststats/selections.php?series=cancer, viewed on November 29, 2013.*)

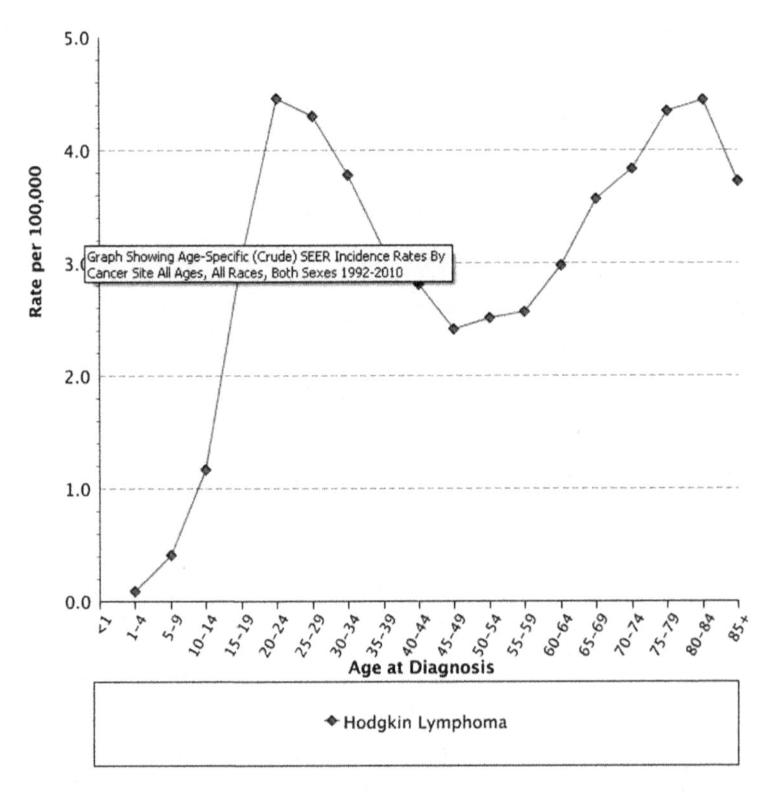

FIGURE 11.2 Graph showing incidence of Hodgkin lymphoma by age of occurrence of disease. There are two peaks in the graph. The first peak occurs in the early 20s. After the first peak, there is a trough, in the mid-40s, after which incidence increases steadily with age, toward a second peak. The graph was generated at the National Cancer Institute's Surveillance, Epidemiology and End Results "Fast Stats" query site. The query input settings were: SEER incidence; Age-specific rates, 1992–2010, All races, Both sexes, All ages, Hodgkin lymphoma. (*Available from: http://seer.cancer.gov/faststats/selections.php?series=cancer, viewed on November 29, 2013.*)

FIGURE 11.3 Centrilobular emphysema, a type of COPD (chronic obstructive pulmonary disease). The cut surface of the lung shows multiple small cavities, each surrounded by black carbon deposits. These are distal airways that have extensive cavitary destruction. (*Source: U.S. Centers for Disease Control and Prevention and entered into the public domain as a U.S. government work.*)

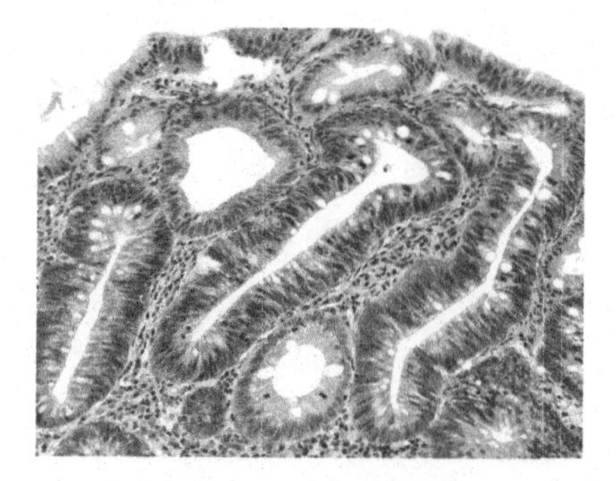

Figure G.1 Histopathologic section of biopsy of a colon adenoma. The specimen has been stained with hematoxylin and eosin, which colors the nuclei blue and the cytoplasm pink. The pathologist can inspect individual cells, as well as the architecture of the tissue. In many cases, the pathologist can render a specific diagnosis based largely on observations of histologic sections. (*Source: Dr. G. William Moore, for the U.S. Department of Veterans Administration, and released into the public domain.*)

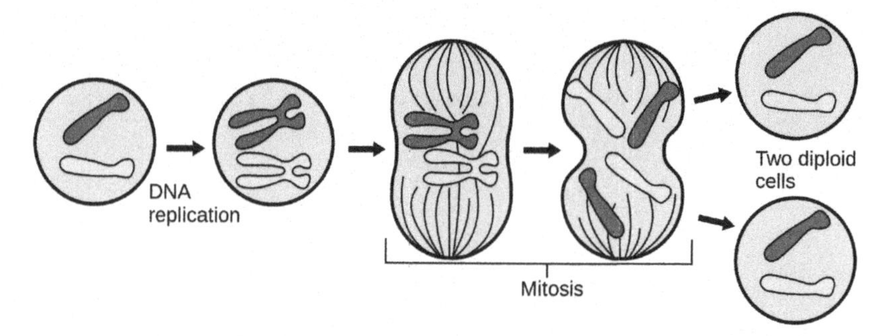

Figure G.3 Schematic of mitosis, indicating how chromosomes replicate and migrate into daughter cells. (*Source: Wikipedia, and created as a U.S. government public domain work by the National Center for Biotechnology Information, part of the National Institutes of Health.*)

Printed in the United States
By Bookmasters